How Students Think When Doing Algebra

How Students Think When Doing Algebra

By
Steve Rhine
Rachel Harrington
Colin Starr

INFORMATION AGE PUBLISHING, INC.
Charlotte, NC • www.infoagepub.com

Library of Congress Cataloging-in-Publication Data

The CIP data for this book can be found on the LIbrary of Congress website (loc.gov).

Paperback: 978-1-64113-411-8
Hardcover: 978-1-64113-412-5
eBook: 978-1-64113-413-2

Copyright © 2019 Information Age Publishing Inc.

All rights reserved. No part of this publication may be reproduced, stored in a
retrieval system, or transmitted, in any form or by any means, electronic, mechanical, photo-
copying, microfilming, recording or otherwise, without written permission
from the publisher.

Printed in the United States of America

CONTENTS

1. Introduction 1
2. Variables and Expressions 35
3. Algebraic Relations 99
4. Analysis of Change (Graphing) 165
5. Patterns & Functions 237
6. Modeling and Word Problems 301

ACKNOWLEDGEMENTS

This book is the result of a multi-year project to synthesize research on algebraic thinking from the past 35 years with primary funding from the Fund for the Improvement of Post-Secondary Education (FIPSE). Thanks are due to the following people for their time and effort in reviewing, discussing, and contributing to this work:

Mike Charles, *Pacific University*
Linda Samek, *George Fox University*
Kasi Allen, *Lewis and Clark University*
Cathy Brown, *Oregon Department of Education, retired*
Krista Heim, *Portland State University*
Barbara Herzberg, *Northwest Christian University*
Estrella Johnson, *Virginia Tech University*
Gulden Karakok, *University of Northern Colorado*
Laura Lethe, *Salem-Keizer School District*
Carolyn McCaffrey James, *Portland State University*
Ron Narode, *Portland State University*
Thelma Perso, *Western Australia, Department of Education*
Nicole Rigelman, *Portland State University*
Jill Summerlin, *Oregon Middle School Teacher, retired*
Darrell Trussell, *McNary High School Teacher, Oregon*
Kris Warloe, *Oregon Middle School Teacher, retired*
Don Watson, *Oregon Middle School Teacher, retired*
John Wilkins, *California State University, Dominguez Hills*

CHAPTER 1

INTRODUCTION

How Students Think When Doing Algebra, pages 1–34.
Copyright © 2019 by Information Age Publishing
All rights of reproduction in any form reserved.

> An error is ...
> a symptom of the nature of the conceptions
> which underlie (a student's) mathematical activity.
> — *Balachef, 1984*

B. F. Skinner thought of a person's mind as a black box: there was no way you could know what was going on inside someone's head. Our educational system was largely built on that premise, resulting in a system where we taught and retaught until we saw students getting correct answers. Yet, as mathematician Sowder (1988) noted, "Correct answers are not a safe indicator of good thinking." One might argue similarly that wrong answers are not a safe indicator of bad thinking. There is lots of gray area between right and wrong. Calvin's experience in the above cartoon suggests that it might be a good idea to explore what is inside that black box now and then ...

"And how did you get seven, Calvin?"

"Well, I was racing towards this alien planet in my spaceship ..."

A classroom is a complex environment in which teachers cannot possibly pay attention to every detail. By necessity, there is a triage process to determine what most needs to be addressed. Teaching in mathematics classrooms requires quick decision making based on a diagnosis of students' thinking, which requires constant assessment of the students' level of understanding.[1] Researchers describe this engagement as *professional vision*.[2] "Teachers' professional vision involves the ability to notice and interpret significant interactions in the classroom."[3] The construct of *significance* is identified in many forms in the research literature, such as "pivotal teaching moments,"[4] "significant mathematical instances,"[5] "the fodder for a content-related conversation,"[6] and Mathematically Significant Opportunities to build on Students' Thinking (MOSTs).[7]

MOSTs are the intersection of important mathematics, pedagogical opportunities, and student thinking. *Important mathematics* is defined by the curriculum, the Common Core State Standards, state and district goals, and the nature of the subject. Content knowledge of important mathematics includes understanding the structures of the subject matter, the rules for determining what is legitimate to say, how content is connected to other concepts, and why it is worth knowing.[8] "In the narrowest sense, this would be a mathematical goal for the lesson in which it occurs, but more broadly, it could also be related to the goals for a unit of instruction, an entire course, or for understanding mathematics as a whole."[9]

Pedagogical opportunities are in the eye of the beholder. Multiple factors—such as how much time is left in class, how many days are available for a unit, mood and atmosphere of a class, and how many students might be influenced by a particular discussion—can be the basis of pedagogical action. Certain problems might be key in facilitating conceptual development or providing prerequisite understanding of a more complex concept for an individual or group. However, opportunities to make use of students' mathematical thinking during instruction can often be missed, particularly by early career teachers.[10] "Novice (teachers') attention is likely to be dominated by the initial traumas of teaching" rather than students' learning.[11]

Finally, the impetus for MOSTs is *students' thinking*. "Observable student actions provide pedagogical openings for working towards mathematical goals for student learning ... student thinking is at the heart."[12] Recognizing a pedagogical opportunity requires teachers to identify a student's thinking as problematic and/or having potential for fruitful discussion. It is important for teachers to have a perspective of students as sense-makers rather than seeing them in black and white as achieving correct or incorrect thinking. Each student pieces together information in a unique way based on experience, prior knowledge, and instruction. Students usually have logical rationale underlying errors that they make. Even Calvin had a path to his response, although the logic of it could be debated. Was he lucky or was he thinking mathematically in a unique context? When teachers have an understanding of the types of struggles or ways students think about mathematical content, they are more prepared to notice MOSTs. Without knowledge of the ways students think about and struggle with a concept, a teacher may believe that a student who incorrectly solved a problem could resolve their misunderstanding simply by witnessing a correct solution, thereby missing opportunities to engage students in discussion and develop their underlying comprehension.

Over time, veteran math teachers develop extensive knowledge of how students engage with concepts—their misconceptions, ways of thinking, and when and how they are challenged to understand—and use that knowledge to anticipate students' struggles with particular lessons and plan accordingly. The art of orchestrating productive mathematics discussions depends on "anticipating likely responses to mathematical tasks."[13] Veteran teachers learn to evaluate whether an incorrect response is a simple error or the symptom of a faulty or naïve understanding of a concept. Novice teachers, on the other hand, lack the experience to anticipate important moments in their students' learning. They often struggle to make sense of what students say in the classroom, and to determine whether the response is useful or can further discussion.[14] They also may assume that students understand, therefore failing to perceive when a student's thinking is problematic.

Learning to foresee and use students' thinking during instruction is complex. Researchers suggest that there are

> two kinds of subject-matter understanding that teachers need to have—knowing that (something is so) and knowing why (it is so)...*Knowing that* refers to research-based and experience-based knowledge about students' common conceptions and ways of thinking about subject matter. *Knowing why* refers to knowledge about possible sources of these conceptions.[15]

The Cognitively Guided Instruction project (CGI) found that in-service teachers who learned about common conceptions and sources of those conceptions could predict students' struggles. The teachers made fundamental changes in their beliefs and practice that ultimately resulted in higher student achievement.[16] Ball, Thames, & Phelps (2008) define this domain of Mathematical Knowledge for Teaching as *Knowledge of Content and Students* (KCS). "KCS includes knowledge about common student conceptions and misconceptions, about what mathematics students find interesting or challenging, and about what students are likely to do with specific mathematics tasks."[17] CGI researchers found that teachers with higher amounts of KCS facilitated increased student achievement and these effects were long-lasting for students and teachers.[18] KCS is traditionally acquired by experience in the classroom—exactly what novice teachers lack.

The purpose of this book is to accelerate early career teachers' experience with how students think when doing algebra as well as to supplement veteran teachers' KCS. The research that this book is based on can provide teachers with insight into the nature of a student's struggles with particular algebraic ideas—to help teachers identify patterns that imply underlying thinking. While this book is useful to help you learn common errors that students make and misconceptions they might have, there is no substitute for exploring a student's thinking with your questions to find out the unique nature of that student's understanding.

ALGEBRA

The National Council for Teachers of Mathematics' (NCTM's) position statement on algebra highlights "the power and usefulness of algebraic thinking and skills—proficiencies that open academic doors and are evident in many professions and careers."[19] Algebra is the foundation for students who want to pursue science, technology, engineering, and mathematics (STEM) careers.[20] The typical 8th- or 9th-grade Algebra I course is considered the gatekeeper to higher education and future employment. More than any other course in K–12 education, algebra predicts students' futures: students who pass algebra are much more likely to go on to college. If students do not pass algebra, many segments of the job market are unavailable to them.[21] However, algebra is a major stumbling block for many students and has an extensive failure rate. Rather than helping students develop mathematical competence and gain access to higher education, it has screened out large numbers of students.[22] For instance, in the Los Angeles Unified School District, 48,000 ninth-graders took beginning algebra, and 44 percent failed. That was nearly twice the failure rate as in English. Seventeen percent finished with Ds. Of those who repeated the class, nearly three-quarters failed again. Algebra triggers more dropouts than any other single subject.[23] Similarly, in Grand Rapids Public Schools, nearly 40 percent of Algebra 1 students received a failing grade and 22 percent of Algebra II students did not pass.[24] Responding to Chicago's increasing algebra failure rate, a school official said, "It's not surprising that you're going to see an increase in [failure] rates [in algebra] if you raise the instructional requirements and you don't raise supports."[25]

Many school districts are responding to the algebra crisis by creating two- or three-year algebra programs or having students take double periods of algebra, but the student success rate is not changing significantly.[26] It is becoming apparent that the amount of time spent studying algebra is not the issue. There is some evidence that new curricula from projects like the Connected Mathematics Program, College Preparatory Mathematics Program, and West Ed. are improving students' success with algebra,[27] but many students still struggle to understand and there is some controversy over reform math's effects.[28]

Research has long documented students' mental hurdles in making the transition from arithmetic to algebra.[29] Over 900 articles spanning the last three decades examine why students struggle in algebra. Teachers need a resource that will easily allow them to understand those common errors and conceptual misunderstandings in algebra, and the paths that students take in coming to understand algebra. Publishing companies are beginning to incorporate some research on students' thinking into teacher editions of textbooks. These often take the form of one or two sentence sidebars that point out a typical struggle that students might have. Although these are helpful, they are not comprehensive.

This book synthesizes those 900 articles into what is most valuable for teachers to understand about how students think when doing algebra. While we break down students' algebraic thinking into the individual elements explored by researchers, it is important to see each exploration as a window into a larger system. The creation of categories and chapters in this book is a matter of convenience in the two-dimensional realm of print that does not do justice to the three- or four-dimensional reality of a student's brain. Ideas about variables are woven throughout equations, functions throughout graphing. They are not distinct. However, by looking at the vast complexities of the development of algebraic thinking through a particular lens, teachers may be able to develop greater insight into students' mental systems overall.

Why is Algebra so Difficult for Students to Understand?

In order to explain students' struggles in algebra, two different arguments are often used.[30] Some people are convinced that inadequate teaching is to blame. "If only we could train teachers well, most students would understand."[31] Mathematician Herscovics argues that this viewpoint presumes all problems that students encounter in algebra can be resolved and the solution is merely improved communication, which is "optimistic and simplistic."[32] A second argument is that society never intended for all students in the general population to learn algebra. However, the NCTM *Principles and Standards for School Mathematics* "challenges the assumption that mathematics is only for the select few with a pervasive argument that everyone needs to understand mathematics."[33] The Mathematical Association of America, in its report on algebra, adds "there is certainly a strong argument to be made that the study of algebra not only 'trains the mind' but is necessary for everyone desirous of participating in our democracy ... we can and should make algebra part of the curriculum for all students."[34]

Perhaps the most plausible perspective on many students' failure in algebra is one that does not blame teachers' skills or students' intellectual potential. Instead, Herscovics suggests we consider the notion of *cognitive obstacles* in which students encounter ideas in algebra that require them to restructure the way that they think about a concept in order to move forward in their thinking. Cognitive obstacles are best understood in terms of Piagetian theory.

Piaget believed that people develop mental structures—or ways of thinking about something—called *schema*. There are two ways that people acquire knowledge: assimilation and accommodation. First, a child may have an idea about something based on experience, such as "I have a dog" (Figure 1.1). When she is faced with new information that is similar to what she knows, she acquires new knowledge by *assimilation*: fitting the new information into a schema she has already developed. "That Great Dane is very different from my dog, but I can put it into my thinking about dogs. There are different kinds of dogs" (Figure 1.2). When she encounters information that does not fit into her existing schema she experiences disequilibrium and her brain strives to make sense and come back into balance. She must *accommodate* her brain structure by creating a new schema. She restructures old ways of thinking in order to integrate the new information. "That pony is not much bigger than the Great Dane, but it is very different, so I need a new category called *horses* and a bigger category to hold them all called *animals*" (Figure 1.3).

Teachers need to decide whether their instructional goal for students is assimilation ("This is just like...") or accommodation ("I need to think about this in a new way"). Accommodation is the restructuring that is often necessary when students experience cognitive obstacles in algebra. When they face ideas that, in order to be fully understood, must be thought about in a radically different way than they have thought about them before, they cannot simply add them to their memories. Learning, therefore, "sometimes requires the significant reorganiza-

FIGURE 1.1   Initial conception

FIGURE 1.2. Assimilation

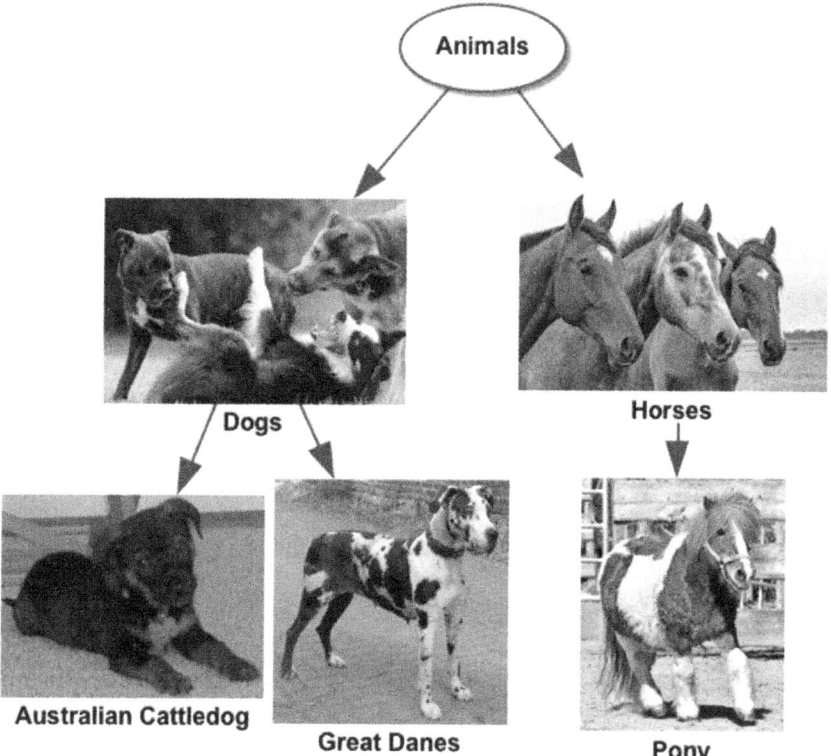

FIGURE 1.3. Accommodation

tion of existing knowledge structures and not just their enrichment."[35] As students move from the concrete world of arithmetic to the abstract world of algebra they often encounter cognitive obstacles that compel them not to simply assimilate but to accommodate the new ideas by restructuring how they think about mathematics. It is the teacher's challenge to figure out ways to facilitate that process.

Another reason algebra may be difficult to understand is based on our struggle to define algebra. Some argue that the traditional emphasis on understanding the structure and manipulation of expressions, and then applying those skills to word problems, is the most useful. From this perspective, algebraic thinking is the ability to transform and represent or recognize forms.[36] This equation-based approach emphasizes the use of symbols to model and explore relationships. Others contend that algebra classes should emphasize the concepts of functions and families of functions. A function-based approach examines relationships between varying quantities, representing those relationships, and using those representations to analyze, generalize, and make predictions. Students should engage with real-world situations and understand relationships that can be described with symbols and models.[37] This latter perspective significantly influenced the *Principles and Standards for School Mathematics*[38] and is somewhat driven by advances in technology, such as graphing calculators and computers that can do algebraic manipulation. There are also those who try to find a middle ground between the two perspectives—that techniques and conceptual understanding in algebra should be complementary components.[39] There is evidence that students' conceptions relate to the type of instruction and curriculum chosen.[40]

We can certainly view algebra from all of these perspectives, but math educators have "found it difficult to connect function- and equation-based views in our instruction ... when both views exist simultaneously in an algebra course, they are not well-integrated."[41] Middle school curricula often approach algebra through a function-based perspective while high schools typically focus on the equation-based approach. Accordingly, students struggle to reconcile these two perspectives themselves. Teachers can help students make sense of algebra by keeping both of these perspectives in mind and integrating multiple representations and approaches where possible.

Regardless of approach, as students move from the concrete world of arithmetic to the abstract world of algebra they must use prior knowledge as a foundation for new ways of thinking. Many arithmetical rules are extended into algebra; for instance, commutativity with numbers is the same with variables. However, students sometimes experience mathematics in elementary school that conflicts with ways of thinking in algebra, such as the functioning of the equal sign as a left to right process rather than a balance of equal sides (see Chapter 3: Algebraic Relations for a more detailed discussion). Educators are currently identifying ways to transform elementary-school mathematics teaching to better establish the foundations of algebraic thinking for study in later grades.[42]

While there is a perception that algebra is difficult for students to understand, we know from research that children at very young ages can use algebraic thinking. Students as young as six years of age can write expressions with letters that represent relationships between attributes.[43] Eight-year-old students can make mathematical statements using letters to represent unknown values[44] and describe algebraic relationships.[45] Nine-year-old students can work with complex algebraic notation with confidence.[46] These research studies certainly raise the question of "whether it is a matter of what barriers are sometimes placed in the way of their learning due to pedagogic approaches taken rather than algebra being an inherently difficult topic to learn."[47] Students are capable of learning algebra. While this book identifies the struggles that students have traditionally had when learning algebra, is important to not focus strictly on what students *can't* do, but also what they *can*.

Algebra is often the course in which students lose their confidence in learning mathematics. Research indicates that this may be partially attributed to a matter of mindsets. Carol Dweck and her colleagues found that students hold different theories about their intelligence. Some believed that their intelligence could not be changed (a *fixed mindset*) while other students believed that their intelligence could be influenced by their effort (a *growth mindset*).[48] They asked students in 5th grade to work on a math quiz. When finished, they either praised the student for how smart they were, or commended them for their effort. Then they asked the student if they wanted to try a more challenging problem. Those that were praised for their effort were willing to try the challenging problem, while those that were praised for how smart they were wanted easier problems.[49] Students with growth mindsets were willing to take risks, because they saw that mistakes helped them learn.

Researchers have found that students with growth mindsets have more brain activity when they make mistakes than students with fixed mindsets.[50] When students make a mistake, their brains are in disequilibrium, having cognitive conflict between what they thought and the correct answer. Students with growth mindsets see mistakes as evidence that they need to work on something and struggle through. They were more likely to go back and correct their errors. When students with fixed mindsets encounter challenges, they see mistakes as evidence that they are not smart, and move on. In another study, researchers found that 7th grade students who had different mindsets had significantly different mathematical success later in school (Figure 1.4).[51]

MISCONCEPTIONS VERSUS ERRORS

Piaget's career as a psychologist began with his work on tests measuring intelligence. He was not particularly interested in the "right or wrong" style of questioning and resulting intelligence quotient, but was more intrigued with the nature of children's incorrect answers. Through interviews pursuing children's reasoning, he saw patterns in their thinking that led them to particular incorrect answers.

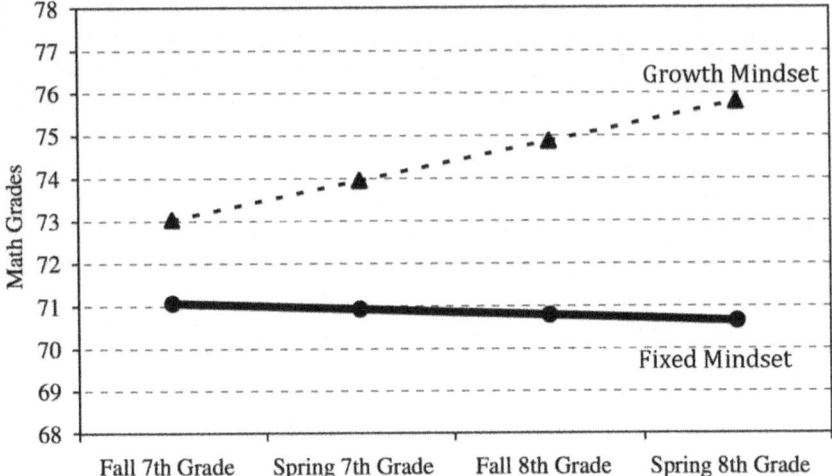

FIGURE 1.4 Math Grades and Mindsets. Adapted from Blackwell, Trzesniewski, & Dweck, 2007, p. 51.

Piaget found that children often had logical reasons for their wrong answers: they based their reasons on some truth they extrapolated incorrectly. They also had tendencies, or similar ways of thinking about a concept. Math education researchers have since found that these tendencies are often influenced by the structure of problems; more specifically, as students try to understand algebra, the nature of the mathematics may lead students to particular incorrect—as well as correct—ways of thinking. For example, students in arithmetic notice that, for the numbers they are working with, multiplication makes quantities bigger faster than addition. While we know that this is not true for all numbers, students can think multiplication makes numbers bigger faster always. This can impact their understanding of an algebraic problem such as "Which is bigger, $2n$ or $n+2$?"

Balachef's quote at the beginning sums up the essence of this book: errors made by a student are often the tip of the iceberg of his or her conceptions of the mathematics beneath. Errors can result from simple carelessness, such as 5 + 3 = 9, or from forgotten rules. The student can quickly identify the mistake in such cases. Researchers have also found that "some kinds of errors are widespread among students of different ages, independent of the course of their previous learning of algebra" and that "a significant number of students did not use the formal methods taught in schools, preferring some intuitive strategies."[52] Similar to Piaget's findings, some students make the same type of errors regardless of their experience with algebra, indicating common struggles that have more complex sources than simple calculation or memory errors.

Errors can be caused by misconceptions that reside deeper in a student's consciousness—a multifaceted structure that has naïve or faulty assumptions inconsistent with accepted mathematical practices. These conceptions can be persistent and resistant to instruction.[53] Such errors are often a result of students' thoughtful but incorrect overgeneralization from correct rules.[54] For instance, an error such as $\frac{a+x}{b+x} = \frac{a}{b}$ can be incorrectly generalized from the correct rule $\frac{ax}{bx} = \frac{a}{b}$. Whether the error is part of a developmental process—that is, self-correcting as the student progresses through school—or a more pervasive type that is only corrected through a student's reflection in concert with teacher intervention, the importance is the immediate impact on the student's understanding of current content.[55] If teachers do not address a student's underlying ways of thinking and align pedagogy with a student's needs, the symptoms will often continue to pop up as the student continues to apply the incorrect rule or variations of that rule in new situations. "One of the greatest talents of teachers is their ability to synthesize an accurate 'picture,' or model, of a student's misconceptions from the meager evidence inherent in his errors."[56]

There is considerable discussion in math and science communities about the idea of misconceptions, or "conceptions that produce systematic errors."[57] When students' conceptions are seen as being in conflict with accepted meanings, the literature uses many different terms to describe what is going on in their minds, including *misconceptions*,[58] *systematic errors*,[59] *preconceptions*,[60] and *alternative conceptions*.[61] Generally, researchers and educators agree that seeing misconceptions as mistakes that impede learning is not a productive orientation for teachers. We want to develop students' growth mindsets, particularly in mathematics where fixed mindsets abound. Errors are opportunities for learning when students have time "to recognize and successfully reflect on their errors" and teachers are "armed with strategies for supporting students in recognizing and reconciling their errors."[62] However, recognizing that patterns of errors may arise from a complex web of related ideas—rather than being one-off mistakes—can be a useful orientation for a teacher's instructional decision-making. Concepts are not distinct, separable, and independent. A student's thinking about algebra is not a battle between abstract and concrete knowledge, nor is it general versus specific. It is not a matter of the correct ideas of experts' "formal" thinking versus the flawed "intuitive" thinking of a student. There is quite a bit of gray area between these dichotomies. Rather, researchers encourage us to think in terms of what thinking is useful and productive in a particular mathematical context.

Getting back to our multiplication example, students experience that multiplication makes numbers bigger.[63] That is true when working with positive integers. However, applying that thinking to fractions or negative numbers results in errors and confusion. Teachers do not want to eliminate the idea that multiplication makes numbers bigger, but to make that thinking more sophisticated and contextual: "It makes numbers bigger sometimes."

MISCONCEPTIONS AND RESEARCH

Incorrect answers are rarely due to guessing, low intelligence, or low mathematical aptitude. They result from systematic strategies or rules which usually have sensible origins and are based on beliefs or misconceptions—they are usually distortions or misinterpretations of sound procedures. (Perso, 1992)

Researchers show us that misconceptions may come from a number of sources, and may be the result of "reasonable, though unsuccessful, attempts to adapt previously acquired knowledge to a new situation."[64] It is natural for students to make sense of new information based on prior experiences. Misconceptions may also develop because of poorly designed and misleading teaching materials.[65] Some of the challenges students face and errors students make do not reflect their cognitive capabilities, but rather the nature of their learning experiences.[66]

When students try to use common sense or generalize from arithmetic rules they know, they can sometimes get confused since algebra does not always follow those norms. Many times, students understand the ideas of algebra, but struggle with the conventions and notation.[67] Mathematical conventions, such as not using 1 as a coefficient for x, can seem arbitrary at times, leading students to believe that algebra is magic and they sometimes just need to memorize things without understanding.[68] Teachers perhaps encourage this sense in students with mantras such as "change sides, change signs," "collect like terms together," "add the same thing to both sides," and "calculate what is in parentheses first." These reminders may result in better short-term performance, but not necessarily better understanding. "To cover their lack of understanding, it appears that students resort to memorizing rules and procedures and they eventually come to believe that this activity represents the essence of algebra."[69] Studies from the National Assessment of Educational Progress (NAEP) support the idea that students consider learning mathematics as mostly memorizing.[70]

At the heart of misconceptions are students' efforts to solve problems. Researchers suggest that there are two components to the process: students' collections of foundational rules in their minds, and a set of extrapolation techniques, or ways of bridging the gap between the rules they know and an unfamiliar problem. Mistakes, therefore, could result from a faulty set of rules, inappropriate use of a rule in a new situation, an incorrect adaptation of a known rule to the new situation, or errors in the execution of a procedure. During instruction, students sometimes create their own rules to make sense of the information given to them,rather than use the formal mathematical methods they were taught in the classroom.

An important distinction for teachers to understand is the difference between students' *rules* and *quasi-rules* that may result in the same error on a test.[71] Rules in students' minds are procedures that they use consistently, so they believe the rule is accurate. For instance, students may believe that the answer to "$13 + 2 = \_\_ + 7$" is 22 because of the "rule" that you add numbers from left to right. Using these rules may help students be successful on one type of problem, which is typically the more simple form. However, when the problem has a variation or becomes more complex, their method often no longer works because they lack the underlying understanding necessary for success. Quasi-rules, on the other hand, are not consistently applied, are often formulated in the moment, and can change haphazardly. For instance, a student may believe that the statement "one less than y" can be interpreted as x rather than "$y - 1$" because x is one letter before y in the alphabet.[72] It is an arbitrary rule applied in a specific context. Misconceptions are generally faulty sets of rules or inappropriate use of rules that guide students' choices. Research indicates that students who are following rules can learn more accurate rules and be successful. Students who are following quasi-rules often have a much more difficult time understanding the mathematics behind the procedures and need more significant intervention by teachers.

Misconceptions that exist in students' minds prior to instruction can be different from students' misinterpretations during the teaching and learning of algebra. For instance, as students enter algebra classrooms, researchers have shown that students bring with them a "natural" tendency to believe that a letter has only one value.[73] The contexts students encounter in algebra require them to restructure and accommodate their conceptions of variables to include the possibility of letters representing multiple quantities or ranges of numbers.[74] Based on their earlier experiences, students may interpret letters in a variety of unexpected ways that can negatively influence their ability to successfully work with problems in algebra. As in Calvin's case at the beginning of this chapter, what may be emphasized in arithmetic is the final numerical answer and not the process, so errors in thinking can go unnoticed. Manipulation of algebraic symbols, however, may require students to make use of processes that they have avoided when dealing with simple arithmetic problems. Students' encounters with algebra, therefore, can bring to light the misconceptions and confusions that they may already have experienced in arithmetic, but have gone unnoticed.[75]

Similar misconceptions appear in spite of differences in age and experience with algebra and can persist in spite of direct instruction. For instance, misconceptions about the nature of variables have been shown to continue into college, in spite of students taking higher-level algebra courses.[76] Themes to these misconceptions can be traced to students' ideas about the focus of algebraic activity and the nature of "answers," the use of notation and convention in algebra, the meaning of letters and variables, the kinds of relationships and methods used in arithmetic, and other notions.[77]

Misconceptions that students have in algebra often relate to its abstract nature. Students have a certain degree of reliance on "reality." Their experience in mathematics prior to algebra often consists of considering what is empirically verifiable, what they can manipulate through objects, or what they can easily connect to experience. As students move from thinking about small numbers to large numbers outside their ability to verify, and then to generalizations, their challenge is to develop a grasp of the abstract structure of things. In dealing with this new, abstract world, when understanding is difficult for students to come by, misconceptions often fill the gaps.

> Of perhaps even more interest than the kind of question that children solve correctly is the nature of the errors that they make, especially when the same error is made by large numbers of children. A study of such errors is important because of the information it provides concerning the ways in which the child views the problem and the procedures that are used in attempting to solve the problem. This information is of interest not only because it might suggest ways of helping children to avoid these errors, but also because it might explain children's apparent lack of progress in attaining higher levels of understanding in Algebra.[78]

Through testing with thousands of students, hundreds of hours of interviews, and observations of classroom interaction, researchers have identified typical cognitive obstacles, how students' misconceptions develop, and particular ways of thinking about ideas in algebra that have tremendous influence on students' potential learning. Research appears in this book for teachers to use in making instructional decisions. This book is meant to help teachers identify potential ways students are thinking about and are confused about algebra. However, the research presented in this book is not an effort to capture the exact thinking of a student in your classroom. With misconceptions research there is a danger that users of that research may misinterpret, over credit, or under credit a student's thinking about an algebraic idea. Nor is this book meant for teachers to memorize an inventory of misconceptions, as there are many more ways of thinking about algebra than can be contained in one book. Rather, by understanding the range of ways students may be thinking about an algebraic problem or idea, perhaps you may have more tools with which to understand your students' thinking, as well as a basis for further questioning. As a student struggles to find success in algebra, this book might help to provide some ideas about why.

ALGEBRAIC THINKING

> Algebraic reasoning in its many forms, and the use of algebraic representations such as graphs, tables, spreadsheets and traditional formulas, are among the most powerful intellectual tools that our civilization has developed. (Yet,) the traditional image of algebra, based in more than a century of school algebra, is one of simplifying algebraic expressions, solving equations, learning the rules for manipulating symbols—the algebra that almost everyone, it seems, loves to hate...

> ...School algebra has traditionally been taught and learned as a set of procedures disconnected both from other mathematical knowledge and from students' real worlds. (Students) memorize procedures that they know only as operations on strings of symbols, solve artificial problems that bear no meaning to their lives, and are graded not on understanding of the mathematical concepts and reasoning involved, but on

their ability to produce the right symbol string. Without some form of symbolic algebra, there could be no higher mathematics and no quantitative science; hence no technology and modern life as we know them. Our challenge then is to find ways to make the power of algebra (indeed, all mathematics) available to all students…[79]

One researcher suggests that an assumption in the traditional teaching of algebra is "if the students spend enough time practicing dull, meaningless, incomprehensible little rituals, sooner or later, something WONDERFUL will happen."[80] If a teacher's approach is for students to learn/memorize a sequence of mechanical activities with corresponding rules, then it is unwise to expect anything but poor understanding.[81] In contrast, the expectation of the Common Core State Standards, particularly in the Mathematical Practices, is that students will develop conceptual understanding that will serve them well when they encounter real-world problems for which mathematics can be a useful tool. Over the past twenty years, there has been quite a bit of argument in the mathematics community about which is the best approach to develop algebraic thinking; that won't be resolved here. However, to understand algebraic thinking, one has to consider the role of arithmetic in its development, the purpose of algebra, how students find meaning in algebra, and finally, how *number sense* transitions into s*ymbol sense* and then to s*tructure sense*.

There can be an assumption that algebra is simply generalized arithmetic; that one flows naturally into the other. Research shows that this is not always the case in students' minds.[82] Students can see arithmetic and algebra as two different worlds in which the rules are not the same. When engaged with a test problem, a student in one study asked the researchers whether they wanted her to answer "in algebra," leading the researchers to believe that students in their study knew the meaning of "$5b$" within the context of algebra, but unless asked to respond "in algebra," students answered based on an arithmetic context, that is, finding a numeric answer without a variable.[83]

Writing and manipulating expressions in algebra and doing so in arithmetic can be significantly different, in spite of the use of common operational signs. Through algebraic manipulation, an expression can take many equivalent forms, but not result in a computed value. In arithmetic, on the other hand, students can directly evaluate an expression with the same form and structure, substituting numbers for letters. For example, substituting 5 for x in "$x + 3 = 10$" can disprove that 5 is a solution. However, no numerical substitutions can prove that "$a^2 + b^2 = c^2$" for right triangles. It must be proved by reasoning. There is much room for confusion in students' minds as they try to reconcile algebra and arithmetic. In order to facilitate the transition from arithmetic to algebra, there has been a concerted effort in the past decade to develop algebraic thinking throughout K–12 classrooms, rather than rely on a single course in algebra around the eighth grade.[84] This has been particularly true in elementary schools where teachers are developing the connection between pattern recognition and generalization, the idea of symbols representing numbers, and the relationship between arithmetic and algebra.[85]

While algebraic thinking develops quite differently—depending on the approach you use and at what ages you introduce algebraic concepts—there are some commonalities of purpose. Mathematical power and flexibility in algebra lie in four themes or activities:

- Generalizing,
- Solving equations and making sense of their solutions,
- Exercising algebraic rules, and
- Building models and learning from them.[86]

Generalizing involves the development of properties or representation of operations based on examination of quantitative problem situations, geometric patterns, or numerical sequences or relationships. It involves determining mathematical symbols that will represent those patterns and ideas in expressions or equations. In solving equations, variables facilitate the transformation of relationships into algebra, which can then be used to find an answer to a problem. Another purpose is understanding the art of using algebraic rules to discover relationships between expressions and to develop the idea of proof or justification. Lastly, algebra is used to better understand real-life situations or to make predictions through the use of symbolic and graphic models, putting words and images into mathematical equations and relationships represented symbolically.

Related to the question of purpose is the search for meaning in algebra. Students trying to understand algebra have four possible sources from which to find meaning:

- The algebraic structure of letters and symbols;
- Other math representations (such as graphs);
- The problem context; and
- The context outside the problem (such as conversation, gestures, or a student's life experience).[87]

The first source is the meaning students find from using letters to represent unknowns, seeing the ideas that are embedded within the symbols,[88] and knowing how the manipulation of algebraic variables and expressions help the student solve a problem. A key to finding this meaning is comprehending the connection between the letters or symbols representing an idea and the numbers that are the foundation of that relationship. Difficulties emerge because of the challenge that arises in relating algebraic symbols to natural language. A student may be able to explain relationships between elements of a situation accurately through language, but be unable to express those same relationships in algebraic symbols.[89]

The second source of meaning is understanding the transformation and relationships between and within three representations in algebra: tables, graphs, and equations.[90] Any of these systems can refer to a mathematical quantity. Part of the meaning derived from these representations is how well they describe the situation: each table, graph, or equation has certain features that highlight particular aspects of the real-life situation. Within each of those systems are also variations that influence the meaning. A line graph expresses different information than a pie chart or box-and-whisker diagram. Varying forms of equations or structures of tables communicate different ideas about the situation. A core of algebraic success is understanding the meaning in how one representation can be interchanged with another and how each *highlights* different aspects of a mathematical situation.

The third source of meaning is the connection between the algebraic representation and its context. Why might an algebraic approach to the situation be more appropriate than another approach? What features of the context are best addressed through representation and manipulation with unknowns? What does an algebraic representation communicate about the situation? Part of acquiring meaning is understanding how the algebraic representation of a context can have a purpose. The classic book *How to Lie with Statistics* demonstrates how the manipulation of axes can convey a significant trend or no trend at all, based on the same data.[91] Decisions made about how to represent the context can reflect the bias of the author or even a political agenda. Understanding the meaning that underlies the context and decisions about its representation lead to fuller comprehension of the mathematics.

The fourth source of meaning is what the students of algebra have access to that surrounds the math and problem situation. Students gain insights from or get confused by incongruities among their lived experiences.[92] For instance, it is very difficult for students to separate the image of a cyclist going up a hill from a graph of the speed versus position along the road (see Figure 1.5).[93] Students often draw a graph mimicking the image of the cyclist on the hill, rather than the relationship between speed and position.

As students learn algebra, they use their foundational understanding of number sense and their arithmetical roots to gradually develop *Symbol Sense,* "an appreciation for the power of symbolic thinking, an understanding of when and why to apply it, and a feel for mathematical structure. Symbol sense is a level of mathematical literacy beyond number sense, which it subsumes."[94] Students with symbol sense understand how symbols can play a dual role in algebra. They can refer to specific objects or situations, or "they can function without continuous reference to the mathematical objects they name."[95] For instance, x can be something that refers to the number 4 in the equation $x + 1 = 5$, or it can simply be something you can multiply times $2y$ to get $2xy$. When there is a close relationship between how the symbols are used and what they refer to, students can find that "the notation is easy to understand and to work with."[96] However, students often

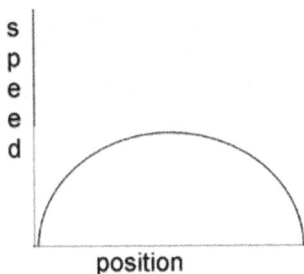

FIGURE 1.5. Speed vs. Position: an incorrect relationship

struggle to see the connections and relationships between the symbols and the situation. Developing an understanding of how symbols can represent real-world relationships—even when the symbols are not closely aligned with the situation or with language—is having symbol sense. Students' mathematical power in algebra depends on how well they ultimately develop that sense.

Structure Sense is an extension of Symbol Sense.[97] Structure Sense includes students' ability to:

> Recognize a familiar structure in its simplest form (as in seeing $81 - x^2$ as the difference of squares and being able to factor it accordingly).
>
> Deal with a compound term as a single entity and, through an appropriate substitution, recognize a familiar structure in a more complex form (as in addressing $(x - 3)^4 - (x + 3)^4$ by seeing $(x - 3)^2$ and $(x + 3)^2$ as single entities and factoring as the difference of squares).
>
> Choose appropriate manipulations to make best use of a structure (as in seeing the possibility of the difference of squares in $24x^6y^4 - 150z^8$, so choosing to extract the common factor of 6 in order to get $4x^6y^4 - 25z^8$ and deal with $2x^3y^2$ and $5z^4$ as single entities, factoring as the difference of two squares).[98]

Underlying Structure Sense is the understanding of the substitution principle, which states that "if a variable or parameter is replaced by a compound term (product or sum), or if a compound term is replaced by a parameter, the structure remains the same."[99] One study found that, of 92 students in 11th grade, only 13% solved problems using structure sense rather than rote procedures.[100] This sense is an aspect of algebraic thinking that needs to be conscientiously developed by teachers. Combined, *Number Sense*, *Symbol Sense*, and *Structure Sense* represent the essence of algebraic understanding, and therefore are appropriate targets for teachers' instruction as the foundation of future mathematical success.

One last way to consider what we mean by the term *algebraic thinking* is the use of algebra as a symbol system and a way of thinking.[101] First, the notations and conventions used in algebra are a significant part of what students must learn to engage in algebraic thinking. Second, algebra is also about a habit of mind: a way of mathematically thinking when you encounter a problem situation.[102] This might include "recognizing and analyzing patterns, investigating and representing relationships, generalizing beyond specifics of an example, analyzing how processes or relationships change, or seeking arguments for how and why rules and procedures work."[103] More specifically, these habits of mind could be described as interactions with functions in three ways:[104]

- *Doing—undoing:* Understanding reversibility so that a student can use a process to achieve a goal as well as work backward from the answer to the starting point. For example, being able to solve the equation $9x^2 - 16 = 0$ and find the solutions 4/3 and -4/3.
- *Building rules to represent functions:* Recognizing patterns and organizing data to represent situations with a rule that represents the relationship of the input to the output. For example, "Take a number and multiply by 4 then subtract 3."
- *Abstracting from computation:* Considering a system, rather than a specific example, by thinking about computations independent of particular numbers being used. For example, given the problem "Compute $1 + 2 + 3 \ldots + 100$," students can regroup numbers into pairs that sum to 101.

As students begin considering algebraic expressions and equations, they must rethink symbols and operations. No longer can they rely on a process perspective or the feeling of a flow of mathematics from left to right, like $5 + 3 = 8$. Instead, students must begin to view symbols with operations as potentially whole objects to be operated on. For instance, the expression $\dfrac{x^2 + 5x + 6}{x + 2}$ can be factored and simplified without even considering an equal sign or the value(s) that x might represent. Students must learn that they can operate on algebraic expressions as well as numbers. Operations on those expressions extend from simply adding, subtracting, multiplying, and dividing to simplifying, factoring, rationalizing the denominator, solving, and so on.[105] The idea of letters, numbers, and operations together as entities rather than left to right calculations is one of the first cognitive obstacles to resolve in the transition from arithmetic to algebraic thinking and, if not comprehensively understood, may lead to mathematical conflicts along the way.

MISCONCEPTIONS AND CONCEPTUAL CHANGE

> Persistent, well retained bodies of knowledge and skill are those which are richly inter-connected and that fresh ideas are often rejected until they become so strong that they force a reorganization of the existing material into a new system, holding together the new idea and the transformed old ones.[106]

"Conviction that a single, very clear general exposition will work is an illusion."[107] Students who passively receive information experience the least effective method of addressing misconceptions. Lecture, or telling a student a new idea, typically does not help the student challenge the existing structure or schema within his or her mind; conceptual change requires students to realize that their current mental structure about a concept is inadequate, and therefore be compelled to restructure their way of thinking. Researchers have found that when students learn to recognize an error, examine the error (rather than simply abandoning strategies that result in errors), consider potential sources for their error, and develop a sense of when they have successfully reconciled an error, they ultimately become more effective problem solvers.[108] Gaining insight into student errors can guide teachers in helping students develop some independence in successfully dealing with their errors and finding meaning in their work.[109]

Sometimes, "we make it too easy for children to understand."[110] Instead, it is insightful sometimes for students to explore invalid logic on their journey to understanding.[111] "The freedom to err is at the heart of developing mathematical knowledge,"[112] so teachers are often faced with the instructional decision of whether to allow students the flexibility to explore their thinking, or to intervene and act as a catalyst for conceptual change. Teachers are encouraged to view misconceptions as part of a student's process of interpreting "phenomena, situations, events, including classroom instruction, through the perspective of the learner's existing knowledge."[113]

Students usually have logic for the mathematical steps that they take. When students make errors, it is important for teachers to understand and honor that logic—and students' efforts—while helping them to understand the faults in that logic. Instead of focusing only on the isolated, mistaken qualities of a student's thinking in a particular moment, teachers should seek understanding of a student's continuous system of thinking: "Why do you think that?" While a misconception may not be a coherently reasoned alternative framework, many threads may weave together to form the way of thinking. Students' difficulty in algebra often arises from their "intuitive assumptions and sensible, pragmatic reasoning about an unfamiliar notation system."[114] They develop understanding of algebra by making use of their experience, comparing results, generalizing and developing rules in their minds, and revising or eliminating previous rules.[115] They can use procedures that have "never been shown during any lesson and were absent in their textbooks."[116] When teachers comprehend a student's framework, they can more effectively support and challenge students' developing understanding.

Upon interviewing hundreds of students, researchers found that what they thought were careless mistakes were really students making sense of algebra in their own ways.[117] Beliefs that can limit the way they understand scientific explanations and physical phenomena may guide that framework.[118] For instance, one misconception is students' tendency to apply the properties of natural numbers to algebraic notation and rules.[119] Knowing how students are constructing meaning can help teachers make appropriate and effective instructional decisions to facilitate conceptual change.

It is not the purpose of this book to advocate for a particular method of addressing misconceptions. Teaching requires "an adequate theory of knowledge, reasoning, and learning [that] must include a range of cognitive and affective structures and processes."[120] However, there are some general principles for achieving cognitive change:

- Given a space of examples, do not choose too narrowly;
- Where possible, anticipate later expansions of meaning;
- Identify the points at which conceptual change is necessary and look for bridging devices;
- Discuss with the students what is going on; and
- Take a long-term perspective.[121]

Bridging devices can include the creation of cognitive conflict or dissonance in students' minds.[122] Based on theories proposed by Piaget, cognitive conflict occurs when students are faced with two different ideas that by their nature are impossible to hold together. Students must resolve the conflict by restructuring their mental framework in order to move forward in their thinking. "The teacher as organizer becomes indispensable in order to create the situations, construct the initial devices which present useful problems to the child. ... [H]e [the teacher] is needed to provide counter-examples, that compel reflection and reconsideration of over-hasty solutions."[123]

Mathematician Bell describes *Diagnostic Teaching* as a method of identifying misconceptions and facilitating development of students' correct thinking.[124] Using research and assessment of students' understanding, teachers present a conflict that students must resolve, or create an environment in which cognitive conflict naturally arises. However, "the construction of knowledge is not only a question of cognitive conflict of an isolated individual, but a collective and social activity joining a culture."[125] Students actively discuss their opinions about the conflict with other students to ensure that they bring all conceptions to the table and build consensus on the most productive, effective, and accurate route to the answer—thereby constructing a socially shared language of algebra. Teaching strategies encountered in the research often take this type of approach to address misconceptions. However, there certainly are other instructional strategies that may prove effective in helping to move students' thinking toward being more mathematically accurate.

STRUCTURE OF THE BOOK

Over the past four decades, a solid research base has been established and is growing, regarding students' algebraic thinking and misconceptions. However, when planning lessons, teachers typically do not have the time in their busy days to find, and then weed through, research articles for pertinent information regarding their students' algebraic thinking. Yet the insights into how students think when engaged with algebra has tremendous potential value to novice and veteran mathematics teachers alike. This book is intended to be a bridge between this research and teachers' practice in the classroom. The goal for the book is to "broaden and support teachers' awareness, judgment, and inquiry" regarding their students' algebraic thinking.[126] Findings from research have been selected and organized into a form that is readily usable as a resource for teachers of students preparing for—and in—classes with algebraic ideas. Based on research into students' thinking over the past three decades, this book provides an easy-to-use library of the development of algebraic ideas, misconceptions, and ways of thinking that are challenging for students to understand.

Four questions form the foundation of the core sections of the book, and guide organization of each section:

1. What is the symbolic representation of the misconception or way of thinking about the algebraic idea? (What does it look like?)
2. How do students think about the algebraic idea or misconception? (What does it sound like?)
3. What are the underlying mathematical issues involved? (Why do they do what they do? When are misconceptions likely to surface?)
4. How might you address this misconception or way of thinking about the algebraic idea? (What strategies could a teacher use to help students understand the concept?)

Answers for each question are organized under five algebraic categories: Variables & Expressions, Algebraic Relations (e.g., equations), Analysis of Change (e.g., graphing), Patterns & Functions, and Modeling & Word Problems. Each category has multiple sections that summarize research on students' algebraic thinking with particular types of algebraic problems. The range of problems presented is dictated by what has been researched. Each section focuses on a particular theme about students' algebraic thinking that has been identified by researchers. Sections include problems that are typical of the algebra curriculum. Each problem raises various mathematical issues based on the unique ways students think and have misconceptions about that type of problem—thereby addressing some core issues in the development of algebraic thinking. Not all research includes student interviews to identify thinking or suggestions of how to address a student's thinking; therefore, when possible, a reasonable possible conversation is given. Interviews directly from research are footnoted.

Each of the categories begins with an introductory section in which some overall ideas are laid out to help you organize the different ways of thinking about each of the algebraic concepts with which students struggle. Words in bold italics indicate a section. The structure of the book unintentionally implies the idea that these categories are distinct and that each of the problems examined can be isolated from the greater realm of algebraic thinking; however, a student's thinking is a complex system of ideas that is continuous and not discrete. In fact, the ideas of variables, equations, functions, graphing, and word problems are thoroughly woven together; the book's organization, therefore, revolves around what struck the authors as the most central theme of a particular idea, with the understanding that other themes are certain to be involved as well. A teacher who honors a student's process of developing thinking and retains that sense of an interrelated web of ideas will help students to feel a sense of growing success, rather than an eternal battle against errors and flawed thinking.

REFERENCES

Ainley, J. (1999). *Doing algebra type stuff: Emergent algebra in the primary school.* Paper presented at the 23rd Conference of the International Group for the Psychology of Mathematics Education, Haifa, Israel.

ENDNOTES

1. Wallach & Even, 2005
2. Sherin, Russ, Sherin, & Colestock, 2008, and Ho & Tan, 2013, among many others
3. Sherin, Russ, Sherin, & Colestock, 2008, p. 28
4. Stockero & Van Zoest, 2012
5. Davies & Walker, 2005, p. 275
6. Schoenfeld, 2008, p. 57
7. Leatham, Peterson, Stockero, & Van Zoest, 2015
8. Shulman, 1986
9. Leatham, Stockero, Peterson, Van Zoest, 2011, p. 839
10. Peterson & Leatham, 2009; Stockero & Van Zoest, 2012
11. Mason, 1998, p. 247
12. Leatham, Stockero, Peterson, Van Zoest, 2011, pg. 840, 842
13. Smith & Stein, 2011
14. Leatham, Stockero, Peterson, & Van Zoest 2011; Peterson & Leatham, 2009
15. Tirosh, Even, & Robinson, 1998, p. 51, based on Shulman, 1986
16. Carpenter, Fennema, Loef Franke, Levi, & Empson, 2000
17. Ball, Bass, Sleep, & Thames, 2005, p. 3
18. Fennema, et al., 1996
19. NCTM, 2014
20. Liston & Donoghue, 2010
21. Katz, 2006, p. 6
22. Ladson-Billings, 1998; Moses & Cobb, 2001
23. Helfand, 2006, p. 1
24. Murray, 2010
25. Viadero, 2009, ¶8
26. Nomi & Allensworth, 2009
27. e.g. Moseley & Brenner, 2009; Riordan & Noyce, 2001
28. e.g. Milgram, 1999
29. e.g. Booth, 1984; MacGregor & Stacey, 1997; and Moseley & Brenner, 2009
30. Herscovics, 1989
31. Herscovics, 1989, p. 60
32. ibid

33 NCTM, 2000, p. 5
34 Katz, 2006, p. 6
35 Christou, Vosniadou, & Vamvakoussi, 2007, p. 283 based on Vosniadou, 1999
36 Pimm, 1995, and Saul, 1998
37 Fey & Good, 1985, and Heid, 1996
38 NCTM, 2000
39 Kieran, 2004, and Lester & Ferrini-Mundy, 2004
40 Payne & Squibb, 1990, Sutherland, 1991, and Tall & Thomas 1991
41 Star, 2009
42 See, for instance, Carpenter, Levi, & Loef Franke, 2003
43 Dougherty & Zilliox, 2003
44 Carraher, Schliemann, & Brizuela, 2001
45 Radford, 2010
46 Hewitt, 2012
47 Hewitt, 2012, p. 158
48 Dweck, 2006
49 Mueller & Dweck, 1998
50 Moser, Schoder, Heeter, Moran, & Lee, 2011
51 Blackwell, Trzesniewski, & Dweck, 2007
52 Demby, 1997, p. 48, discussing Hart, 1981, Booth, 1984, and Kuchemann, 1981
53 Anderson & Smith, 1987
54 Matz, 1980
55 Based on Capraro, Capraro, Parker, Kulm, & Raulerson, 2005
56 Brown and Burton, 1978, pp. 155–156
57 See Smith, diSessa, & Roschelle, 1993
58 e.g., Leinhardt et al., 1990; Nesher, 1987
59 e.g. Brown, J. S. & VanLehn, K. (1980
60 e.g. Novak, 1985
61 e.g., Confrey, 1990, based on Mevarech & Kramarsky, 1997
62 Lannin, Arbaugh, Barker, & Townsend, 2006
63 For a discussion of 'phenomenological primitives' such as "multiplication makes numbers bigger" and a counter-argument to the misconceptions perspective, see Hammer, 1996, and diSessa, 1993
64 Matz, 1982, pp. 25–6
65 Kajander & Lovric, 2009
66 MacGregor & Stacey, 1997, p. 3 based on Cohors-Fresenborg, 1993, Sutherland, 1991, and Tall & Thomas, 1991
67 Ainley, 1999
68 Proulx, 2007
69 Kieran, 1992
70 Brown et al., 1988
71 Demby, 1997
72 Macgregor & Stacey, 1997
73 Kuchemann, 1981
74 Kuchemann, 1978, 1981 and Ursini & Trigueros' work on their 3UV model (2001, 2004, 2008).
75 Bills, Wilson, & Ainley, 2006; Carraher & Schliemann, 2000; Cerulli & Mariotti, 2001; Cooper, Boulton-Lewis, Atweh, Pillay, Wilss, & Mutch, 1997; Filloy & Rojano, 1984; Filloy & Rojano, 1989; Gonzalez, Ambrose, & Castro Martinez, 2004; Herscovics & Linchevski, 1994; Linchevski & Herscovics, 1996; Nathan & Koellner, 2007; Russell, Schifter, & Bastable, 2011; Warren, 2003
76 Trigueros & Ursini, 1999

77 Booth, 1988, pp. 20–21
78 Booth, 1984, pp. 2–3
79 Kaput, 1999, pp. 133–4
80 Davis, 1989, pp. 118
81 Tall & Thomas (1991), p. 127
82 Lee & Wheeler, 1989
83 Chalouh & Hercovics, 1988
84 Kaput, 1995
85 Carpenter, Franke, & Levi, 2003
86 Arcavi, 2008, p. 47. See also Bell, 1995, Kieran, 2004, and Saul, 2008
87 Kieran, 2007
88 Sfard & Linchevski, 1994
89 Bazzini, 1999, and Arzarello, Bazzini, Chiappini, 1994
90 Kaput, 1989
91 Huff, 1993
92 Kieran, 2007, p. 712
93 Monk, 1992
94 Picciotto & Wah, 1993, p. 42
95 Resnick, Cauzinille-Marmeche, & Mathieu, 1987, p. 171
96 Bergsten, 1999, pp. 123–4
97 Hoch, 2003; Hoch & Dreyfus, 2004; Novotna & Hoch, 2008
98 Hoch, 2007
99 Novotna & Hoch, 2008, p. 95
100 Hoch & Dreyfus, 2004
101 Sfard, 1991, and Sierpinska, 1995
102 Cuoco, Goldenberg, & Mark, 1996
103 Magiera, Van Den Kieboom, & Moyer, 2010, p. 2, based on Driscoll, 1999
104 Driscoll, 1999, p. 1
105 Kieran (1992), p. 393
106 Bell & Purdy, 1986, p. 39
107 Demby, 1997, p. 67
108 Heemsoth, 2014; Lannin, Arbaugh, Barker, & Townsend, 2006; McCann, 2014
109 Lannin, Arbaugh, Barker, & Townsend, 2006, p. 82
110 David Tall in Greer, 2006, pp. 1–176
111 Durkin & Rittle-Johnson, 2012, Heemsoth, 2014, McCann, 2014
112 Movshovitz-Hadar & Hadass, 1990, p. 266
113 Smith, diSessa, Roschelle, 1993, p. 116
114 MacGregor & Stacey, 1997, p. 1
115 Demby, 1997
116 Demby, 1997, p. 61
117 MacGregor & Stacey, 1993, p. 28
118 Christou & Vosniadou, 2005, p. 453
119 Kieran, 1990
120 Hammer, 1996, p. 122
121 Greer, 2006, p. 176
122 See, for instance, Chun-Yi & Ming-Puu, 2008; Muldoon, Lewis, & Francis, 2007; Rea-Ramirez & Clement, 1998; and Watson 2002 & 2007
123 Piaget, 1975, p. 16
124 Bell, 1995
125 Haspekian, 2003, p. 2
126 Hammer, 1996, p. 123

Anderson, C. W., & Smith, E. (1987). Teaching science. In Richardson-Koehler, V. (Ed.), *The educator's handbook: A research perspective* (pp. 84–111). New York, NY: Longman, Inc.

Arcavi, A. (2008). Modeling with graphical representations. *For the Learning of Mathematics, 28*(2), 2–10.

Arzarello, F., Bazzini, L., & Chiappini, G. (1994). *The process of naming in algebraic problem solving*. Paper presented at the 18th Annual Conference of the International Group for the Psychology of Mathematics Education (PME), Lisbon, Portugal.

Ball, D. L., Bass, H., Sleep, L., & Thames, M. (2005). *A theory of mathematical knowledge for teaching. The fifteenth ICMI Study: The professional education and development of teachers of mathematics.* State University of Sao Paolo at Rio Claro, Brazil, 15–21 May 2005. Retrieved from http://stwww.weizmann.ac.il/G-math/ICMI/log_in.html.

Bazzini, L. (1999). *On the construction and interpretation of symbolic expressions.* Paper presented at the Proceedings of the First Conference of the European Society for Research in Mathematics Education.

Bell, A. (1995). Purpose in school algebra. *The Journal of Mathematical Behavior, 14*(1), 41–73.

Bell, A., & Purdy, D. (1986). Diagnostic teaching. *Mathematics Teaching, 115*, 39–41.

Bergsten, C. (1999). *From sense to symbol sense.* Paper presented at the Proceedings of the First Conference of the European Society for Research in Mathematics Education.

Bills, L., Wilson, K., & Ainley, J. (2006). Making links between arithmetic and algebraic thinking. *Research in Mathematic Education, 7*, 67–82.

Blackwell, L. S., Trzesniewski, K. H., & Dweck, C. S. (2007). Implicit theories of intelligence predict achievement across an adolescent transition: A longitudinal study and an intervention. *Child Development, 78*(1), 246–263.

Booth, L. (1984). *Algebra: Children's strategies and errors.* Windsor, UK: NFER-Nelson.

Booth, L. R. (1988). Children's difficulties in beginning algebra. *The ideas of algebra, K–12* (1988 Yearbook, pp. 20–32). Reston, VA: National Council of Teachers of Mathematics.

Brown, C. A., et al. (1988). Secondary school results for the Fourth NAEP Mathematics Assessment: Algebra, geometry, mathematical methods, and attitudes. *Mathematics Teacher, 81*(5), 337–347,397.

Brown, J. S., & Burton, R. (1978). Diagnostic models for procedural bugs in basic mathematical skills. *Cognitive Science, 2*, 155–192.

Brown, J. S., & VanLehn, K. (1980). Repair theory: A generative theory of bugs in procedural skills. *Cognitive Science, 4*, 379–426.

Capraro, R. M., Capraro, M. M., Parker, D., Kulm, G., & Raulerson, T. (2005). The mathematics content knowledge role in developing preservice teachers' pedagogical content knowledge. *Journal of Research in Childhood Education, 20*(2), 102–118.

Carpenter, T. P., Fennema, E., Loef Franke, M., Levi, L., & Empson, S. (2000). *Cognitively guided instruction: A research-based teacher professional development program for elementary school mathematics.* Retrieved from Madison, WI: http://ncisla.wceruw.org/publications/reports/RR00-3.PDF.

Carpenter, T., Franke, M., & Levi, L. (2003). *Thinking mathematically: Integrating arithmetic and algebra in the elementary school.* Portsmouth, NH: Heinemann.

Carraher, D., & Schliemann, A. (2000). *Bringing out the algebraic character of arithmetic: Instantiating variables in addition and subtraction.* Paper presented at the 24th annual Psychology of Mathematics Education (PME) Conference, Hiroshima, Japan.

Carraher, D., Schliemann, A., & Brizuela, B. (2001). *Can young students operate on unknowns?* Paper presented at the 25th Conference of the International Group for the Psychology of Mathematics Education., Utrecht, Netherlands.

Cerulli, M., & Mariotti, M. A. (2001). *Arithmetic and algebra, continuity or cognitive break? The case of Francesca.* Paper presented at the 25th Conference of the International Group for the Psychology of Mathematics Education, Utrecht, Netherlands.

Chalouh, L., & Herscovics, N. (1988). Teaching algebraic expressions in a meaningful way. In A. F. Coxford (Ed.), *The ideas of algebra, K–12* (1988 Yearbook) (pp. 33–42). Reston, VA: National Council of Teachers of Mathematics.

Christou, K. P. & Vosniadou, S. (2005). How students interpret literal symbols in algebra: A conceptual change approach. In B. G. Bara, L. Barsalou, & M. Bucciarelli (Eds.), *Proceedings of the XXVII Annual Conference of the Cognitive Science Society* (pp. 453–458). Italy.

Christou, K., Vosniadou, S., & Vamvakoussi, X. (2007). Students' interpretations of literal symbols in algebra. In S. Vaosniadou, Baltas, A., & Vamvakoussi, X. (Ed.), *Reframing the conceptual change approach in learning and instruction.* New York, NY: Elsevier.

Chun-Yi, L., & Ming-Puu, C. (2008). Bridging the gap between mathematical conjecture and proof through computer-supported cognitive conflicts. *Teaching Mathematics & its Applications, 27*(1), 1–1.

Cohors-Fresenborg, E. (1993). Integrating algorithmic and axiomatic ways of thinking in mathematics lessons in secondary schools. In the *Proceedings of Southeast Asia Conference on Mathematics Education (SEACME-6) and the Seventh National Conference on Mathematics* (pp. 74–81). Kampus Sukolilo, Surabaya.

Confrey, J. 1990. A review of the research on student conceptions in mathematics, science, and programming, In C. E. Cazden (Ed.), *Review of Research in Education, 16,* 3–56.

Cooper, T. J., Boulton-Lewis, G., Atweh, B., Pillay, H., Wilss, L., & Mutch, S. (1997). *The transition from arithmetic to algebra: Initial understanding of equals, operations, and variables.* Paper presented at the 21st Conference of the International Group for the Psychology of Mathematics Education, Lahti, Finland.

Cuoco, A., P. Goldenberg, E., & Mark, J. (1996). Habits of mind: An organizing principle for mathematics curricula. *The Journal of Mathematical Behavior, 15*(4), 375–402.

Davies, N., & Walker, K. (2005, July). Learning to notice: One aspect of teachers' content knowledge in the numeracy classroom. In *Building connections: Theory, research and practice (Proceedings of the 28th annual conference of the Mathematics Education Research Group of Australasia)*(pp. 273–280). [AU: City, state?]

Davis, R. B. (1989). Three ways of improving cognitive studies in algebra. In S. Wagner & C. Kieran (Eds.), *Research in the learning and teaching of algebra* (pp. 115–119). Reston, VA: National Council of Teachers of Mathematics.

Demby, A. (1997). Algebraic procedures used by 13-to-15-year-olds. *Educational Studies in Mathematics, 33*(1), 45–70.

diSessa, A. A. (1993). Toward an epistemology of physics. *Cognition & Instruction, 10*(2/3), 105.

Dougherty, B. & Zilliox, J. (2003) Voyaging from theory and practice in teaching and learning: A view from Hawai'i. In N. Pateman, B. Dougherty, & J. Zilliox (Eds.), *Proceedings of the 27th conference of the international group for the psychology of mathematics education* (vol. 1, pp. 31–46). College of Education, Honolulu: University of Hawaii.

Driscoll, M. (1999). *Fostering algebraic thinking: A guide for teachers grades 6–10.* Portsmouth, NH: Heinemann.

Durkin, K., & Rittle-Johnson, B. (2012). The effectiveness of using incorrect examples to support learning about decimal magnitude. *Learning and Instruction, 22*(3), 206–214.

Dweck, C. S. (2006). *Mindset.* New York, NY: Random House.

Fennema, E., Carpenter, T., Franke, M. L., Levi, L., Jacobs, V., & Empson, S. (1996). A longitudinal study of learning to use children's thinking in mathematics instruction. *Journal for Research in Mathematics Education, 27*(4), 403–434.

Fey, J. T., & Good, R. A. (1985). Rethinking the sequence and priorities of high school mathematics curricula. In C. R. H. M. J. Zweng (Ed.), *The secondary school mathematics curriculum* (Yearbook of the National Council of Teachers of Mathematics). (pp. 43–52). Reston, VA: National Council of Teachers of Mathematics.

Filloy, B., & Rojano, T. (1989). Solving equations: The Transition from arithmetic to algebra. *For the Learning of Mathematics: An International Journal of Mathematics Education, 9(*2), 19–25.

Filloy, E., & Rojano, T. (1984). *From an arithmetical to an algebraic thought: A clinical study with 12–13 years old.* Paper presented at the Sixth Annual Meeting of the North American Chapter of the International Group for the Psychology of Mathematics Education, Madison: University of Wisconsin.

Gonzalez, M., Ambrose, R., & Castro Martinez, E. (2004). *In the transition from arithmetic to Algebra: Misconceptions of the equal sign.* Paper presented at the Psychology of Mathematics Education 28.

Greer, B. (2006). *Designing for conceptual change.* Paper presented at the the 30th Conference of the International Group for the Psychology of Mathematics Education, Prague, Czech Republic.

Hammer, D. (1996). Misconceptions or P-Prims: How may alternative perspectives of cognitive structure influence instructional perceptions and intentions. *Journal of the Learning Sciences, 5(*2), 97.

Hart, K. (1981). *Children's understanding of mathematics: 11–16.* London, UK: John Murray.

Haspekian, M. (2003). *Between arithmetic and algebra: A space for the spreadsheet? Contribution to an instrumental approach.* Paper presented at the Third Conference of the European Society for Research in Mathematics Education, Pisa: University of Pisa.

Heemsoth, T. (2014). *Learning fractions from reflecting the rationale behind one's own errors.* Paper presented at the American Educational Research Association, Philadelphia, PA.

Heid, M. K. (1996). A technology-intensive functional approach to the emergence of algebraic thinking. In C. K. N. Bednarz, & L. Lee (Ed.), *Approaches to algebra: Perspectives for research and teaching* (pp. 239–255). Dordrecht, The Netherlands: Kluwer.

Helfand, D. (2006, January 30). A formula for failure in L.A. schools. *Los Angeles Times*. Retrieved from http://www.latimes.com/news/education/la-me-dropout-30jan30,1,2605555.story.

Herscovics, N. (1989). Cognitive obstacles encountered in the learning of algebra. In S. W. C. Kieran (Ed.), *Research issues in the learning and teaching of algebra* (Vol. 4, pp. 60–86). Reston, VA: National Council of Teachers of Mathematics.

Herscovics, N., & Linchevski, L. (1994). A cognitive gap between arithmetic and algebra. *Educational Studies in Mathematics, 27*(1), 59–78.

Hewitt, D. (2012). Young students learning formal algebraic notation and solving linear equations: Are commonly experienced difficulties avoidable? *Educational Studies in Mathematics, 81*(2), 139–159.

Ho, K. F., & Tan, P. (2013). Developing a professional vision of classroom practices of a mathematics teacher: Views from a researcher and a teacher. *Teaching Education, 24*(4), 415–426.

Hoch, M. (2003). Structure sense. In M. A. Mariotti (Ed.), *Proceedings of the 3rd Conference for European Research in Mathematics Education.* Bellaria, Italy: CERME.

Hoch, M. (2007). *Structure sense in high school algebra.* Unpublished doctoral dissertation, Tel Aviv University, Israel.

Hoch, M., & Dreyfus, T. (2004). *Structure sense in high school algebra: The effect of brackets.* Paper presented at the 28th Annual Meeting of the International Group for the Psychology of Mathematics Education, Bergen, Norway.

Huff, D. (1993). *How to lie with statistics.* New York, NY: W.W. Norton & Company.

Kajander, A., & Lovric, M. (2009). Mathematics textbooks and their potential role in supporting misconceptions. *International Journal of Mathematical Education in Science and Technology, 40*(2), 173–181.

Kaput, J. (1989). Linking representations in the symbol systems of algebra. In S. W. C. Kieran (Ed.), *Research issues in the learning and teaching of algebra* (Vol. 4 of Research agenda for mathematics education, pp. 167–194). Reston, VA: National Council of Teachers of Mathematics.

Kaput, J. (1995). *A research base supporting long-term algebra reform?* Paper presented at the Annual Meeting of the North American Chapter of the International Group for the Psychology of Mathematics Education, Columbus, OH.

Kaput, J. (1999). Teaching and learning a new algebra with understanding. In E. Fennema & T. A. Romberg (Eds.), *Mathematics classrooms that promote understanding* (pp. 133–155). Mahwah, NJ: Lawrence Erlbaum.

Katz, V. (2006). *Algebra: Gateway to a technological future. Mathematical Association of America.* Retrieved from: http://www.maa.org/news/new-report-algebra-gateway-technological-future.

Kieran, C. (1990). Cognitive processes involved in learning school algebra. In P. N. J. Kilpatrick (Ed.), *Mathematics and cognition: A research synthesis by the International Group for the Psychology of Mathematics Education* (pp. 96–112). Cambridge: Cambridge University Press.

Kieran, C. (1992). The learning and teaching of school algebra. In D. A. Grouws (Ed.), *Handbook of research on mathematics teaching and learning* (pp. 390–419). New York, NY: Macmillan.

Kieran, C. (2004). Algebraic thinking in the early grades: What is it? *The Mathematics Educator, 8*(1), 139–151.

Kieran, C. (2007). Learning and teaching algebra at the middle school through college levels. In F. Lester (Ed.), *Second handbook of research on mathematics teaching and learning* (pp. 707–762): Information Age Publishing.

Kuchemann, D. (1978). Children's understanding of numerical variables. *Mathematics in School, 7*(4), 23–25.

Kuchemann, D. (1981). Algebra. In K. M. Hart (Ed.), *Children's understanding of mathematics* (pp. 102–119). London: John Murray.

Ladson-Billings, G. (1998). Teaching in dangerous times: Culturally relevant approaches to teacher assessment. *Journal of Negro Education, 67*(3), 255–267.

Lannin, J., Arbaugh, F., Barker, D., & Townsend, B. (2006). Making the most of student errors. *Teaching Children Mathematics, 13*(3), 182–186.

Leatham, K. R., Peterson, B. E., Stockero, S. L., & Van Zoest, L. R. (2015). Conceptualizing mathematically significant pedagogical opportunities to build on student thinking. *Journal for Research in Mathematics Education, 46*(1), 88–124.

Leatham, K., Stockero, S., Peterson, B., & Van Zoest, L. (2011). *Mathematically important pedagogical opportunities.* Paper presented at the 33rd Annual Meeting of the North American Chapter of the International Group for the Psychology of Mathematics Education, Reno, NV.

Lee, L., & Wheeler, D. (1989). The arithmetic connection. *Educational Studies in Mathematics, 20*(1), 41–54.

Leinhardt, G., Zaslavsky, O., & Stein, M. K. (1990). Functions, graphs, and graphing: Tasks, learning, and teaching. *Review of Educational Research, 60*(1), 1–64.

Lester, F. K., & Ferrini-Mundy, J. (2004). *Proceedings of the NCTM Research Catalyst Conference.* Reston, VA: National Council of Teachers of Mathematics.

Linchevski, L., & Herscovics, N. (1996). Crossing the cognitive gap between arithmetic and algebra: Operating on the unknown in the context of equations. *Educational Studies in Mathematics, 30*(1), 39–65.

Liston, M., & O'Donoghue, J. (2010). Factors influencing the transition to university service mathematics: Part 2. A qualitative study. *Teaching Mathematics and Its Applications, 29*, 53–68.

MacGregor, M., & Stacey, K. (1993). Cognitive models underlying students' formulation of simple linear equations. *Journal for Research in Mathematics Education, 24*(3), 217–232.

MacGregor, M., & Stacey, K. (1997). Students' understanding of algebraic notation: 11–15. *Educational Studies in Mathematics, 33*(1), 1–19.

Magiera, M., Van Den Kieboom, L., & Moyer, J. (2010). *An extensive analysis of preservice middle school teachers' knowledge of algebraic thinking.* Paper presented at the American Educational Research Association, Denver, CO.

Mason, J. (1998). Enabling teachers to be real teachers: Necessary levels of awareness and structure of attention. *Journal of Mathematics Teacher Education, 1*(3), 243–267.

Matz, M. (1980). Toward a computational theory of algebraic competence. *Journal of Mathematical Behavior, 3*(1), 93–166.

Matz, M. (1982). Towards a process model for high school algebra errors. In D. S. J. S. Brown (Ed.), *Intelligent tutoring systems.* (pp. 25–50). New York, NY: Academic Press.

McCann, N. (2014). *What could go wrong? Error anticipation relates to conceptual and procedural knowledge in algebra students.* Paper presented at the American Educational Research Association, Philadelphia, PA.

Mevarech, Z. R., & Kramarski, B. (1997). From verbal descriptions to graphic representations: Stability and change in students' alternative conceptions. *Educational Studies in Mathematics, 32*(3), 229–263.

Milgram, R. J. (1999). *An evaluation of CMP*. Retrieved from ftp://math.stanford.edu/pub/papers/milgram/report-on-cmp.html.

Monk, S. (1992). Students' understanding of a function given by a physical model. In G. H. E. Dubinsky (Ed.), *The concept of function: Aspects of epistemology and pedagogy.* USA: Mathematical Association of America.

Moseley, B., & Brenner, M. E. (2009). A comparison of curricular effects on the integration of arithmetic and algebraic schemata in pre-algebra students. *Instructional Science: An International Journal of the Learning Sciences, 37*(1), 1–20.

Moser, J., Schoder, H. S., Heeter, C., Moran, T. P., & Lee, Y. H. (2011). Mind your errors: Evidence for a neural mechanism linking growth mindset to adaptive post error adjustments. *Psychological Science, 22,* 1484–1489.

Moses, R. P., & Cobb Jr, C. (2001). Organizing algebra: The need to voice a demand. *Social Policy, 31*(4), 4–12.

Movshovitz-Hadar, N., & Hadass, R. (1990). Preservice education of math teachers using paradoxes. *Educational Studies in Mathematics, 21*(3), 265–287.

Muldoon, K. P., Lewis, C., & Francis, B. (2007). Using cardinality to compare quantities: The role of social-cognitive conflict in early numeracy. *Developmental Science, 10*(5), 694–711.

Murray, D. (December 20, 2010). Algebra in elementary school? As demands change, educators look for new ways to teach math. *The Grand Rapids Press.* Retrieved from: http://www.mlive.com/news/grand-rapids/index.ssf/2010/12/algebra_in_elementary_school_a.html

Nathan, M. J., & Koellner, K. (2007). A framework for understanding and cultivating the transition from arithmetic to algebraic reasoning. *Mathematical Thinking and Learning: An International Journal, 9*(3), 179–192.

National Council of Teachers of Mathematics. (2000). *Principals and standards for school mathematics.* Reston, VA: NCTM.

National Council of Teachers of Mathematics. (2014). *Algebra as a strand of school mathematics for all students: A position of the National Council of Teachers of Mathematics.* Retrieved from: http://www.nctm.org/Standards-and-Positions/Position-Statements/Algebra-as-a-Strand-of-School-Mathematics-for-All-Students/.

Nesher, P. (1987). Towards an instructional theory: The role of student's misconceptions. *For the Learning of Mathematics, 7*(3), 33–40.

Nomi, T., & Allensworth, E. (2009). "Double-dose" algebra as an alternative strategy to remediation: Effects on students' academic outcomes. *Journal of Research on Educational Effectiveness, 2*(2), 111–148.

Novak, J. D. (1985). Metalearning and metaknowledge strategies to help students learn how to learn. In L. West & A. Pines (Eds.), *Cognitive structure and conceptual change,*(pp. 189–207). New York, NY: Academic Press.

Novotna, J., & Hoch, M. (2008). How structure sense for algebraic expressions or equations is related to structure sense for abstract algebra. *Mathematics Education Research Journal, 20*(2), 93–104.

Payne, S. J., & Squibb, H. R. (1990). Algebra mal-rules and cognitive accounts of error. *Cognitive Science, 14*(3), 445–481.

Peterson, B., & Leatham, K. (2009). Learning to use students' mathematical thinking to orchestrate a class discussion. In L. Knott (Ed.), *The role of mathematics discourse in producing leaders of discourse* (pp. 99–128). Charlotte, NC: Information Age Publishing.

Piaget, J. (1975). *The origin of the idea of chance in children.* London: Routledge and Kegan Paul Ltd.

Picciotto, H., & Wah., A. (1993). A new algebra: Tools, themes, concepts. *Journal of Mathematical Behavior, 12,* 19–42.

Pimm, D. (1995). *Symbols and meanings in school mathematics.* London, UK: Routledge.

Proulx, J. (2007). *Addressing the issue of the mathematical knowledge of secondary mathematics teachers.* Paper presented at the 31st Conference of the International Group for the Psychology of Mathematics Education, Seoul, Korea.

Radford, L. (2010). Layers of generality and types of generalization in pattern activities. *PNA, 4*(2), 37–62.

Rea-Ramirez, M. A., & Clement, J. (1998). *In search of dissonance: The evolution of dissonance in conceptual change theory.* Paper presented at the Annual Meeting of the National Association for Research in Science Teaching, San Diego, CA.

Resnick, L. B., Cauzinille-Marmeche, E., & Mathieu, J. (1987). Understanding algebra. In J. A. Sloboda & D. Rogers (Eds.), *Cognitive processes in mathematics* (pp. 169–203). Oxford, UK: Clarendon Press.

Riordan, J. E., & Noyce, P. E. (2001). The impact of two standards-based mathematics curricula on student achievement in Massachusetts. *Journal for Research in Mathematics Education, 32*(4), 368–398.

Russell, S. J., Schifter, D., & Bastable, V. (2011). Developing algebraic thinking in the context of arithmetic. In J. Cai & E. Knuth (Eds.), *Early algebraization* (pp. 43–69). Heidelberg: Springer Berlin.

Saul, M. (1998). *Algebra, technology, and a remark of I. M. Gelfand.* Paper presented at The nature and role of algebra in the K–14 curriculum: Proceedings of a National Symposium organized by the National Council of Teachers of Mathematics, the Mathematical Sciences Education Board, and the National Research Council, Washington, D.C.

Saul, M. (2008). Algebra: The mathematics and the pedagogy. In C. Greenes & R. Rubenstein (Eds.), *Algebra and algebraic thinking in school mathematics* (pp. 63–79). Reston, VA: The National Council of Teachers of Mathematics.

Schoenfeld, A. H. (2008). Chapter 2: On modeling teachers' in-the-moment decision making. *Journal for Research in Mathematics Education. Monograph, 14,* 45–96.

Sfard, A. (1991). On the dual nature of mathematical conceptions: Reflections on processes and objects as different sides of the same coin. *Educational Studies in Mathematics, 22*(1), 1–36.

Sfard, A., & Linchevski, L. (1994). The gains and the pitfalls of reification—The case of algebra. *Educational Studies in Mathematics, 26,* 191–228.

Sherin, M. G., Russ, R. S., Sherin, B. L., & Colestock, A. (2008). Professional vision in action: An exploratory study. *Issues in Teacher Education, 17*(2), 27–46.

Shulman, L. S. (1986). Those who understand: A conception of teacher knowledge. *American Educator, 10*(1), 9–15,43–44.

Sierpinska, A. (1995). *Understanding mathematics.* London: Falmer Press.

Smith, J., diSessa, A., & Roschelle, J. (1993). Misconceptions reconceived: A constructivist analysis of knowledge in transition. *Journal of the Learning Sciences, 3*(2), 115.

Smith, M., & Stein, M. K. (2011). *5 practices for orchestrating productive mathematics discussions.* Reston, VA: NCTM.

Sowder, L. (1988). Children's solution of story problems. *Journal of Mathematical Behavior, 7,* 227–238.

Stockero, S., & Van Zoest, L. (2012). Characterizing pivotal teaching moments in beginning mathematics teachers practice. *Journal of Mathematics Teacher Education, 16*(2), 125–147.

Star, J. R. (Producer). (2009, February 10, 2009). *Why students struggle with algebra and how schools are helping. Making algebra easier.* Retrieved from http://www.edweek.org/ew/marketplace/webinars/webinars.html.

Sutherland, R. (1991). Some unanswered research questions on the teaching and learning of algebra. *For the Learning of Mathematics, 11*(3), 40–46.

Tall, D., & Thomas, M. (1991). Encouraging versatile thinking in algebra using the computer. *Educational Studies in Mathematics, 22*(2), 125–147.

Tirosh, D., Even, R., & Robinson, N. (1998). Simplifying algebraic expressions: teacher awareness and teaching approaches. *Educational Studies in Mathematics, 35*(1), 51–64.

Trigueros, M., & Ursini, S. (1999). *Does the understanding of variable evolve through schooling?* Paper presented at the Conference of the International Group for the Psychology of Mathematics Education., Haifa, Israel.

Trigueros, M., & Ursini, S. (2008). *Structure sense and the use of variable.* Paper presented at the Proceedings of the 32rd Conference of The International Group for the Psychology of Mathematics Education.

Ursini, S., & Trigueros, M. (2001). *A model for the uses of variable in elementary algebra.* Paper presented at the 25th Conference of The International Group for the Psychology of Mathematics Education., Utrecht, Netherlands.

Ursini, S., & Trigueros, M. (2004). *How do high school students interpret parameters in algebra?* Paper presented at the 28th Conference of the International Group for the Psychology of Mathematics Education, Bergen, Norway. http://search.ebscohost.com/login.aspx?direct=true&db=eric&AN=ED489663&site=ehost-live.

Viadero, D. (2009, March 11). Algebra-for-all policy found to raise rates of failure in Chicago. *Education Week.* Retrieved from http://www.edweek.org/ew/articles/2009/03/11/24algebra.h28.html?tmp=1587018900.

Vosniadou, S. (1999). Conceptual change research: State of the art and future directions. In W. Schnotz, S. Vosniadou, & M. Carretero (Eds.), *New perspectives on conceptual change* (pp. 3–13). Bingley, United Kingdom: Emerald Group Publishing.

Wallach, T., & Even, R. (2006). Hearing students: The complexity of understanding what they are saying, showing, and doing. *Journal of Mathematics Teacher Education, 8*(5), 393–417.

Warren, E. (2003). The role of arithmetic structure in the transition from arithmetic to algebra. *Mathematics Education Research Journal, 15*(2), 122–137.
Watson, J. M. (2002). Inferential reasoning and the influence of cognitive conflict. *Educational Studies in Mathematics, 51*(3), 225.
Watson, J. M. (2007). The role of cognitive conflict in developing students' understanding of average. *Educational Studies in Mathematics, 65*(1), 21–47.

CHAPTER 2

VARIABLES AND EXPRESSIONS

> There are a number of conceptual obstacles to progress in algebra
> and one of the most important of these
> is the failure to understand the concept of variable.
> —*Graham and Thomas, 2012*

INTRODUCTION

We can often trace students' struggles in algebra back to their understanding of variables.[1] There are many uses of the term *variable*, which is often the basis for students' difficulties in understanding.[2] Research has shown that some students struggle with the differences between the name of an object (e.g., the person Michael), the name of an attribute (e.g., Michael's height), and the name of a measurement or quantity (h units).[3] They can often capably manipulate variables without really understanding the power and flexibility of the symbols.[4] Students may interpret letters or algebraic expressions based on intuition, simple guessing, or comparison with other symbol systems they know.[5] Moreover, there is evidence to show that students' understanding of variables can remain largely unchanged through many years of schooling, all the way to the college level.[6] Yet variables

How Students Think When Doing Algebra, pages 35–97.
Copyright © 2019 by Information Age Publishing
All rights of reproduction in any form reserved.

are the core of algebra, and understanding of their nature and behavior is critical to students' success.

Some of the confusion over the nature of variables reflects how our views of the concept of variables have changed over time. When researchers asked mathematicians, educators, linguists, computer scientists, and logicians for a one-word definition of variable, they heard "symbol, placeholder, pronoun, parameter, argument, pointer, name, identifier, empty space, void, reference, and instance."[7] Each person uniquely defined the word based on his or her experience, prior knowledge, and context. A variable has historically been understood as "a changing number," "representing numbers," able to have "two or more values," "a symbol for which one substitutes names for some objects, usually a number," and more recently, "a symbol for which things . . . can be substituted."[8] A letter can represent a constant, a specific unknown, and a variable.

Moreover, how we use variables in algebra can take the same form, but have a different sense.[9]

1. $A = LW$
2. $40 = 5x$
3. $\sin x = \cos x \tan x$
4. $1 = n(1/n)$
5. $y = kx$

Although each of the above equations has a term equal to a product of two other terms, the way the variables are used in each case communicates a different idea about the mathematics involved. In the first case—a formula—a relationship is a mathematical tool. Each letter has a distinct connection to a physical reality, the dimension of a rectangle. In the second, x is an unknown quantity that students can solve. The third use of the variable x is more abstract, in that students can algebraically manipulate this trigonometric identity without finding a specific value; students can manipulate the identity to learn about mathematical systems and structure. Fourth, n is in a property that defines mathematical operations, a value based on a generalized pattern. Finally, in the fifth use, y and x are truly variable in the full sense of the word, communicating an independent/dependent relationship that takes its shape with k representing some constant value.

In the above discussion on algebraic thinking, it was proposed that there are four purposes or themes to algebra: generalizing, solving equations and making sense of their solutions, exercising algebraic rules, and building models and learning from them. Each of these four approaches to learning algebra leads to a different conception of the idea of variable.[10] When you plug 2 into a calculator and repeatedly press the + sign, you get the pattern expressed by $2x$. The use of x helps students translate the mathematical ideas into a generalized form. There are no unknowns to seek; the variable simply helps in representation and students are "finished" when they find an appropriate algebraic form. In contrast, working with an equation such as $40 = 5x$ leads students to simplify and solve for the unknown, represented by the variable. Solving problems can also involve variables treated as constants, such as k in the equation $y = kx$. Algebra can work on a more abstract level, as well: one in which students manipulate letters based on algebraic rules. In this use, there is no connection between the variable and a specific value or values. Letters can be "arbitrary marks on paper" as students study structures and systems of variables and operations.[11] Finally, students use variables to represent and study situations in the real world with mathematical models. Within this use, variables relate to each other in important ways. In the area formula $A = LW$, each variable refers not only to an aspect of a rectangle, but also to the relationships among the aspects. Students do not need to put specific values into the model to consider the relationships of the variables. Variables stand for numbers on which other numbers depend, thereby acting as a parameter of the real-world situation, explaining dependent and independent relations. Students need to understand the many meanings within the name of *variable* in order to access the powerful mathematics behind the word.[12]

The way we conceive of variables is also parallel to the way we use language. For instance, consider how we use the word *hat*. When we refer to "Lincoln's hat," it has a specific value—a black top hat—as might the equation $40 = 5x$ in which x has the specific value of 8. We also might say "Michael always wears a hat." Knowing that Michael wears a range of hats, it could be a baseball cap, a beret, or a fedora. Our understanding of hat, given the constraint that it is Michael's, lets us know that it could be a few specific hats. Given the constraints that x is a positive integer and $x < 4$, we know x could be 1, 2, or 3. We could also refer to "a lady's hat," which is constrained to the type of hats worn by ladies, but an unlimited range of possibilities, in a similar way that $|x| \leq 5$. The variable x in this case has an infinite number of possibilities, but is also constrained so that it cannot be just any number. We might even say, "a woman's choice of hat depends on what she wears." There is a relationship between the hat and clothing, similar to the relationship between the two variables in $y = kx$. The parameter k quantifies the relationship between x and y. If she chooses a particular dress, that corresponds to a particular hat. There is some kind of relationship between the two, such as color coordination or style. The variable y similarly depends on what you choose for x, but k defines how the two are connected. It is the words that surround *hat* that help define what we are thinking when we say "hat," just as the numbers, variables, functions, etc. surrounding a letter define how a variable is understood. We have interesting research that explores **What Can Variables Stand for?**[1] and **Can Variables Change?** that is explored in more detail later.

Research on students' understanding of variables has also evolved over the past thirty years. There are many ways that researchers have categorized students' developing understanding of variables, but the following categories can be helpful to teachers. Based on tests and interviews of hundreds of students, researchers

---

[1] **Bold and italicized** print indicates a reference to a section in the book

have identified different ways that students think about letters representing unknowns.[13] The first three categories describe students' struggles to develop understanding of what a letter can represent. Each of the next three categories indicates different uses of *variable* in algebra, as well as increasing levels of sophistication of thought about the use of letters.

- *Letter evaluated*
  Students assign the letter a numerical value from the outset.
- *Letter not used*
  Students ignore the letter, or at best acknowledge its existence, but without giving it a meaning.
- *Letter used as an object*
  Students regard the letter as shorthand for an object or as an object in its own right.
- *Letter as an unknown number*
  Students regard the letter as a specific but unknown number, and can operate upon it directly.
- *Letter used as a general number*
  Students see the letter as representing, or at least as being able to take, several values rather than just one.
- *Letter used as a functional relationship*
  Students see the letter as representing a range of unspecified values, and a systematic relationship is seen to exist between two such sets of values.

These categories help us consider how students' thinking about letters evolves into thinking about variables. Initially, these categories were considered to be a Piagetian hierarchy of levels of increasingly sophisticated thinking about variables.[14] However, research has demonstrated that students' understanding of variables is deeply impacted by their prior knowledge and experience and does not necessarily have a linear developmental path.[15] Each of these categories is described in more detail later in this chapter.

The last three categories of specific unknown, general number, and functional relationship are broken down in Table 2.1 as a decomposition of the concept of variable. The authors of this research suggest that difficulties with understanding the multiple perspectives of variables are reinforced when each of the different aspects are taught separately, particularly when teachers stress manipulation and transformation rules.[16] Research indicates that Algebra teachers typically focus on the use of letters as specific unknowns rather than variables.[17] Instead, teachers should emphasize how the same symbolism, syntax, or rules for manipulation apply to the different uses of a variable, while having different purposes and meaning. Understanding how context can influence the use and meaning of letters is an important aspect of algebraic thinking. Again, helping students understand how letters mean different things in different situations might involve examining

TABLE 2.1. Decomposition of the Concept of Variable

	Conceptualization and Symbolization	Interpretation	Manipulation
Generalized number	The ability to see symbols as able to represent general methods or rules that are deduced from patterns or families of similar problems.	The ability to find meaning in symbols as general objects in algebraic expressions or in general methods.	The ability to manipulate symbols by factoring, simplifying, and expanding to rearrange expressions.
Specific unknown	The ability to see a symbol as representing an unknown in a particular situation.	The ability to find meaning in a symbol representing a specific unknown in equations in which it appears one or more times.	The ability to manipulate symbols by factoring, simplifying, and expanding in order to transpose or balance an equation to achieve a solution.
Variable in a functional relationship	The ability to see symbols as able to represent functional relationships based on a table or graph or word problem.	The ability to find meaning in symbols regarding their connection between expressions, tables, and graphs.	The ability to manipulate symbols by factoring, simplifying, and expanding to rearrange an expression and substitute values to determine intervals of variation, maximum or minimum values, and behavior of the relationship.

Adapted from Trigueros, Ursini, & Reyes (1996, p. 4–317)

language. Avid crossword puzzle doers, for example, realize that the art of a good puzzle is in the ability of words to have multiple meanings. The word *train*, for instance, could mean a locomotive, preparation for a baseball game, or the end of a bridal dress. These are very different uses of one word, depending on how and when it is used. Similarly, symbols take on different purposes in different mathematical contexts.

Students of arithmetic typically believe mathematics flows from left to right, to a numeric conclusion. For instance, in one study, 145 sixth-grade students were given the problem $8 + 4 = ? + 5$. All 145 students answered either 12 or 17.[18] Children as young as kindergarten come up with the same conclusion. This tendency leads to problems in algebra when students encounter expressions, such as $x + 5$. They often struggle to understand that an "answer" can be an expression rather than a number. They want to "complete" the problem by continuing the flow from left to right. Researchers describe the ability to see an expression as an answer (and not need a single number) as an *Acceptance of Lack of Closure*, or the ability to overcome the *Expected answer obstacle* (expecting an answer to be a single entity without operations).[19] Arithmetic leads many students to believe that math requires specific numerical answers, so they may be reluctant to give an algebraic answer, assuming that something else must have been intended. Students struggle to find meaning when given expressions because they lack an equal sign and something on the other side.[20]

Algebra requires a shift in thinking so that an expression such as $x + 5$ can be seen as an instruction for a procedure to add 5 to the variable x; or as an answer, the number that is 5 more than x; or as a function, mapping the values achieved for values of x onto another. This notion that an expression can represent a procedure and an answer simultaneously is called the *Process-Product Dilemma*[21] and is challenging for students in early algebra to accept.[22] Students must be able to see that 13 can be written as $5 + 8$, which can represent the number of items in two sets, 5 and 8, as $a + b$ can represent the number of items in two sets containing a and b items.[23] The ability to see $a + b$ as an object as well as a process can be initiated as students learn arithmetic. Some of the difficulty for students can be traced back to their understanding of *Algebraic Notation*, as learning notation in mathematics can be as challenging as learning a new language. The rules of notation can be based on a particular rationale or simply mathematical convention, and it is difficult for students to differentiate.

Finally, variables are used as a tool to help solve problems. While the roles of variables in modeling and word problems are discussed in a later chapter, it is useful to explore some research in regard to using variables in context. We have organized the research on variables in these contexts into the final two topics in this chapter: *Area, Perimeter, and Variables* and *Translation Difficulties in Changing Words to Algebraic Sentences.*

In the following quote, researchers describe our primary goal for this chapter:

> As soon as children are unable to give meaning to concepts, they hide their difficulties by resorting to routine activities to obtain correct answers and gain approval. Once committed to such a course, it easily degenerates into a never ending downward spiral of instrumental activity: learning the "trick of the week" to survive, soon leading to a collection of disconnected activities that become more and more difficult to coordinate, even at a purely mechanistic level. Therefore, the beginning phase of the subject—giving meaning to the variable concept and devising ways of overcoming the cognitive obstacles—is fundamental to laying a foundation for meaningful algebraic thinking.[24]

This chapter on *Variables and Expressions* can be used as a basis to help explain some of the mathematical issues that may underlie possible answers that students give when encountering algebraic problems. A clear and comprehensive understanding of variables is at the heart of algebraic thinking.

REPRESENTATION: WHAT CAN VARIABLES STAND FOR?

Common Core Standards: 6.EE.2a, 6.EE.2c

One of the first obstacles for students to overcome when learning algebra is the idea that letters mean different things in different contexts. Students use their prior experience to make sense of variables and generalize their thinking; however, depending on the situation, the representation can either confuse them or help them. They know, for instance, that p6 means page six because p represents a word. In other examples, 3*a* could mean the first section of item number 3, and 7*m* can mean 7 meters, 7 times the variable *m*, or an abbreviation for 7 mangos. In algebra, *m* all by itself could mean 1 meter, an unknown number of meters, or have nothing to do with meters at all.

Therefore, in an expression such as 10*m* – 1, students might think, "ten meters minus one is nine meters." Students have a difficult time knowing when letters mean words and when they represent single numbers, multiple numbers, or a range of numbers. Research identifies this struggle for students as the ***Letter Used as an Object*** way of thinking about variables.[25] Students may struggle to understand when letters represent things rather than quantities.

Teachers often say that a variable "can be anything" to encourage a broader interpretation of letters, but in mathematics, there are rules that students are trying to discern about what a letter can represent. For example, sixth-, seventh-, and eighth-grade students were asked what the letter *n* could stand for in the expression 2*n* + 3.[26] When asked if *n* could stand for the number 4 in this expression, 56% of sixth-grade students, 77% of seventh-grade students, and 87% of eighth-grade students said "Yes." Other responses included:

> Can '*n*' stand for 37?   Yes (30%, 67%, 81%).
>
> Can '*n*' stand for 3*r* + 2?   Yes (26%, 30%, 40%).

As students had more experience with variables, their success on these questions increased. However, at only 40%, the percentage of eighth-grade students who thought *n* could be 3*r* + 2 was still quite low after instruction about variables. A student gave explanations for some of his thinking in a follow-up interview: in the case of *n* being 4, the student explained, "letters are also used as variables and so *n* could be any number." However, when asked, "Could *n* stand for 15 + 27?" the student responded:

> I think so. Well, actually, I do not think so really because variables just stand for one number. You could have *n* plus another letter or variable and *n* could be 15 and the other variable could be 27. *n* = 15 and *p* can equal 27 so then if you did *n* + *p* it would equal 42.[27]

The student could not accept that the addition sign could be part of what n represented—that it could represent an expression.

> This student sees a distinction between an expression that represents a process—such as 27 + 15—and the result of that process. The distinction the student makes between r and (r)—what r equals—indicates that this student conceives of r as representing an action or process of replacing the symbol r with a value, and not the result of that process.[28]

He had developed the rule in his mind, likely from his experience, that variables can only represent single numbers. This is an example of the *Process-Product Dilemma* described in detail later in this chapter.

Upon further questioning, the student also had developed some erroneous rules about the impact of parentheses on the potential for what n could represent (see Table 2.2). In the interviews, when parentheses isolated expressions, students responded differently. For instance, writing the expression as $(3r + 2)$ instead of $(3r) + 2$ or simply putting r into parentheses altered students' perspectives so that they were more comfortable thinking about what that expression might represent as an object, as opposed to thinking of it as a process. However, the student in this case struggled with the idea that an expression (with or without parentheses) could be one thing. He seems to have understood that you typically evaluate what is inside parentheses, which influenced his thinking about what n could represent: if it was evaluated first, then it would work. Researchers found that using parentheses in different ways might be an effective approach that could help bridge the Process-Product Dilemma for students.

TABLE 2.2. Interviews for Representation Problems

Interviewer:	Can n stand for $(3r + 2)$?
Student:	I do not think so because it couldn't really stand for, well, actually yes it could because it's all in parentheses so the r could stand for another number and then it would all be one number. Inside the parentheses you'd have to come up with an answer.
Interviewer:	What about $(3r) + 2$? Could n stand for that?
Student:	No. Because what is in the parentheses stands for a separate number and it's basically what you did up here (points to 15 + 27). It's as if you were doing 10 plus 5 in parentheses and then plus 27 out of the parentheses.
Interviewer:	Could n stand for r?
Student:	I do not think it could because variables stand for numbers and if n were a variable then it would have to stand for a number, not another letter.
Interviewer:	Would it make a difference if I put parentheses around it? (r)?
Student:	I think it might because then n is standing for what r equals, not just r but I'm not quite sure.

Source: Weinberg, et al. (2004, pp. 7–8)

In a different study, 14- to 15-year-old students were asked to "Write down numerical values that you think *cannot* be assigned to the algebraic objects a or -b."[29] The following were their responses:

Category of response: $(a, -b)$
No answer: (4%, 6%)
Positive whole numbers: (1, 2, 3, etc.) (3%, 50%)
Negative whole numbers: (-1, -2, -3, etc.) (46%, 4%)
Positive numbers: (2.33, etc.) (8%, 14%)
Negative numbers: (-2.33, etc.) (10%, 8%)

Students have quite a bit of difficulty thinking about a variable being a negative number as well as thinking about the negative or opposite of the value of a variable. Research indicates that students' experience with natural numbers strongly influences their interpretation of symbols in algebra. Specifically, when thinking about the values of variables in algebraic expressions, students tend to believe that the letters can only represent natural numbers. As students move from working with natural numbers to real numbers, their understanding of what variables can represent must change similarly.

As the -b part of this problem implies, some students interpret the sign in front of the variable as the sign of the numbers that they represent. When engaged with expressions such as $k + 3$ or $d + d + d$, some students believed that the variables stood for positive numbers and a negative number could not be substituted. Fifty percent of students believed that a positive number could not be put in for b in a term such as -b. See **Algebraic Relations: Negatives** for further discussion on this topic.

In one study, the interviewer asked four different students to explain their understanding of the variable x when given the problem (see Table 2.3): What is the value of $10 - x$, when $x = 6$?

The mathematical convention of considering $1x$ to be the same as x may be confusing to students. For some, their prior experience with numbers next to each other may negatively influence their interpretation of this convention of leaving

TABLE 2.3. Problem 1: What is the Value of 10 – x, When x = 6?

	Student Responses
Student #1:	9. Because x is just like 1. Like having one number. And so you take one of the xs out of the tens and you get 9.
Student #2:	Well, x equals 1. By itself it is 1, the x.
Student #3:	x is just one single thing, so like x times x is just like 1 times 1.
Student #4:	x is one because it hasn't got no number.

Source: MacGregor & Stacey (1993c.)

out the multiplication symbol. For instance, students' experience with compound fractions (e.g., 2 ½ = 2 + ½) and place value (43 = 4 tens + 3 ones) is contrary to the convention of 1x meaning multiplication.[30] Others may get confused as they try to make sense of what they are learning about mathematical practice in algebra, with the use of letters in places such as textbook exercises labeled 1a, 1b, 1c, etc.[31] Asked, "When I show you something like 3a, what does it mean to you?" five of six students in one study interpreted 3a as a subdivision label, saying: "third problem, first part" or "like 1, a, b, c."[32] In addition, students may misunderstand

> what teachers mean when they say "x" without a coefficient means 1"x" . . . the power of x is 1 if no index is written and that x with zero as the (power) equals 1 (e.g., $x = x^1$ and $x^0 = 1$). . . . The student gets a vague message that the letter x by itself is something to do with 1.[33]

As a result, researchers have noted that some students assume that $x = 1$, as the students explain above.

Another common misconception regarding what letters represent is thinking there is some relationship between the choice of letter used and its value. In particular, many students believe that there is a correspondence between the letter's position in the alphabet and its value. Problems such as "Find three consecutive integers that add to 63" may be set up initially as $x + y + z = 63$, in which x is the first integer, y is the second, and z is the third. This strategy makes organizational sense but can have the unintended effect of teaching students that there is some connection between a letter's position in the alphabet and the answer. Students may bring this type of understanding to the learning of algebra, so it is difficult for them to believe that the use of letters is arbitrary and not related to other letters. This interpretation can also be influenced by decoding puzzles or other contexts that connect numbers to particular letters, such as in Table 2.4. This misconception might be reinforced by letters used in close association with specific numbers, such as **α = 1, β = 2**, etc.[34]

In another research problem, students were asked to "Simplify $e + 2 + 6$." Some students responded "13" and explained, "I added 2 and 6 to get 8, but I didn't really know what to do next so I guessed that because e is the fifth letter of

TABLE 2.4. Word Code

Using the substitution cipher below, find the following word: 1 12 7 5 2 18 1												
1	2	3	4	5	6	7	8	9	10	11	12	13
A	B	C	D	E	F	G	H	I	J	K	L	M
14	15	16	17	18	19	20	21	22	23	24	25	26
N	O	P	Q	R	S	T	U	V	W	X	Y	Z

the alphabet, the answer must be 13."[35] Students answering in this manner did not have ***Acceptance of a Lack of Closure***. They may have wanted to "complete" the problem by combining the numbers and letter. Not having enough information about the letter *e* to "finish" the problem, the student assumed that *e* represented 5 because it is the fifth letter of the alphabet. In another problem, when asked if they could simplify $10 + h$, some students answered "$10 + h = r$" because ten letters after *h* is *r*.[36] When researchers posed the question, "if $e = 3$ and $g = 5$, what can you say about the value of *f*?" a large proportion of students answered "4" because *f* is between *e* and *g*, so the answer must be between 3 and 5. Students may not understand that what a letter represents is independent of the letter used. This is an important point that teachers will want to make in the classroom.

Some students may confuse what the variable can represent with place value.[37] For instance, one interview asked: "If $x = 6$, what is the value of $2x$?" The student responded: "26. You put the 6 in for the *x* because it is supposed to be something more than 20." In this case, the 6 is a digit in the number 26 representing the ones column. Students may assume that the *x* in $2x$ is the second digit in a number of twenty something. There is implicit addition in arithmetic that may lead to this confusion.[38] For instance, 26 represents $20 + 6$ and $5\ ½$ represents one half more than five. This type of thinking, called *concatenation* in the research, may also be encouraged with *digit problems*, in which *x* might stand for the tens digit and *y* might stand for the ones digit.

A significant part of the movement from concrete to abstract thinking is understanding what a variable can represent. There are significant differences between how numbers work and how letters work in mathematics. In Table 2.5, researchers have summarized the important differences between numbers and letters that stu-

TABLE 2.5. Natural Numbers vs. Letters

	Natural Numbers in Arithmetic	**Literal Symbols as Variables in Algebra**
Form sign	1, 2, 3, … Actual sign (positive)	a, b, x, y, … Phenomenal sign (positive or negative)
Symbolic representation	Each number in the natural number set has a unique symbolic representation; different symbols correspond to different numbers.	Each literal symbol corresponds to a range of real numbers; different symbols could stand for the same number.
Ordering—density	Natural numbers can be ordered by means of their position on the count list. There is always a successor or a preceding number. There are no numbers between two subsequent numbers.	Literal symbols in algebra cannot be ordered by means of their position on the alphabet. There is no such thing as a successor or preceding literal symbols.
Relationship to the unit	The unit is the smallest natural number.	There is no smallest number that can be substituted for a variable, unless otherwise specified.

Source: Christou, Vosniadou, & Vamvakoussi (2007)

TABLE 2.6.

Problem 2: What numbers might work for d in the problem $5d > \frac{4}{d}$?

	Interview
Interviewer:	Could we try with a negative number; let's say -2?
Student:	Well, yes, of course. Mmm! That is -10 > -2, which is not valid. Well, that changes everything.
Interviewer:	So, now do you think we could find a number which wouldn't hold for the inequality?
Student:	Certainly, the negative numbers.
Interviewer:	Any other kind of number?
Student:	We have tried everything, positives and negatives.
Interviewer:	Could we try with a fraction; let's say 1/2?
Student:	Yes, we could [he tries it]. This also wouldn't hold.
Interviewer:	Why didn't you try with such kind of numbers before?
Student:	I don't know, I should have thought both of negatives or fractions.

Source: Christou & Vosniadou (2012)

dents need to understand. Researchers suggest that students' difficulties in algebra may relate to students' difficulty in understanding arithmetic,[39] or in applying ideas from arithmetic to algebra inappropriately.[40] It is important to differentiate between when students don't understand something and when they just don't think of something. Consider the exchange in Table 2.6. With appropriate prompting, the student did understand the concepts. The issue for this student was not a misconception, but a bias toward only considering natural numbers. Researchers found that students have a bias toward thinking about variables as representing positive integers.[41]

Students' experience inside and outside the classroom can influence how they develop rules in their minds for what letters can mean in different contexts. The problems they experience typically do not include a variable representing a non-integer or negative number. One of the critical responsibilities of a teacher is to help students develop a comprehensive understanding of the way letters are used in mathematics.

Teaching Strategy

It often helps to tap into students' prior knowledge and build on it toward an algebraic idea. In regard to the idea of representation, consider the approach in Table 2.7 that builds an understanding of variable on students' understanding of

TABLE 2.7. Teaching Strategy: What do Variables Represent?

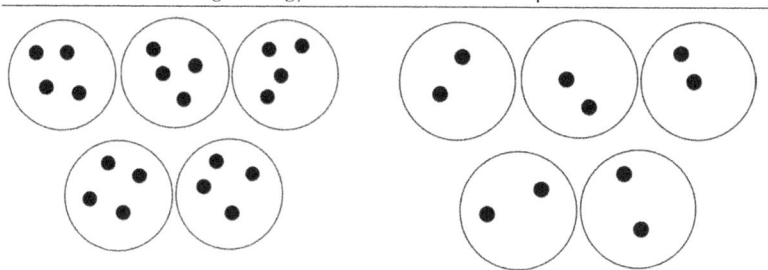

Teacher: What is 3 sets of 4 plus 2 sets of 4?

> Teacher shows physical objects, such as circle chips, in groups of 4. Students may respond "5 sets of 4" or "20." Teacher helps them to understand the equivalence of either response by showing the equation: $3(4) + 2(4) = 5(4)$

Teacher: What is 3 sets of 2 plus 2 sets of 2?

> Teacher shows physical objects, such as circle chips, in groups of 2. Students may respond "5 sets of 2" or "10." Teacher helps them to understand the equivalence of either response by showing the equation: $3(2) + 2(2) = 5(2)$

> Teacher continues by offering additional examples with larger, two- or three-digit numbers, fractions, and decimals. When students seem ready, teacher introduces questions about the pattern they see with the equation. Teacher suggests the idea of a letter to take the place of the 4, 2, etc., and asks questions about what that letter represents: "Should we use one letter or more than one letter? Why?" Either approach could work. "Which is better? Why?" Teacher should encourage as much flexibility as possible with the idea of variable.

---

number. Similar strategies are found in math curriculum such as Connected Math with "coins and bags." Coins represent numbers and bags represent variables.

Teachers

- Help students understand how letters mean different things in different contexts. How does a particular context lead to a particular interpretation of the use of a letter?
- Help students understand the role of parentheses and how they effect what a variable can represent.
- Help students understand how negatives affect the value of variables.
- Consider how arithmetic rules may influence students' thinking about the use of variables.
- Help students understand that a letter's place in the alphabet or the choice of a particular letter has no impact on its value.
- Sometimes use random letters for variables, rather than always using consecutive letters in the alphabet.

CONSERVATION OF VARIABLES: CAN VARIABLES CHANGE?

Common Core Standards: 6.EE.4

Students struggle to understand when variables can change and when they stay the same. The problems in Table 2.8 explore this concept. Table 2.9 shows the results from a study of Japanese and American students in grades 5–11, including the percentages of students who answered both problems correctly. Other studies of students find similar results.[42] The very low percentages of students across the grade levels who answered both problems correctly indicate how students struggle to understand the stability of variables, what letters can mean, and when. They need to understand when a letter can represent a specific unknown *or* a range of values, as well as if the value of letters can change within or across problems. They also need to understand when variables represent quantities and when they represent units or entities. Students were also interviewed about their thinking after solving the problems (see Table 2.10).

Based on the results of the study in Table 2.10, researchers developed a categorization of their responses.[43] The first level of students "had a vague conception of literal symbols" and had "no rules to interpret literal symbols, or no rules for substituting numbers into literal symbols."[44] The second level of students circled all answers for both problems. These students "appear to treat the letters x and y as they might treat empty boxes (i.e., interpret Problem 3 as □ + □ + □ = 12) in a primary school exercise."[45] As mentioned earlier, when developing students'

TABLE 2.8. Problems for Changing Variables

Problem 3	Problem 4
Mary has the following problem:	Jon has the following problem:
"Find value(s) for x in the expression:	"Find value(s) for x and y in the expression:
$x + x + x = 12$"	$x + y = 16$"
She answered in the following manner:	He answered in the following manner:
a. 2, 5, 5 b. 10, 1, 1 c. 4, 4, 4	a. 6, 10 b. 9, 7 c. 8, 8
Which of her answers is (are) correct?	Which of his answers is (are) correct?

Source: Fujii (2003)

TABLE 2.9. Percent of Students Correct on Both Changing Variables Problems, by Ethnicity and Grade Level

Grade	5	6	7	8	9	10	11
Japanese	0%	3.7%	9.5%	10.8%	—	18.1%	24.8%
American	—	11.5%	—	11.5%	5.7%	—	—

Source: Fujii (2003)

TABLE 2.10. Problem 3 Interviews

Student 1:	(2,5,5) and (10,1,1) are acceptable because x is unknown so it could be anything.
Student 2:	I think that since in this sentence there are $3xs$, all of the xs have to be the same number, even though they are unknown, so that would have to be just the three numbers that add up to 12.
Student 3:	It depends on what x equals, which, because x can equal 10, the first x, and then second x can equal 2.
Student 4:	I think that all the xs are the same number and so you can write $3x$. I will say that x is a variable and if it is in the same problem with another x then it has to be the same number.
Interviewer:	Is (2, 5, 5) acceptable for $x + x + x = 12$?
Student:	x is unknown so it could be anything.

Source: Fujii (2003)

sense of variables, teachers sometimes say, "x can be anything," leading students to have difficulty understanding that variables within a problem are all the same value. The use of empty boxes—instead of letters—in primary school may have the same effect.

In Problem 3, the first student interviewed had a misconception that *the same letter in an expression does not necessarily stand for the same number* (x could be 2, 5, and 5 simultaneously because "it is a variable"). Similarly, the next level of students answered Problem 3 correctly, but circled "c" on Problem 4, thinking it didn't work because different letters must mean different values. They believed that the reason a different letter was chosen by the problem maker was to identify a different value:

Interviewer: Is (8, 8) acceptable for $x + y = 16$?
Student: They have to be different numbers because they are different variables.

Researchers describe this way of student thinking as "Changing the variable symbol as changing the referent."[46] For example, the expression $3n$ is not considered the same concept as $3x$ because n and x would never be equal because they are different letters.[47]

Finally, the top level of students could solve both problems correctly and understood that variables within a problem must stay the same value and that different letters could still have the same value. One of the important lessons from this research is that

a student with the *Empty Box* misconception, who interprets $x + x + x = 12$ as $\square + \square + \square = 12$ and believes that $10 + 1 + 1$ is an acceptable answer, is unlikely to make sense of the explanation that $x + x + x = 12$ is equivalent to $3x = 12$ and then $x = 4$.[48]

TABLE 2.11. Problems 5-7: Variable Changing

Problem	Response
5. $L + M + N = L + P + N$ is Always Sometimes (when?) Never true	Never (51%[a], 40%[b], 50%[c], 74%[b]) Incorrect (89%)[a] P must be different than M (74%)[c]
6. Which is larger, W or N? $7 \times W + 22 = 109$ or $7 \times N + 22 = 109$	Students saying either N or W is larger (45%) "You can't tell" (17%) "Neither" (38%)[d]
7. David is 10 cm taller than Con. Con is h cm tall. What can you write for David's height?	Correct: $h + 10$, 36% - 64% (Year 7 – Year 10)[e] Conjoined: $10h$ or $h10$, 14% - 5% (Year 7 – Year 10)

[a]Percent of 3,000 13- to 15-year-old students in Hart and Kuchemann (2005).
[b]Percent of 3,550 13- to 15-year-old students in Booth (1984).
[c]Percent of 278 students in Coady and Pegg (1993).
[d]Percent of 30 10- to 18-year-old students in Wagner (1981).
[e]Percent of Year 7–Year 10 students in Stacey and MacGregor (1994).

It would make no sense to condense $x + x + x = 12$ into $3x = 12$ if all the xs were different numbers. The difficulty with accurate understanding is also apparent with other problems (see Tables 2.11 and 2.12).

Similarly, in the answer for Problem 7, the student believed the letter h generally referred to height, so it could mean both David's height and Con's height. In another study, a student was given the equation $x + 5 = x + x$ and responded that the second x on the right side had to be 5, but the other xs could be anything.[49] Hearing that "x can be any number" repeatedly in class, students may logically take that principle to the extreme.

Similar to Problem 4 above, a student may respond "Never true" in Problem 5 because he or she may be aware that the same letter stands for the same number, but think that the converse is also correct, that M and P each represent specific unknowns and do not represent the same unknown because they are different letters.[50] In the same problem, some students crossed off the Ls and Ns from both sides, leaving $M = P$. However, they did not necessarily understand the reasoning behind the conclusion and may simply have been manipulating the terms. In the interview for Problem 5, Student #4 believed that the position in the alphabet had some influence on the value of the variable. These responses also indicate the misconception that different letters must mean different values.[51]

One way to discover why students struggle with algebra is to identify the kinds of errors that are common to students and then see if the *why* behind the errors can be found. Four areas to concentrate on are described as:

- The focus of algebraic activity and the nature of "answers";
- The use of notation and convention in algebra;
- The meaning of letters and variables; and
- The kinds of relationships and methods used in arithmetic.

TABLE 2.12. Interviews for Problem 5

Interviewer:	Why do you think that equation would never be true?
Student 1:	It'll have different values…because P has to have a different value from M and the other values, so that'll never be true.
Interviewer:	Why do you say that?
Student 1:	Well, if it didn't have a different value, then you wouldn't put P, you'd put M. You see, you put a different letter for every different value.
Student 2:	You wouldn't have down two letters for one number.
Interviewer:	Oh, I see, so using different letters…
Student 3:	Means they're different amounts.
Interviewer:	Oh, I see. And are they always different amounts?
Student 3:	Well, I've always found they're different. I've never come across one where they're the same.
Interviewer:	Does P have to be a larger number than A?
Student 4:	Yes because A starts off as 1 or something.
Interviewer:	What made you think of that?
Student 4:	Because when we were little we used to do a code like that…in junior school.…A would equal 1, B equals 2, C equals 3. There were possibilities of A being 5 and B being 10 and that lot but it would come up too high a number to do it. It was always in some order.

Source: Kuchemann (2005); Booth (1984); Coady and Pegg (1993); Stacey and MacGregor (1994); and Wagner (1981)

Students need to understand when letters can vary and when they can't. Letters can vary across problems but not within a problem. For example, in the problem $x + y = 2x - 3$, x represents the same number whenever it is used on both sides of the equation. This is confusing in that x is actually a different number depending on what y is, so it does change based on y. Possibly adding more confusion for students, in a system of equations, x could be the same across equations (at the point of intersection) but could also be different. However, in two different problems, x could be a different number or range of numbers. It could also be the same number, but for different reasons. Conservation of relations is just one common misconception students might develop when working with equations, functions, and variables. Teachers can address this possibility by purposefully designing experiences focused on changing the alphabetic variable and nothing else about a relation.

An understanding of when variables can change and when they stay the same is critical when students get to the complex situation of systems of equations. When solving systems of equations, students might have the following two equations: $x + y = 3x - 5$ and $2x + 3y = x + 2$. Within those two equations, the xs have the same value. Across the two equations, the *x*s are different values except at their point of intersection (3, 1), where they are the same values. This can be very complicated for students to understand. Success with systems of equations begins with the foundation of understanding the stability of variables in single equations.

Teachers

- Help students understand that variables *within* an equation have the same value (e.g., $x + 9 = 2x - 8$).
- Help students understand that letters chosen to represent values do not have any influence on their value. For instance, when m is chosen to represent a value and n is chosen to represent another value, the value of m is not necessarily less than n. Also, m and n could represent the same value.

ACCEPTANCE OF LACK OF CLOSURE

Common Core Standards: 6.EE.2, 6.EE.4

Students can struggle with the difference between expressions and equations: they may not understand when variables can and cannot be evaluated. The problems in Table 2.13 provide some context for this difficulty. Researchers describe the ability to see the expression as an answer—and not need a single number—as an "Acceptance of Lack of Closure"[52] (see also **Patterns and Functions: Process versus Object**). Arithmetic leads many students to believe that math requires specific numerical answers, so they may be reluctant to give an algebraic answer in the form of an expression, assuming that something else must have been intended. They can be uncomfortable having an answer that includes an operation sign (i.e., $8 + g$), as with the interview of Student 1 (see Table 2.14) who notes, "$8g$ sounds more like math." Students struggle to find meaning when given expressions because they lack an equal sign and something "on the right side."[53] Some students even add "= 0" to an expression when they are asked to simplify, in order to create meaning and make it appear closer to their experience with equations.[54]

TABLE 2.13. Problems on Lack of Closure

Problem	Response
6. If $e + f = 8$, what is $e + f + g$?	Correct: $8 + g$ (24%[a], 41%[b])
	$8g$ (4%[a], 3%[b])
	12 (13%[a], 26%[b])
	9 (2%[a], 6%[b])
	some other number (10%[a])
7. Simplify $2a + 5b + a$.	Correct: $3a + 5b$ (60%)[b]
	$8ab$ (20%)
8. Add 4 onto $3n$.	Correct: $3n + 4$ (36%)[b]
	$7n$ (31%)
9. Peter's age is represented by x. Alan is 2 years older than Peter. How can we represent Alan's age?	Correct: $2 + x$ (49%)[c]
	$2x$ (10%)
10. I have x pence and you have y pence. How many pence do we have altogether?	Correct: $x + y$ (48%)[c]
	xy (34%)
11. Add 5 to an unknown number n, then multiply the result by 3.	Correct: $(n + 5)3$ (14%, 47%)[d]
	e.g., $15n$, $5n*3$ (14%, 12%)
12. David is 10 cm taller than Con. Con is h cm tall. What can you write for David's height?	Correct: $10 + h$ (39%, 74%)[d]
	$10h$ (14%, 5%)
13. Add 12 to x.	Correct: $12 + x$ (68%, 93%)
	$12x$ (2%, 2%)

[a]Percent of 5475 12- to 14-year-old students from Hodgen, et al. (2008)
[b]Percent of 3,000 13- to 15-year-old students in Hart and Kuchemann (2005).
[c]Percent of British 15-year-old students in the Assessment of Performance Unit (1985).
[d]Percent of 1806 12- or 15-year-old students, respectively, in Stacey and MacGregor (1994).

TABLE 2.14. Interviews on Lack of Closure Problems: Problem 6

Student 1:	8g sounds more like math . . . than 8 + g . . . which sounds (like a) bit of a sum which you have to work out; but 8g just seems like an answer . . . in itself . . . but 8 + g, you still think, "Oh, what will it equal?"
Student 2:	It has to be 8 + g, because you're not told what g is.
Student 3:	You need to find out what g is and then you can add g to 8 to give you the answer. In the meantime, the best you can do is just guess.
Student 4:	e + f could be anything . . . 2 + 6, 4 + 4 . . .
Interviewer:	What is the best we can do for an answer, since we don't know g?
Student 4:	Perhaps it could be 12, because e + f + g = 4 + 4 + 4.

Source: Adapted from Hodgen, Küchemann, Brown, & Coe, (2008).

Sometimes, students have more difficulty with the representation than with the concepts, such as Student 3 (see Table 2.14), who can describe his thinking but struggles with what to write.

Other researchers describe this struggle with closure as the *Process-Product Dilemma*.[55] Algebra requires a shift in thinking so that an expression such as "$x + 5$" can be seen as an instruction for a procedure to add 5 to the variable "x," or as an answer, the number that is 5 more than "x," or as a function giving the output values resulting from the possible values of x. Whether "$x + 5$" is to be understood as a process or as an object/product depends on the situation. This notion is challenging for students in early algebra to accept. The ability to see $a + b$ as an object as well as a process can be initiated as students learn arithmetic. Students must be able to see that 13 can be written as $5 + 8$, which can represent the number of items in two sets, 5 and 8, as $a + b$ can represent the number of items in two sets containing a and b items.

"Many features of an item determine students' responses, one being the size of the number and the ease of automatic responses."[56] For instance, students correctly answered Problem 13 (add 12 to x) at a high rate but erred more often with Problem 8 (add 4 onto $3n$), perhaps because of the coefficient in front of the variable. For some students, mathematical convention is confusing, particularly the elimination of the multiplication operation using symbolism such as $10 \times h = 10h$.[57] Students working on Problem 12 may generalize this procedure by writing $10h$ and intending $10 + h$. Students may struggle with the representation of the mathematical idea rather than the concept. Some suggest that students' struggles are due to conventions in natural language. For instance, the similarity between the words *and* and *plus* may lead students to consider ab to mean the same as $a + b$.[58] It is important for teachers to explore students' rationale for why they conjoin numbers and variables in problems such as these, in order to determine if students accept lack of closure.

Researchers find that exploration of examples and non-examples by plugging in values for variables helps students overcome their inability to accept lack of closure and become comfortable with expressions as processes as well as products. In particular, computer programs, such as spreadsheets, can "encourage individuals to manipulate examples, to predict and test, and to develop experiences on which higher level abstractions may be built."[59] A cell in a spreadsheet can represent many different types of quantities and formulas, helping students to expand the potential role of variables in their minds.[60]

Teachers

- Help students understand how an expression can be an answer.
- Help students understand when they can evaluate a variable and when they cannot.
- Have students test numbers in the original expression and conjoined expression to see if they are equivalent.

LETTER EVALUATED

Common Core Standards: 6.EE.2, 7.EE.2, A-SSE.1, A-SSE.2, A-SSE.3

When students have very little understanding of variables, they may assign the letter a numerical value from the outset. The problems in Table 2.15 make this way of thinking evident. In Problem 14, the student may struggle to understand that an "answer" can be an expression. The interview (see Table 2.16) indicates that the students may not be comfortable with leaving an expression such as "$8 + g$" and feel compelled to come up with a numerical answer (see **Acceptance of Lack of Closure**). When students have the *Letter Evaluated* way of thinking, they figure out numbers for the letters using any means possible. This *numerical thinking* indicates the student's belief that the variable is a single number.[61] The first student interviewed assumed that the value of letters is related to their position in the alphabet, and evaluated the letter accordingly. The student in the second interview for Problem 14, as well as the student in Problem 15 (see Table 2.17),

TABLE 2.15. Problems on Evaluating Letters in Expressions and Equations

Problem	Response
14. If $e + f = 8$, $e + f + g =$	Correct: $8 + g$ (24%[a], 41%[b])
	$8g$ (4%[a], 3%[b])
	12 (13%[a], 26%[b])
	9 (2%[a], 6%[b])
	some other number (10%[a])
15. What can you say about r if $r = s + t$ and $r + s + t = 30$?	$r = 10$ (21%)
16. Describe the expression $2 + 3x$	5
17. David is 10 cm taller than Con. Con is h cm tall. What can you write for David's height?	Correct: $10 + h$ (39%, 74%)[c]
	$10h$ (14%, 5%)
18. What values can a take in the expression $7 + a + a + a + 10$?	(Over 50% solved for a)[c]
	$7 + a + a + a + 10 = 0$?
	$17 + 3a = 0$
	$3a = -17$
	$a = -17/3$

[a] Percent of 5475 12- to 14-year-old students from Hodgen, Küchemann, Brown, & Coe, (2008).
[b] Percent of 3,000 13- to 15-year-old students in Hart and Kuchemann (2005).
[c] Percent of 164 freshman college students in Trigueros, Ursini, and Reyes (1996).

TABLE 2.16. Interviews for Problem 14

Interviewer:	Why 15?
Student 1:	Because g is the seventh letter of the alphabet and $7 + 8 = 15$.
Interviewer:	Why 12?
Student 2:	Because $4 + 4 + 4 = 12$.

TABLE 2.17. Interview for Problem 15

Interviewer:	How did you get your answer?
Student:	I saw that there was supposed to be three things that added up to 30 so I figured they were each 10.
Interviewer:	What can you say about r?
Student:	It is where one of the 10s go.

TABLE 2.18. Interview for Problem 16

Interviewer:	How did you get your answer?
Student:	x is 1, so $2 + 3x = 2 + 3 * 1 = 5$.

assumed that there are three letters, so three equal numbers would be the answer. This again may be an example of the Letter Evaluated way of thinking, in which the student does not consider the letters as representing unknowns—or that they have any relationship to each other—but simply puts numbers in their places.

In Problems 16 and 17 (see Tables 2.18 and 2.19), the student believes x can be evaluated as the number 1 (see ***Representation: What Can Variables Stand for?***), so evaluates the expressions accordingly. Sometimes students are confused when teachers say, "$1x$ is just like x," and they interpret this as $x = 1$. Students solving Problem 18 for a may be a bit more advanced in algebra, but still are challenged in thinking that a must be evaluated, so they figure out a way to find that value. In the interview for Problem 18, the student appears to understand the problem and can describe how he or she might get the answer, but struggles to understand that the answer can be something with a letter—challenged by the symbolism in the answer, rather than the thinking behind it. The student wants to evaluate the problem into a specific number, and is therefore confused as to how to represent the answer. The tendency to evaluate the letter in the expression is based on many faulty assumptions such as these and can persist into college.

In order to help students not feel the need to evaluate letters in expressions into numbers, research suggests that teachers might consider having students use more language in algebra, such as having students write out words to describe

TABLE 2.19. Interviews for Problem 17

Interviewer:	How did you get 11? ($10 + h = 11$)
Student 1:	h is just like one, like having one number.
Student 2:	By itself it is 1, the h.
Student 3:	Because $h = 1$.

their thinking, and then help students understand how to symbolize that thinking in algebraic terms.

Teachers

- Help students understand that the letters chosen to represent a value(s) do not have any influence on the value(s) that letter represents.
- Help students become comfortable with answers that are expressions, such as $h + 10$.
- Help students understand when variables can be evaluated and when they cannot.

Variables and Expressions • 59

LETTER NOT USED

Common Core Standards: 6.EE.2, 6.EE.3, 6.EE.4, 7.EE.1, 7.EE.2, A-SSE.1, A-SSE.2, A-SSE.3

Another approach for students who do not understand the meaning of variables is simply to ignore the letter, or at best acknowledge its existence, but without giving it a meaning. The problems in Table 2.20 demonstrate that way of thinking.

The answers to these problems may represent the *letter not used* way of thinking. In the first problem, the student may acknowledge the existence of the letter, but not give it meaning or simply ignore it (see interviews in Table 2.21). The student can ignore the letter and just deal with the numbers in both of the first two problems, adding 4 and 5 or 4 and 3.[62] In the third problem, a student may correctly respond 45, but ignore the letters and just deal with the numbers

TABLE 2.20. Problems on Ignoring Letters

Problem	Response
18. Add 4 onto $n + 5$	Correct: $n + 9$
	9 (20%)[a]
19. Add 4 onto $3n$	Correct: $3n + 4$
	$7n$ (31%)[a] also $4 + 3 = n$[b]
20. If $a + b = 43$, $a + b + 2 = ?$	Correct: 45 (97%)[a]
21. Simplify $2a + 5b + a$	Correct: $3a + 5b$ (40%, 60%, 66%)[c]
	8ab (20%)[a]

[a]Percent of 3,000 13- to 15-year-old students in Hart and Kuchemann (2005).
[b]One of the six students in Chalouh and Herscovics (1988).
[c]Percent of 3,000 13-, 14-, and 15-year-old students (respectively) in Hart and Kuchemann (2005).

TABLE 2.21. Interviews for Problems on Ignoring Letters

	Problem 18
Interviewer:	How did you get 9?
Student:	I just added 4 plus 5.
	Problem 19
Interviewer:	How did you get $7n$?
Student:	I just added 4 plus 3.
Interviewer:	How did you get $4 + 3 = n$?
Student:	In math you add the numbers to get the answer.
	Problem 20
Interviewer:	How did you get 45?
Student:	I added 43 and 2.

43 and 2, thereby not really showing understanding of the equivalency meant in the problem. Finally, in Problem 21, students are aware of the letters, but simply collect everything into one term. When students think this way, they are willing to have **Acceptance of Lack of Closure**, in regard to the ambiguous existence of the letters in the final answer, but do not assign any meaning or significance to the letters.

In order to help *students* address this inaccurate way of thinking about variables, it may be helpful to consider how reading mathematics is analogous to reading English. When reading English, it doesn't make sense to skip over words. In mathematics, it doesn't make sense to skip over symbols. Seeing distinctions can help, such as through the following examples:

1. Ask students to describe the difference between 3 and $3 + n$.
2. Ask students to describe the difference between $3 + n$ and $n + n$.
3. Ask students to describe the difference between $3 + n$ and $3n$.
4. Explicitly identifying n as an unknown number may also be useful: "I don't know what n is, but can you tell me what number is 3 more than n?"
5. Translating between verbal and mathematical descriptions may help students understand the idea of a variable (see **Translation Difficulties in Changing Words to Algebraic Sentences** and **Modeling: Translating Word Problems into Equations**). "How could I calculate the age I will be in 12 years? I could write it as (my age now) + 12, but that's a lot to write! Let's shorten it to $a + 12$, where I've put "a" in place of "my age now.""

Teachers

- Help students understand that variables represent values.
- Help students understand how variables influence expressions and equations.
- Help students understand the impact of eliminating variables from expressions and equations when it is mathematically incorrect, versus when it is mathematically correct. For instance, the difference between subtracting an x from both sides of the equation $2x + 3 = x - 7$ versus ignoring the variable in the expression $4n + 7$ to get 11.
- Help students understand mathematical convention with variables. For instance, $12a$ meaning 12 multiplied times a—rather than added—and why $1x$ is the same as x.

LETTER USED AS AN OBJECT

Common Core Standards: 6.EE.2, 6.EE.3, 6.EE.4, 7.EE.1, 7.EE.2, A-SSE.1, A-SSE.2, A-SSE.3

In this third category, describing students' thinking when they struggle to attach meaning to variables, students regard the letter as shorthand for an object or as an object in its own right (see Table 2.22). Answers to the problems in Table 2.22 may represent the *letter used as an object* way of thinking. The interviews in Table 2.23 demonstrate how students might talk about these problems when they think about the letter as referring to an object rather than a value. The letter is per-

TABLE 2.22. Problems on Letters as Objects

Problem	Response
22. Simplify $2a + 5b + a$	$3a + 5b$ (60%)[a]
23. If $x + y = 10$, circle all the meanings that y could have: 3 10 12 7.4 the number of apples in a box an object like a cabbage	3 (76.5%)[b] 10 (51.5%) 12 (11.9%) 7.4 (66.5%) number of apples in a box (40.4%) object like a cabbage (22.7%)
24. The following question is about this expression: $2n + 3$ ↑ The arrow above points to a symbol. What does the symbol stand for?	An object or word beginning with n (4%, 11%, 9%)[c]
25. Simplify $2a + 5a$	$7a$ (100%)[d]

[a] Percent of 3,000 13- to 15-year-old students in Hart and Kuchemann (2005).
[b] Percent of 379 12- to 15-year-old students in Warren (1999).
[c] Percent of 110 sixth-grade students in McNeil et al. (2010). McNeil et al., 2010, differentiated between using c and b versus other letters in a similar problem.
[d] Percent of 24 12-year-old students in Bills, Ainley, and Wilson (2003).

TABLE 2.23. Interviews for Problems on Letters as Objects

	Problem 22
Interviewer:	How did you get $3a + 5b$?
Student:	I collected all the a's together
	Problem 25
Interviewer:	Why did you put $7a$?
Student:	Anything can be an a, so you can put two apples and five apples, so then you can add the five and the two, which is seven, and then you can put seven a equals seven apples.

haps seen as an abbreviation for something, or a label. For instance, in a problem about the perimeter of a quadrilateral in which a side is labeled "x," the student may believe that x is the name of the side rather than the unknown measurement of that side (see **Area, Perimeter, and Variables**). A student may respond correctly "$3a + 5b$" for Problem 22 or "$7a$" for Problem 25, but have a very different understanding of the letters. A student could treat the a as an object in itself, literally as the letter a or as an object such as apples, rather than as representing an unknown number or range of unknowns. In these cases, the letters' abstract meaning is ignored and they are treated as concrete, actual objects.

In a test of 3,000 students, researchers found that 73% of 13-year-olds, 59% of 14-year-olds, and 53% of 15-year-olds treated the letters in equations and expressions as objects.[63] In research with Problem 23, interviews with students indicated that some were thinking y represented "a number of cabbages," rather than a label of an object, even though they referred to the variable as an object.[64] It is often unclear from the language teachers or students use whether they are referring to unknown quantities or objects, which can make this issue difficult to diagnose. Some students expect to find single answers, leading to the common error of "simplifying" terms such as $2a + 5b$ to $7ab$. Researchers refer to this misconception as *non-numerical* thinking in regard to variables.[65] In another study of 1000 14-year-olds, less than half were able to think of a letter as a number.[66]

Using letters as labels can be confusing for students who see formulas such as "$A = L \times W$" and are taught to think of "area equals length times width." Rather than understanding the letter as representing a quantity or quantities, the student could think of the letter as a label, which may contribute to struggles in reaching the sophisticated understanding of letters as variables. "Students notice that concepts in applied mathematics are usually denoted by the initial letters of their names (A for area, m for mass, t for time, etc.). It is likely that this use of letters reinforces the belief that letters in mathematical expressions and formulas stand for words or objects rather than for numbers."[67]

Researchers suggest that literal symbols as variables are difficult for students to understand until they reach Piaget's highest stage of development, *formal operational thought*, around age 15. Before then, students tend to think of literal symbols as shorthand labels for objects or in place of a specific number.[68] However, age alone does not explain all aspects of students' misconceptions,[69] and some studies show that students can understand variables at early ages.[70] Researchers also found that the idea of "letter as abbreviated word" may be caused by certain teaching materials or teachers' explanations. Some materials and textbooks used by teachers explicitly presented letters as abbreviated words (e.g., c could stand for cat, so $5c$ could mean five cats), and this misconception is particularly resilient through schooling.[71]

Algebra teachers sometimes instruct students to think of the a as representing apples, so they can "just combine the apples together" and successfully solve the problem "Add $3a$ and $4a$." Further, in a problem such as Problem 22, teachers sometimes instruct students to treat the *a* as apples and *b* as bananas so they don't add 3*a* and 5*b* to get 8*ab* because "You can't combine apples and bananas." This type of thinking is described by researchers as *fruit salad algebra*,[72] which encourages students to think of letters as names or labels of things. While this approach might help students in the moment in learning to combine like terms, this can lead to long-term confusion and "the belief that one cannot multiply expressions such as 2*a* and 5*b* because one cannot multiply apples and bananas."[73] Instead, teachers should clearly distinguish between the names and numbers of objects, as in "*a* represents the unknown number of apples." Teachers understand the shortcut language "apples" as implying "unknown number of apples" but students do not necessarily understand it: in early algebra, they are struggling to comprehend how letters are used in different ways, and in this context, letters do not signify objects, but unknown quantities of those objects—an important distinction.

In Problem 26 (Table 2.24), many students understood the letters to represent the objects "cakes and buns" rather than the cost of cakes and the cost of buns. Those students who answered incorrectly were found to make poor progress in their understanding of literal symbols as variables. Research indicates that teachers' choice of letter used to represent the quantity may influence students' tendency to think about the letters as objects instead of variables (see Figure 2.1[74]). In one study, only 37% of Grades 6–8 students answered Problem 26 correctly when *c* and *b* were used as the variables while 57% answered correctly when *x* and *y* or Ψ and ϕ were used to represent the quantities instead.[75] While students' performance in this study changed depending on the letters used to represent the quantities, other researchers caution against the assumption students treat the letters as objects and encourage teachers to have students explain their rationale.[76]

In order to have a comprehensive understanding of the interaction between numbers and variables and operations in a term such as "4*c*" students need to understand the value of the number, the potential value of the letter, and that the representation of 4 next to *c* implies multiplication. Students may rely on their previous experiences and interpret 4*c* as "four of some object that is represented by *c*." On the other hand, if the student can understand 4*c* as that combination of number, variable, and operation, in Problem 26 they must understand 3*b* in the same way. Finally, the student must understand 4*c* + 3*b* as a similar encapsulated quantity that represents the total cost of cakes and buns, with each variable signifying potential values.[77] What might seem simple to those of us who teach math—

TABLE 2.24. Problem 26: Letters as Objects

Problem	Response
Cakes cost *c* pence each and buns cost *b* pence each. If I buy 4 cakes and 3 buns, what does 4*c* + 3*b* stand for?	Correct: total cost of cakes and buns (22%)[a] "4 cakes and 3 buns" (39%)[a]

[a]Percent of 3,000 13- to 15-year-old students in Hart and Kuchemann (2005).

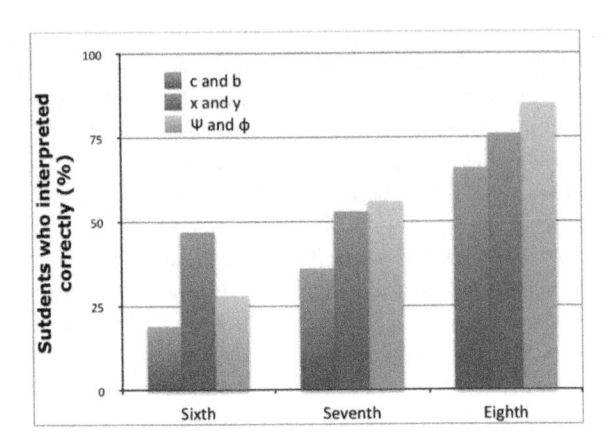

FIGURE 2.1. Use of Variable in Cakes and Buns Problem.

and have been successful in many math classes beyond algebra—is actually quite involved, with multiple pieces of understanding coming together to represent the concept of the cost of cakes and buns.

Researchers have shown that when students were asked to use their intuition and self-generate ways of representing a problem instead of relying on the teacher or instructional materials, it led to better understanding of what the variable represented (see ***Algebraic Relations: Student Intuition and Informal Procedures***). Although, in the short term, the use of mnemonic devices such as "think of a as apple(s) and b as banana(s)" may be beneficial, it may reinforce students' naïve conception that letters stand for labels instead of variables. We can't teach that variables don't ever stand for objects. In day-to-day life, we use letters to stand for objects all the time. Recipes call for 4c, or 4 cups, of flour. A grocery list might have shorthand for items. To counteract students' experiences, teaching variables should include when letters stand for labels and when they stand for quantities; this will help students understand the difference. It is important to help students understand the range of uses of letters and when each use is appropriate, in what context, and how the meaning of letters changes in each context.

Teachers
- Help students understand the many different uses of letters in algebra.
- Help students differentiate between letters as signifying objects and letters as variables.
- Be aware of how your use of language in the classroom may encourage or discourage students' thinking of letters as objects.
- In initial work with story problems, not always using the first letter of the item indicated when naming variables might help students understand what variables represent.

LETTER USED AS AN UNKNOWN NUMBER

Common Core Standards: 6.EE.2, 6.EE.3, 6.EE.4, 7.EE.1, 2; A-SSE.1, A-SSE.2, A-SSE.3

In the most basic understanding of variable, students regard a letter as a specific but unknown number, and can operate upon it directly.[78] Consider the problems in Table 2.25. The answer to the problems in Table 2.25 may represent the *letter used as an unknown number* way of thinking about variables. At this lowest level of understanding variables, the student thinks of the letter as representing a specific number. The student can operate on that letter as if it were the specific number. In Problem 27, for instance, thinking that c can only have one possible value, the student may believe it must be 4.

The struggle for understanding that is evident in the interviews in Tables 2.26 and 2.27 may also be based on the students' experience with natural numbers. The symbols 5 and 6 have direct reference to certain quantities, and students may be applying this type of thinking to letters.[79] There can be frustration for students because, as students learn that letters are "variables," they may get the sense that they really can't ever say anything about them unless they know the specific value.

Students' answers to these questions can be divided into two categories. First, students have no awareness of conditions that might restrict the variable. They confine their manipulation to terms in the given problem. Students might substitute one number into the problem to test, or simply manipulate the terms to no end. Second, students are able to take into account the various conditions, can understand that the possibilities for the answer are determined by the limitations given, and can reason how the variables vary in coordination with each other. As with Problem 27 in Table 2.25, a pre-complete understanding may involve students providing a systematic list for c, such as "0, 1, 2, 3, 4," indicating a pos-

TABLE 2.25. Problems Regarding Letters Used as an Unknown

Problem	Response
27. What can you say about "c" if $c + d = 10$ and c is less than d?	• One number (e.g., $c = 4$) (39%[a], 4.5%[b]) • Only integer responses (e.g., 1, 2, 3, 4)(42.9%[b]) • $c = 10 - d$; so $c = 4$ and $d = 6$ (64%[c]) • $c < 5$ (can't be $= d$ and can't be $> d$) (2.1%[b])

28. Given $a = 28 + b$, determine whether
  i. a is greater than b
  ii. b is greater than a
  iii. $a = 28$
  iv. Cannot tell which is greater
	iv. Cannot tell which is greater (25%[d])

[a]Percent of 3,000 13- to 15-year-old students in Hart and Kuchemann (2005).
[b]Percent of 379 12- to 15-year-old students in Warren (1999).
[c]Percent of 278 students in Coady and Pegg (1993).
[d]Percent of students in Stacey and MacGregor (1997).

TABLE 2.26. Interviews for Problem 27

Interviewer:	How did you get 4?
Student 1:	c is smaller than d, so it must be 4.
Interviewer:	How would we work out an expression involving c?
Student 2:	c plus d equals 10 [writes $c + d = 10$], c is less than d [writes $c < d$] ... could we write it like that? [writes $c = 10 - d$, $c < 10 - d$]
Interviewer:	c is less than d and c is less than $10 - d$. Are they the same statement?
Student 2:	No.
Interviewer:	What do we have to do now?
Student 2:	I think maybe we should plot some numbers in.
Interviewer:	Okay, let's try that.
Student 2:	Because it equals 10, we have to choose numbers less than 10 ... I really don't have any idea.
Interviewer:	What can you write down about c?
Student 3:	First of all, because c is less than d, I actually write it down graphically [writes $c < d$] Now because you have $c + d = 10$, that [points to $c < d$] means d must be greater than c. So for any number c, c is less than d so therefore c is less than 5 [writes $c < 5$].
Interviewer:	Why 5?
Student 3:	Because you have 10 as a number and because d is greater than c, then c cannot be greater than 5 because 5 is the half way mark, so you have $5 + 5$, that will be 10, but it says c is less than d, then it cannot be 5 but it could be 4.9 plus 5.1 as d.

TABLE 2.27. Interviews for Problem 28

Interviewer:	Why do you think you can't tell which is greater?
Student 1:	The letters stand for unknown numbers, so you can't tell. They could be anything.
Student 2:	b is greater because it has 28 added to it.
Student 3:	a is 28 because the equation says "a equals 28 then add b."

TABLE 2.28. Problem 29: Letters as Unknown Issues

Problem	Response
29. Can you tell which is larger $3n$ or $n + 6$? Why?	Justification: One number (3%, 5%, 4%)[a]

[a] Percent of 373 sixth-, seventh-, and eighth-grade students' answers, respectively, in Knuth et al. (2005).

sible natural number bias—rather than considering fractional answers or negative numbers. This way of thinking about variables is considered in the **Letter Used as a General Number** section.

As Table 2.28 indicates, Problem 29 is a little bit different, in that students only tested one number to determine which side was larger. It is unclear from the research whether they might have considered using additional numbers, but they were satisfied that their test was valid with only one trial. They did not understand that different numbers might result in different answers for the question. Given the very low percentage of students answering these types of problems correctly throughout schooling, it is clear that students find it difficult to understand the full nature of variables.

A comprehensive understanding of a letter used as an unknown number requires that students:[80]

> recognize and identify in a problem situation the presence of something unknown that can be determined by considering the restrictions of the problem; interpret the symbols that appear in an equation as representing specific values; substitute a value for the variable that makes the equation a true statement; determine the unknown quantity that appears in problems by performing the required algebraic and/or arithmetic operations; and interpret a specific situation to symbolize the unknown quantities and use them to design equations representing the situation mathematically.

Teachers

- Create opportunities for students to understand how letters can represent more than one value, such as $x + y = 10$.
- Help students understand that testing one number in a situation is not sufficient justification for an argument.

LETTER USED AS A GENERAL NUMBER

Common Core Standards: 6.EE.2, 6.EE.3, 6.EE.4, 7.EE.1, 7.EE.2, A-SSE.1, A-SSE.2, A-SSE.3

In the *letter used as a general number* level of developing an understanding of variables, students see the letter as potentially representing several values, rather than just one. Consider the problems in Table 2.29. A student's answer to a problem in Table 2.29 may represent the letter used as a general number level of thinking. At this level, the student can think of the letter as representing many potential values.[81] Some research indicates that "a prerequisite in order to develop an understanding of variable as a general number is the ability to recognize patterns and to find or deduce general rules and methods describing them."[82] From this perspective, students need to understand that patterns can extend indefinitely before they can understand that a letter can help represent the general relationship indicated by the pattern.

When asked about the meaning of a variable, students who understand that letters can represent multiple values have responded "The symbol is a variable, it can stand for anything," and "a number, it could be 7, 59, or even 363.0285."[83] Researchers found that less than 50% of sixth-grade students understood that letters could represent multiple values, but more than 75% of eighth-grade students had that conception.[84] For Problem 30 in Table 2.29, ($c + d = 10$), the interviewer asked "Why do you think the answer is 1, 2, 3, or 4?" and the student responded "If c is less than d, then d has to be 6, 7, 8, or 9." The explanation the student provides demonstrates an understanding that n could potentially be multiple different values, thereby providing different answers to the question of which expression would be larger.

With the problem $c + d = 10$, a student may respond with either $c < 5$ or $c = 1, 2, 3, 4$—indicating an understanding of c representing many numerical values. However, researchers found that the constant c in an expression of the form $x + y = c$

TABLE 2.29. Problems for Letters Used as a General Number

Problem	Response
30. If $c + d = 10$, and c is always less than d, what values may c have?	• Correct: $c < 5$ (included 0, negatives, and fractions) (5%[a]) • no response/incorrect list of numbers (29.4%[a]) • One number (e.g., $c = 4$) (4.5%[a], 39%[b]) • 1, 2, 3, 4 (only integers) (42.9%[a], 19%[b]) • included 0, negatives, and/or fractions (20.6%[a]) • < 5 (2.1%[a], 11%[b])
31. Can you tell which is larger, $3n$ or $n + 6$?	"Can't tell" (with multiple values justification) (11%, 51%, 60%)[c]

[a]Percent of 379 12- to 15-year-old students in Warren (1999).
[b]Percent of 3,000 13- to 15-year-old students in Hart and Kuchemann (2005).
[c]Percent of 373 sixth- through eighth-grade students, respectively, in Knuth et al. (2005).

influenced students' thinking about the possibilities for x and y. Students were reluctant to pick values for either x or y that were greater than c.[85] As in this problem, students typically did not pick numbers greater than 10 for either x or y. Few students considered the possibility that c could be a negative number, constrained by their thinking about the composition of the number 10. While many students in the research had an understanding of the letter possibly representing more than one number, few students in the research understood the infinite possibilities of that relationship. A student with a sophisticated understanding of **Letter Used as a Functional Relationship** is able to comprehend the full range of possibilities for the letter in a mathematical context. A student with the letter used as a general number level of thinking, on the other hand, may have some self-imposed constraints on what those numbers might be.

A comprehensive understanding of a letter used as a general number requires that students:[86]

1. Recognize patterns and perceive rules and methods in sequences and in families of problems;
2. Interpret a symbol as representing a general, indeterminate entity that can assume any value;
3. Deduce general rules and general methods in sequences and families of problems;
4. Manipulate (simplify, develop) the variable; and
5. Symbolize general statements, rules, or methods.

Teachers

- Help students understand a multiple-values interpretation of letters.
- Probe students' answers to determine the extent of their understanding of the range of possibilities. Would 3.3 work for c? Would -11? Would numbers between 4 and 11?

LETTER USED AS A FUNCTIONAL RELATIONSHIP

Common Core Standards: 6.EE.2, 6.EE.3, 6.EE.4, 6.EE.6, 7.EE.1, 7.EE.2, A-SSE.1, A-SSE.2, A-SSE.3

In the highest level of developing understanding of variables, students see the letter as representing a range of unspecified values, and a systematic relationship is seen to exist between those sets of values.[87] The problems in Table 2.30 help differentiate whether students see the functional relationships with variables. The answers that students #3 – #6 gave to the interviewer in Table 2.31 may represent the *letter used as a functional relationship* way of thinking. At this level, the student understands that the letter can represent a range of potential values and that there is a relationship that exists between the values. One of the keys to this level of thinking is to comprehend *how* the values of the unknown(s) change. In the "Which is Larger?" problem, a student with this level of thinking would be able to explain the relationship between the two expressions and how the size of "$2n$" or "$n + 2$" depended on the value of n, as student #4 did. For Problem 32 in Table 2.30, the number of students who say "can't tell" steadily increases with grade level. This may be a good indication of students' movement toward understanding variables in a functional relationship, but it is important that students are able to articulate that relationship, which is the difference between the responses of Student #3 and Student #6.

The answer "$2n$" makes sense with the rationale that multiplication generally grows faster than addition. Students learn that multiplication makes things bigger faster.[88] However, being able to comprehend that the relationship changes for $n < 2$ is evidence of a higher-order understanding of the complex relationship between

TABLE 2.30. Problems for Letter Used as a Functional Relationship

Problem	Response
Problem 32	
Which is larger, $2n$ or $n + 2$?	$2n$ (71%[a]) (79%[b])
	$2n$, because multiplication makes things bigger (48%[d])
	$n + 2$ (16%[a])
	Can't tell (18%, 54%, 64%[c])
	vs.
	It depends upon the value of n; if $n < 1$ then $2n < n + 2$ ... if $n > 2$, $2n > n + 2$ (1%[d])
Problem 33	
A friend gives you some money. Can you tell which is larger: the amount of money your friend gives you plus six more dollars, OR three times the amount of money your friend gives you?	Can't tell (43%, 57%, 68%[b])

[a] Percent of 3,000 13- to 15-year-old students in Hart and Kuchemann (2005).
[b] Percent of 278 students in Coady and Pegg (1993).
[c] Percent of 373 sixth-, eighth-, and eighth-grade students (respectively) in Weinberg, et. al (2004).
[d] Percent of 5475 12- to 14-year-old students in Hodgen, Brown, Coe, and Küchemann (2012).

TABLE 2.31. Interviews for Problem 32

Interviewer:	Can you tell which is larger, $2n$ or $n + 2$?
Student 1:	$2n$ because it is multiplying.
Student 2:	Yes, $n + 2$ is bigger because they $+$.
Student 3:	No, because you do not know what n is (sixth-grade student).
Student 4:	$2n$, because it is multiply, where the other is only plus ... for example if $n = 3$, then $2(3) = 6$, but $2 + 3$ is only 5; and 6 is bigger than 5.[a]
Student 5:	No, because n is not a definite number. If n was 1, $2n$ would be 2 and $n + 2$ would be 4, but if n was 100, $2n$ would be 200 and $n + 2$ would be 102. This proves that you cannot tell which is larger unless you know the value of n (eighth-grade student).[b]
Student 6:	If n is less than 1, $2n$ is less than $n + 2$ (writes down $2n < n + 2$, $n < 1$); if $n = 2$, $2n$ equals $n + 2$, (writes down if $n = 2$, $2n = n + 2$); if n is greater than 2, $2n$ is greater than $n + 2$ (writes down if $n > 2$, $2n > n + 2$).[a]

[a]From Coady and Pegg (1993).
[b]From Knuth et al. (2005).

the two expressions. A student at the level of letter used as a functional relationship recognizes how one set of values ($2n$ and $n + 2$) changes as another set (n) varies in a dependent relationship. A sophisticated understanding of this problem requires students to view the variable as "something that can not only take on multiple values, but must have those static values compared to each other."[89]

This level of understanding is perhaps best seen in the problem noted in the previous section on **Letters Used as General Number**: "If $c + d = 10$, and c is always less than d, what values may c have?" It is not that c can be anything, but that there is a dependent relationship. If c is 8.7, then d must be 1.3. The choice of the independent variable dictates what the dependent variable will be. Students at this level are able to take into account the "if . . . then" relationship and consider constraints, understanding how the answer might change, depending on the value of c. While Students #3 and #4 may be at that level of thinking with Problem 32, it is not clear from their answer the extent of their reasoning, which would require further probing.

TABLE 2.32. Justification of Responses to Which is Larger, $2n$ or $n + 2$?

Basis of Justification	6th	7th	8th
Variable: the value depends on what number you plug in	11%	51%	60%
Singular value: the letter has a single numeric value	3%	4%	5%
Operation: the operation (add or multiply) determines the value	0%	5%	9%
Other: incorrect explanation	42%	15%	16%
No response/do not know	45%	23%	11%

Source: Based on Coady & Pegg (1993)

Table 2.32 provides information as to how students justified their response to Problem 32. Sixth-grade students have the least experience with variables, so it makes sense that they are generally not able to justify their answer with explanations incorporating the idea that letters can represent multiple values, while the majority of seventh- and eighth-grade students can. Other students justified their response with the idea that the letter could only represent a single unknown or that the operation dictated the size of the expression (as in the comments of Students #1 and #2, above).

One interesting additional finding from research is that students have different responses when the problem is given verbally rather than abstractly, particularly with sixth-grade students. Most of these students have not had experience with using letters to represent quantities, and yet, when the situation is grounded within a context, they are more able to understand the idea of varying quantities. Use of verbal forms may, therefore, be a way to help students develop a full sense of the idea of variable. Students need to practice creating contexts to match algebraic language (e.g., a reason for $6x - 3y = 18$ to make sense).

A comprehensive understanding of the letter used as a functional relationship way of thinking about variables requires that students:[90]

- Recognize the correspondence between related variables, independent of the representation used;
- Determine the value of the dependent variable, given a value of the independent one;
- Determine the values of the independent variable, given a value of the dependent one;
- Recognize the joint variation of the variables involved in a relation, independent of the representation used;
- Determine the interval of variation of one variable, given the interval of variation of the other one; and
- Symbolize a functional relationship, based on the analysis of the data of a problem.

"An understanding of variable implies the comprehension of all these aspects and the possibility to shift between them depending on the problem to be solved."[91] Given that there are so many ways to interpret variables, it is important that students develop the flexibility to adapt, depending on the specific situation. A rich understanding of variable requires students to have facility in manipulating the symbols, while maintaining a sense of what the variable represents in the problem.

Teachers

- Help students understand how changes in one variable affect the other variable(s).

Variables and Expressions • 73

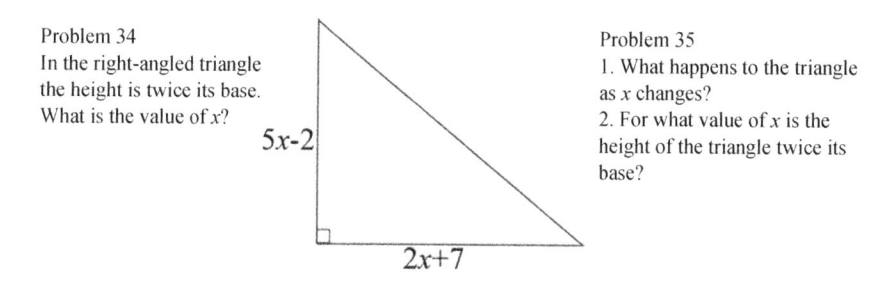

Problem 34
In the right-angled triangle the height is twice its base.
What is the value of x?

5x-2

2x+7

Problem 35
1. What happens to the triangle as x changes?
2. For what value of x is the height of the triangle twice its base?

FIGURE 2.2. Triangle Problem.

- Help students understand the difference between variables that represent a single value, a range of values, and infinitely many values—and the circumstances that create those differences.
- Help students understand the circumstances in which variables can represent an infinite number of values, but within a particular range.
- Consider ways to alter textbook problems to extend students' thinking about variables. For example, in a textbook, you might find something such as Problem 34 (Figure 2.2).[92] When you take the same image, but change the questions as shown in Problem 35, the questions transform "the task to one where the learners have to think of x as a variable."[93]

ALGEBRAIC NOTATION

Common Core Standards: 6.EE.1, 6.EE.2, 6.EE.3, 6.EE.4, 7.EE.4, A-SSE.1, A-SSE.2, A-SSE.3, F-IF.2

The ability to think algebraically comes before the understanding of notation.[94] Sometimes students' struggles with algebraic notation are not based on the mathematical concepts, but the mathematical conventions. "For students learning formal notation, all they are given is a collection of someone else's symbols."[95] For instance, it can be quite confusing for students that $4\frac{1}{2}$ means $4 + \frac{1}{2}$ and 89 means 8 tens + 9 ones, but $4x$ means 4 multiplied times x.[96] The symbol "3" can have multiple meanings depending on its location, such as 34, 43, 4^3, $\sqrt[3]{}$, and 4/3. "Words gather meanings from the company they keep,"[97] and so it is with the symbols of mathematics.

There are specific ways of symbolizing in algebra that students need to learn. Algebraic notation exists because we needed a way to be consistent, and to be able to communicate mathematical ideas clearly. While many uses of notation can seem second nature to teachers, they can be considered a similar challenge as learning a new language for many students.[98]

> Students' interpretations of algebraic notation are based on intuitive assumptions and sensible, pragmatic reasoning about an unfamiliar notation system; analogies with symbol systems used in everyday life, in other parts of mathematics, or in other school subjects; interference from other new learning in mathematics; and poorly designed and misleading teaching materials.[99]

TABLE 2.33. Problems for Algebraic Notation

Problem	Response
36. David is 10 cm taller than Con. Con is h cm tall. What can you write for David's height?	Correct: ($h + 10$) (39%, 52%, 63%, 73%)[a] Other responses: $h10$, Uh (unknown height), 110 (a reasonable height), 18 (10 + 8th letter of the alphabet)
37. Add 12 to x	Correct: ($12 + x$ or $x + 12$) (68%, 93%)[b] $12x$ (2%)
38. Simplify $(n + 5) \times 3$	Correct: ($3n + 15$) (14%, 47%)[c] (9%)[d] $15n$ or $5nx3$ (14%, 12%) $n + 15$ (20%)[d] $3n + 5$ (25%)[d] 15 (8%)[d]
39. Is $t + t$ ever larger than $t + 4$? If so, when? Are $t + t$ and $t + 4$ ever equal? If so, when?	Correct: (46.4%, valid reason 28.8%)[c] Correct: (43.0%, valid reason 36.1%)

[a]Percent of 1463 seventh- through tenth-year students in MacGregor and Stacey (1997), respectively.
[b]Percent of 678 seventh- and tenth-year students in Stacey and MacGregor (1994), respectively.
[c]Percent of 379 12- to 15-year-old students in Warren (1999).
[d]Percent of 5475 12- to 14-year-old students in Hodgen, Brown, Coe, and Küchemann (2012).

Yet, as mentioned earlier, students as young as eight years of age can learn to express algebraic relationships[100] and operate on unknown values with letters.[101] Consider the problems in Table 2.33.

Basic Operations

All students understand the concept of one person being taller than another by a certain amount. The issue for students is how to express that algebraically. In Problem 36, students who write $h10$, for instance, can intend the notation to mean $h + 10$. This type of thinking is apparent in the interview for Problem 39. The interviewer asks: "Are $t + t$ and $t + 4$ ever equal? If so, when?" The student responds: "$t + t$ is $2t$ and $t + 4$ is $4t$ so $t + 4$ would be bigger." In the next section on *Area, Perimeter, and Variables* there are multiple examples of this type of thinking—writing a collection of variables and numbers with the operation missing, but intended. As students begin formal algebra, they may just put terms together because they don't know what else to do, or they may generalize the mathematical convention that $3y$ means 3 multiplied times y to $3y$ also being able to symbolize $3 + y$. They may collapse $h + 10$ into $h10$ because they don't have *Acceptance of Lack of Closure* and need to come up with a single-term answer as described previously with Problem 39.

Lastly, the sequence of natural English language is reading from left to right. This may lead students to

> read the symbol ab as *a and b*, and interpret it as $a + b$. Or the student may read the expression $2 + 3a$ from left to right as $2 + 3$ giving 5, and consider the full expression to be the same as $5a$.[102]

While this type of difficulty appears to diminish over time, as the data from Problem 37 suggests, with 93% of students answering correctly by 15 years of age, students still had significant difficulty with Problem 38 in the same study. Students in early algebra also have considerable difficulty interpreting the equal sign as a balance of equal quantities, rather than a "get the answer" sign (see *Algebraic Relations: Variables on Both Sides*). Ultimately, how problems are structured may lead to the types of confusion with notation described here.

Students learning algebra may not understand when there is mathematical rationale for notation being the way it is, and when mathematicians just agreed on a particular practice. Much of notation in algebra is arbitrary, and simply agreed upon by mathematicians for the sake of convenience. This can lead students to the perception that math is just a bunch of things to memorize without meaning. Math teachers may contribute to this perception by using tricks or mnemonics to help students learn mathematics, such as FOIL (first, outer, inner, last for multiplying two binomials) or PEMDAS (parentheses, exponents, multiplication, division, addition, subtraction for order of operations). These also can include sayings for students to memorize, such as "collect together like terms," "of means multiply," "add the same thing to both sides," "change sides, change sign," and "to divide, turn upside down and multiply."[103] While helpful in that moment, it may contribute to the feeling some students have: that the nature of mathematics is procedures you do by rote, rather than understanding. Furthermore, those strategies may be limiting (e.g., what do you do when you have to multiply a binomial with a trinomial?) or misleading (e.g., multiplication does not always precede division in the order of operations).[104] It is important for teachers to make distinctions between what is *arbitrary*—agreed-upon notational conventions that must be memorized—and what is *necessary*—procedures that have mathematical rationale.[105]

Generalization

Perhaps the primary reason for algebraic notation is the need to generalize. In arithmetic, mathematical symbols indicate concrete ideas with numbers. In algebra, variables are introduced in order to represent mathematical ideas such as unknown solutions ($x + 9 = 13$) and the character of possibilities or the relationship in the pattern ($5m = n$). In the first case, a single answer is represented by the letter. In the second, a general relationship is described that for every case of m, n is five times as large.

In arithmetic as well as algebra, symbols typically refer to some reality. The number 5 can refer to five apples in a basket, and a letter can refer to the solution of the equation ($x + 9 = 13$) or the dependent part of the relationship (n depends upon what m is). Understanding the reference[106] and representing generalization[107] are concepts that are typically very difficult for novice algebra students to understand. One of the mental shifts in algebra is for students to be able to manipulate symbols as a goal in itself—without any connection to a reference reality—but based on rules regarding relationships between symbols and the notation. In other sections of the book we discuss generalization in further detail, such as **Letter Used as General Number** and **Patterns and Functions: Promoting Generalization.** The main issue addressed here is not the act of generalization, but rather developing students' understanding of how generalization is represented in the form of letters.

Exponents and Subscripts

Researchers found that there is widespread misunderstanding of subscript and superscript notation. Consider the problem in Table 2.34. The interview for Problem 40 in Table 2.34 reveals that students can have unique understandings of notation. For instance, a simple reading of the answer would not have uncovered the level of conceptual understanding but lack of notational understanding behind the number "25."[108] Upon questioning, teachers may find that students understand the concepts but lack the means to represent that concept in the way that is acceptable to mathematicians. In particular, when the student writes 2_5 to represent two

TABLE 2.34. Problem 40: Find the Perimeter

Interview

Find the perimeter:

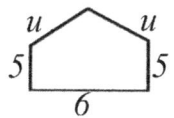

[Figure: pentagon (house shape) with bottom side 6, two slanted top sides labeled u and u, two vertical sides labeled 5 and 5]

Student:	I'd do two fives, two u's, and one six (writes $25 + 2u + 16$).
Interviewer:	You've written two fives there. Now what do you mean by that?
Student:	Oh! It should be smaller! Because otherwise you'll think ... (writes 2_5 with the five as a subscript).
Interviewer:	What do you mean by that?
Student:	Two fives, multiply it. Two multiplied by five, two multiplied by u ...
Interviewer:	You do not mean twenty-five?
Student:	No! Two lots of five. Oh ... (writes 1_6 instead of 16).
Interviewer:	Why did you write $162u$?
Student:	5 plus 5 plus 6 is 16 and two "u's"

fives, it is evident that, in the absence of clarity, students can be creative in using such mathematical representations as subscripts. This confusion may come from other experiences with symbols, such as chemical notation, in which CO_2 means a carbon atom plus two oxygen atoms. Students often struggle to understand the meaning and purpose of subscripts such as y_1 or F_{n-1}.[109] For instance, in one study, students would write $F_{n-1} + 2 = F_n$ as $Fn - 1 + 2 = Fn$ or $Fn_{-1} + 2 = Fn$, indicating that they didn't really understand the difference between the notation and the terms in the equation. In particular, using this type of notation for recursive equations is challenging, as the following interview suggests:[110]

Interviewer:	[Writing $F_n = 18$] Does this mean something to you?
Student:	Yes, it means the answer is eighteen.
Interviewer:	If I write this [F_{n+1}] what does that mean?
Student:	The answer after it ... no, the answer plus one, eighteen plus one is nineteen.

The researchers found that only 12% of 34 seventh-grade students were able to use the notation correctly as they engaged in recursive problems.[111] Students viewed F_n as a single symbol for the answer and interpreted F_{n-1} as "back up one from the answer," that is, subtract one from the answer, rather than move to the previous case before the nth case.

Another common wrong answer is when students express powers instead of products (e.g., n^2 instead of $2n$). In one study, the misconception of using powers instead of products increased from 5% at Year 7 (11–12 years old) to 18% at Year 10 (14–15 years old), indicating that new learning often pushes students to reinterpret past learning and contributes to misconceptions.

Finally, there are many factors that may influence students' interpretation of algebraic notation:[112]

- What the student is able to perceive and prepared to notice;[113]
- Difficulties with operation signs;[114]
- The nature of the questions asked and the medium in which they are asked;[115]
- The presence of multiple referents and shifts in the meaning of unknowns;[116] and
- The nature of the instructional activity—that is, whether teachers emphasize procedures or concepts.[117]

The bottom line is that students may have understanding of the algebraic concepts, but struggle to determine how to represent that understanding in a way that is deemed appropriate by the mathematical community. Rather than immediately considering students' incorrect answers as representative of incorrect thinking, teachers may want to question students about the rationale behind their use of algebraic notation to differentiate between understanding of concepts versus representation of those concepts.

Teachers

- Help students understand the difference between mathematical convention and mathematical necessity.
- Help students convey the meaning behind their use of notation, in order to assess whether they struggle with representation of the answer/idea or with mathematical concepts.
- Help students understand the utility of letters to represent generalization not simply as a placeholder for a solution.
- Help students understand the difference between uses of notation in mathematics that are different than uses of similar notation in other subject areas, like science (e.g., CO_2 versus $x_2 - x_1$).

AREA, PERIMETER, AND VARIABLES

Common Core Standards: 7.EE.4, A-SSE

When students work with problems with figures in algebra, they need to understand the role of variables in a new way. In algebra, we are interested in the situation in which letters are assigned to indicate the measurement of physical characteristics of shapes. For instance, in the case of perimeter, the letter represents the length of a side. Researchers have examined the development of students' understanding of the concept of perimeter and area with identified lengths, and have found some unique struggles for students as they engage in these types of problems.[118] Consider the problems in Table 2.35:

Students who are successful with the types of problems in Table 2.35 are ones who have *Acceptance of a Lack of Closure*. That is, they understand that an answer can include a variable and need not be evaluated to a single number. Those students who struggle with this sense of "incompleteness" can be at two levels. In the first level, students may not accept any answer with a letter, so will evaluate the problem into a number at all costs, putting a number in for the variable based on some rationale (see *Letter Evaluated*). For instance, in Problem 42, students gave number answers between 32 and 42; or, in Problem 41, students might simply give a value to h. Their sense of "complete" is that the final answer must be a number.

In the second level, students may be comfortable with a variable in the answer, but not comfortable with having an operation remaining, such as the answer "$4ht$"

TABLE 2.35. Problems on Finding Perimeter

Problem	Response
41. Find the perimeter: (pentagon figure with sides labeled h, h, h, h, t)	• $hhhht$, $4ht$, or $5ht$ (27%[a], 20%[b]) Correct: $4h + t$ (57%[a], 68%[b]) – a correct verbal description of how to get the answer, but no symbolic representation (14%) – with the unsimplified $h + h + h + h + t$ (29%)
42. Part of this figure is not drawn. There are n sides altogether, all of length 2. (star-shaped figure with sides labeled 2, 2, 2)	• A number between 32 and 42 (25%)[a] – correctly with $2n$ (9%) – 33% gave a correct verbal description of how to get an answer (such as "Two times however many sides there are") but were unable to give a symbolic representation – the rest provided no answer • A number between 32 and 42 (18%)[b] – correctly with $2n$ (38%) – the rest provided no answer

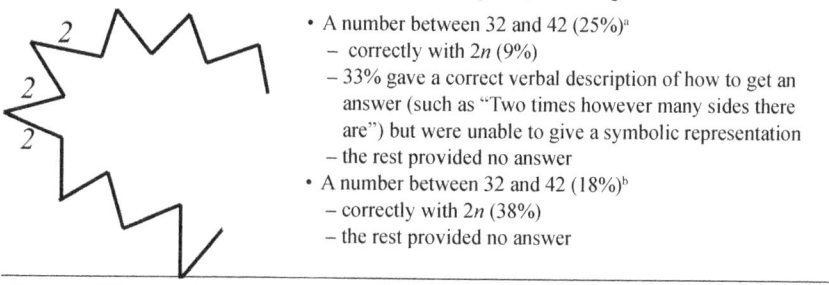

[a] Percent of 13-year-old students in Booth (1984).
[b] Percent of 14-year-old students in Hart and Kuchemann (2005).

in Problem 41, or 10e in Problem 44 in Table 2.37. Their sense of "complete" is that all operations must be finished (i.e., there are no operation symbols remaining) in order to get a final answer. At this second level of understanding, it would be important to differentiate between students who are simply collecting the letters (see **Letter Used as an Object**), and those adding lengths of sides to find the perimeter. In the first case, students do not associate "h" with a specific length of the side but consider it a label or name for the side. For example, in the interview for Problem 41 in Table 2.36, the student seems to be simply collecting things (for example, answering "4ht" or "$hhhht$"). Even students who respond with the correct answer, "4$h + t$," may similarly be collecting objects. It is not necessarily clear whether they understand the concept of perimeter as the addition of unknown lengths. They may simply be putting all the letters together in a "pile."

On the other hand, students who answer "4ht" or "$hhhht$" for Problem 1 may understand the concept of perimeter and the variables representing the lengths of the sides, but struggle with the algebraic representation (see **Algebraic Notation**). The student who answers 4ht may intend 4$h + t$ but simply does not know the convention for notation. In the interview for Problem 41 in Table 2.36, the student understands how to get the perimeter with numbers, but struggles with variables representing numbers. Further questioning is necessary to reveal whether a student is regarding the letter as the unknown measurement of the side. When learning addition, students often associate length with numbers as they combine blocks, etc. For instance, a student "combines" a length of 2 and a length of 3 to get a length of 5. This may lead to thinking that a length of 2 added to a length of x results in a length of 2x (a "combination" of 2 and x), or in Problem 43 writing "am" as the total of the two sides. **Algebraic Notation** may be the primary struggle that students are having with these types of problems, rather than the concepts of area or perimeter.

TABLE 2.36. Interview on Problem 41

Interviewer:	How did you get *hhhht*?
Student:	I just added them all up.
Interviewer:	I see you couldn't get an answer. Do you know how to find it for *this* shape (a new shape B provided with sides 5, 5, 4, 3, and 2)? What do you do there?
Student:	Counted all the numbers up. But I can't here because it's got no numbers. I do not know what "h" means and what "t" is.
Interviewer:	Suppose I tell you that "h" just means some number and "t" means a number, but it's a game and you do not know what the numbers are. Could you tell me anything about how you'd find the perimeter, what you'd do to the "hs" and "t" to get the perimeter?
Student:	I suppose I'd have to measure them or something.
Interviewer:	And then what?
Student:	Count them up.

TABLE 2.37. Problems on Area

Problem	Response
43. Find the area of the figure. (rectangle with sides p, and base a + m)	• p x a + m (36%) • am x p or pam (25%) • No algebraic answer (21%) • Correct, p(a + m) (7%)[a]
44. Find the area of the figure. (rectangle with side 5, and base e + 2)	• 5e2, e10, 10e, or e + 10 (42%) • 41% of students thought that 5 x e + 2 was the same as 5(e + 2) • Correct, 5(e + 2) (7%)[a]

[a]Percent of 13-year-old students in Booth's (1984) study.

Problems 43 and 44 in Table 2.37 focus on finding the area. Research, including the interviews in Tables 2.38 and 2.39, suggests that there are four possible reasons for students' answers. The student:

- interpreted the letter as a "thing" which could be merely collected up with the numbers (an interpretation or "input" error);
- did not know how to interpret the letter *or* did not know how to operate with it and so performed the numerical calculation and wrote down the letter afterwards (a process or "input" error);
- knew that "e" represented a number expressing part of the length of the base, but thought that area meant multiplying everything together (a process error); or
- interpreted the letter correctly and applied the correct method, but recorded the answer to "e added to 2" as "e2" or "2e" (an "output" or representation error).

TABLE 2.38. Interview on Problem 43

Interviewer:	How would you do this problem?
Student:	I suppose I could put it down like "*am* times *p*" (writes *am* x *p*)
Interviewer:	What's that *am* part?
Student:	That is the joining together of the two lengths.
Interviewer:	How would you get that joining together, what would you do?
Student:	First I'd have to measure what the length of *a* and *m* is, and then I'd add them together.

TABLE 2.39. Interview on Problem 44

Interviewer:	Why did you answer "5e2"?
Student 1:	Well, I just put all the numbers and letters together.
Interviewer:	Why did you do that?
Student 1:	I didn't know what to do about the e.
Interviewer:	Why did you answer "10e"?
Student 2:	I just took length times width and I took 2 times 5 is 10 and I just put 10e whatever e is for the variable e.
Interviewer:	Suppose that e is equal to 4. What would your answer give if e is 4?
Student 2:	I'd put 30 because 4 and 2 is 6 and 6 times 5 is 30.
Interviewer:	Does that agree with what you would give if you gave that answer with e equal to 4? Forget the figure. What would that (points to 10e) be if e was equal to 4?
Student 2:	It would be 40.
Interviewer:	So one way you get 30, and the other way you get 40.
Student 2:	Yeah.

Errors in thinking may occur at one of several points in the process of solving the problem. Students may have misread the question or part of the problem, or, in spite of a correct interpretation of the problem, students may have used an incorrect method in solving the problem. They also may have understood the problem, but had difficulty putting the answer into a format that is understood by mathematicians. Considering an input-process-output model of students' engagement with problems may help gain a clearer picture of the points at which students' understanding of the problem breaks down.

Problem 42 in Table 2.35 is a different style of perimeter problem that demands an understanding of variables used for the purpose of generalization (see **Letter Used as a General Number**). In order to deal with the unknown situation, students may simply substitute a number or complete the figure in their minds. The first student interviewed for Problem 42 (Table 2.40) simply ignored the letters and incompleteness of the problem. The student simply counted the number of sides, rather than consider the figure as going on "n" times, getting an answer of "32." Other students physically or mentally continued the shape, closed the figure, and then counted, getting an answer such as "42." These students may have lacked the ability to generalize, or simply did not consider it as an option. Students may have difficulty visualizing a figure in which there are many more sides than what they can see.

The interview with the second student for Problem 42 demonstrates that students are learning not only the mathematical procedures, but also the mathematical expectations. Similar to our earlier discussion of algebraic representation with perimeter problems, the student here may simply need to learn what is "okay" in mathematics and mathematical conventions. A comprehensive understanding of

Variables and Expressions • 83

TABLE 2.40. Interview for Perimeter Problem 42 From Table 2.35

Interviewer:	How did you get 32 for your answer?
Student 1:	I just added them all up.
Student 2:	There's a letter there ("n," in the problem).
Interviewer:	What does the letter mean?
Student 2:	It's telling you how many stages.
Interviewer:	Right. Can you write anything for the perimeter?
Student 2:	Times it by that number.
Interviewer:	Could you write something down, to say how much that is?
Student 2:	What, shall I write down what I would do? (writes "if n was a number, I would times it by 2")
Interviewer:	Now can you write that without using words, using math instead?
Student 2:	What, how'd you mean, like 2 times n?
Interviewer:	Yes, OK.
Student 2:	(writes "2 x n") Is that it?
Interviewer:	Fine.
Student 2:	What, is that all it was? Why didn't you say so, I thought you wanted the answer.
Interviewer:	Do you mean a particular number answer?
Student 2:	Yes!
Interviewer:	Well, is there a particular number answer?
Student 2:	No! Not unless you know what n is.
Interviewer:	Well, then, how could you give me a particular number answer?
Student 2:	Well, you can't, but I didn't know you only had to put *that*.

perimeter problems with variables involves an understanding that it is acceptable to have an expression as an answer, the appropriate mathematical notation for that answer, and a sense that the variable represents the unknown length of the side, or, in the case of Problem 42, the unknown number of sides.

Teachers

- Have students explain their thinking so you can determine their understanding of variable as collecting things *or* their understanding that variables represent values.
- Differentiate between students who are struggling with the concept of perimeter or area versus those who have conceptual understanding, but lack the understanding of mathematical notation or conventions, and those who struggle with what variables represent.
- If students have difficulty with letters in perimeter or area problems, consider walking through a problem with just numbers to help them get the overall understanding.

TRANSLATION DIFFICULTIES IN CHANGING WORDS TO ALGEBRAIC SENTENCES

Common Core Standards: 6.EE.2a, 6.EE.2c, 6.EE.5, 6.EE.6, 6.EE.9, 7.EE.3, 7.EE.4, A-CED.1, A-CED.2

Students struggle for multiple reasons when translating word problems into algebraic equations (also see **Modeling: Translating Word Problems into Equations**). Their understanding of the use of letters in mathematics may influence their thinking about translation. Students may use letters to represent labels or objects when the letters are needed as variables with values (also see **Variables and Expressions: Letter Used as an Object**); as a result, they can make a *reversal error* and translate problems in the wrong order. Consider the problems in Table 2.41. In one study, 37% of 150 college calculus students answered the Student/Professor problem incorrectly.[119] In another study, 35% of college faculty and 47% of high school teachers got it incorrect.[120] The reversal error "is a resilient one that is not easily taught away."[121] These problems provide evidence of a challenging concept for students in algebra to understand.

There are essentially two ways for students to address these types of problems: translate key words to symbols (syntactic translation) or try to express the meaning of the problem algebraically (semantic translation).[122] Semantic translation is where a student understands the entire context and can represent the mathematics in the problem accurately. In regard to syntactic translation, some research suggests that students may be overly focused on *word order matching*, wherein students assume that the order of the words in the problem will be the order of the corresponding variables in the equation. This can be a successful strategy in the majority of school problems; for instance "the number of students is six times the number of professors" can be represented as $S = 6P$.[123] A variation of this approach is *within-phrase adjustment*, in which students are able to make small adjustments to the phrasing and then proceed with word order matching. For instance, "the number of disks is four less than the number of notebooks" can be rephrased "the number of disks is equal to the number of notebooks, subtract four" without necessarily understanding the meaning of the whole context.[124]

TABLE 2.41. Problems for Translation Difficulties

Problem	Response
44. There are six times as many students as professors.	(Incorrect and common response) $6S = P$
45. Write an equation using the variables C and E to represent the following statement: There are 8 times as many people in China as there are in England.	(Incorrect and common response) $8C = E$

Students may also be thinking in terms of a *static comparison* process, wherein students think of variables as objects or labels instead of numbers.[125] Specifically, in the students and professors problem, students do not understand that S is a variable that represents the number of students. Instead, they treat S like a label or a unit attached to the number 6. For instance, there are 6 students (6S) for every 1 professor (P). This is a similar equivalence comparison to the conversion 12 inches = 1 foot.[126] They often understand that there are more students than professors, but struggle with representing that concept algebraically.

The interview in Table 2.42 focuses on a student working on the China/England problem, and demonstrates that the mistake by the student is not simply carelessness based on quick guessing, but rather a set of beliefs about the problem.[127] The first comment by the student indicates difficulty understanding the meaning of the equal sign (see also ***Algebraic Relations***). Students may understand the problem as a rate (see also ***Analysis of Change: Understanding Rate of Change***); for example, 8:1 Chinese to English, and use an equal sign to connect the two. In the student's mind, the equal sign indicates a "correspondence or association

TABLE 2.42. Problem 45 Interview

Interviewer:	This is one where you have to write an equation for the English sentence.
Student:	"There are 8 times as many people in China as there are in England." (Writes "1 per E = 8 per C")
Interviewer:	So let's use the letters C and E.
Student:	OK—for every 1 person in England there are 8 people in China so—8E equals 1C as an equation. (Writes 8E = 1C; Expression 2.) ... Well, if I put—there are more people in China than there are in England, so if I put 8 Chinese equals one Englishman—(Writes 8C = 1E; Expression 3.)
Interviewer:	8C equals 1E—uh-huh.
Student:	8 Englishmen equals one Chinese—um—
Interviewer:	What are you thinking about now?
Student:	I'm thinking about how stupid it is that I just put this wrong. No—this second equation [8E = 1C] here is wrong—the third one—I don't know what I just—
Interviewer:	The third one looks good?
Student:	Yeah, the third one—for every 8 Chinamen—but then again, saying for every 8 Chinese—yeah, that's right—for every 8 Chinese there's one Englishman.
Interviewer:	So just explain how the third one looks good.
Student:	All right—it means that there is a larger number (points to 8C) of Chinese for every Englishman (points to 1E).
Interviewer:	So you're pointing to the 8C and the 1E?
Student:	Yeah, and there is a larger number of Chinese than there are Englishmen; therefore the number of Chinese to Englishmen should be larger... 8C.

Source: Clement (1982)

rather than equality"[128] and functions as a separator symbol, rather than indicating an equivalent relationship: 8 Chinese *for every* 1 English.[129] Reviewing the meaning of the equal sign may be helpful here. The Interviewer suggests using the letters C and E, but neither designates the meaning of those variables.

The student's struggles from that point forward imply that he is using *static comparison* and considers the letters to represent labels for Chinese (*C*) and English (*E*), rather than the number of Chinese or English people. Being careful to define the variables and continually refer to the variables as "the number of" rather than simply "Chinese" or "English" may help students understand the role of the variable in the problem. This mistake may have its roots in the English language, in which adjectives precede nouns. Students who are used to writing "red car" and "six students" in English may use the same logic in mathematical representation. Teachers might unintentionally reinforce this tendency by switching between labels, units, and variables readily when working on word problems. While these are perfectly reasonable uses of letters in mathematics, these problems indicate that students are not as adept at differentiating between the different uses. Research also suggests that this same type of error (following a familiar pattern for direct translation) may be due to students' familiarity with the standard form of an equation as $ax = y$.[130]

Some research suggests that using different letters than those that represent the initial letters of words such as x and y or N_C and N_E may encourage a more accurate use of variable, but the results are inconclusive.[131] One strategy that appears to help students not make the reversal error appears to be placing a multiplication sign between the number and variable.[132] Writing the problem out as $8 \times E = 1 \times C$ may help students not think of the letters as labels.

Other research suggests that the reversal error is not solely evident with multiplication problems, but also with problems involving addition, and that syntactic translation only describes a small subset of the reasons for students' difficulty. More broadly,

> reversed equations and expressions are produced when students attempt to represent on paper cognitive models of compared unequal quantities. . . . Students need experience in recognizing when a problem sentence needs to be reorganized before it can be translated syntactically and instruction in how the reorganizing can be done.[133]

One study found that requiring students to alter their equation into a division format ($y/a = x$) significantly increased their success with the Student/Professor problem.[134]

In contrast, the interview in Table 2.43 demonstrates a student's semantic understanding of the Students/Professors problem. Researchers have found that there is a significant connection between language and mathematics; in particular, high language scores tend to be associated with high algebra scores.[135] In order for students to understand language, they need to understand symbols, syntax, and

TABLE 2.43. Problem 44 Interview

Student:	(Writes: S = # Students; P = # Prof.) S equals the number of students and P equals the number of professors. OK, and there are 6 times as many students as professors, so it would be S would equal 6 times the number of professors. (Writes $S = 6P$)
Interviewer:	How does it fit with the sentence—just if you can, say a little more about how you knew to write it that way.
Student:	Well, I just looked at the sentence and it says 6 times as many students as professors, so the number of students has to equal the number of professors times 6, and plus your—it would make sense that there would be more students than professors.
Interviewer:	Can you see that in the equation?
Student:	Yes, because if you assigned a number to that (P) and multiplied it by 6 you'd have S equals some number larger than what you assigned for that one (P).

Source: Clement (1982)

ambiguity in the words used that need context for understanding. Similar skills are needed to understand algebra.

One strategy suggested by researchers to address students' difficulty with this type of problem is to teach the concept of variables using visual representations of data, such as tables or images, rather than reliance on problem-solving patterns. "Generalizing from tables is significantly correlated with understanding the variable concept."[136] For instance, assist students in drawing pictures or building models for the problem. A chart that is developed prior to writing an equation may be useful. For example:

```
Students ------- Professors
6   ------------------------- 1
12  ------------------------- 2
18  ------------------------- 3
Etc.
```

Students should then translate the problem statement into an algebraic sentence. Following this, they may substitute their numbers from the chart into the equation. Another study found that visual cues (i.e., boxes representing containers for variables) as well as feedback based on test cases helped some students overcome reversal errors.[137]

A strategy to address students' tendency to simply use the same order as the words for the equation —*word order matching*—is to give them a more familiar context. For example, have students consider how to write the expression 5 less than 11 as 11 - 5. In this case, as in the problems above, the representation of the words does not follow the order of the words.

Finally, there should be explicit discussion to help students discover the shortcomings of the incorrect equation(s) that were generated by students. Creating a hypothetical situation and plugging in numbers for S and P makes it clear that the representation $6S = P$ does not accurately convey the sentiment that there are more students than teachers.[138] Teachers should also consider emphasizing the meaning of a variable (as opposed to a label) and of the equal sign during this activity.

Teachers

- Have students test their constructed equations from word problems with different values to ensure the equations accurately reflect the situation.
- Help students identify key words and phrases that help them write equations while simultaneously having them consider broader ideas in the problem, and whether their translation makes sense with the broader problem.
- Encourage students to use multiple representations to help them accurately construct equations from word problems, such as drawings, tables, and graphs.

ENDNOTES

1. Kieran, 1992, and Stacey & MacGregor, 1997.
2. Schoenfeld & Arcavi, 1988.
3. MacGregor & Stacey, 1997, p. 16.
4. Wagner & Parker, 1993, p. 330.
5. MacGregor & Stacey, 1997, p. 15.
6. Trigueros & Ursini, 1999, and Trigueros, Ursini, & Reyes, 1996.
7. Schoenfeld & Arcavi, 1988, p. 421.
8. Based on Hart & Kuchemann, 2005, and May & Van Engen, 1959, from Usiskin, 1988, p. 9.
9. Usiskin, 1988, p. 9.
10. Usiskin, 1988
11. Usiskin, 1988, p. 17.
12. Bazzini, 1999, p. 113.
13. These six categories integrate the work of Küchemann, 1981, and Ursini & Trigueros' work on their 3UV model (2001, 2004, 2008).
14. Hart & Kuchemann, 2005 and Trigueros & Ursini, 2003
15. Stacey & MacGregor, 1997, and Weinberg, Stephens, McNeil, Krill, Knuth & Alibali, 2004.
16. Trigueros, Ursini, & Reyes, 1996, pp. 4–321.
17. Küchemann, Hodgen, & Brown, 2011
18. Falkner, Levi, & Carpenter, 1999.
19. Collis, 1974, 1978; Tall & Thomas, 1991, p. 126.
20. Chalouh & Herscovics, 1988, and Kieran, 1981.
21. Davis, 1975. Sfard, 1991, and Tall, 1994, frame this similarly, explaining that abstract math can be seen as structural (objects) or operational (processes).
22. Booth, 1988.
23. Kieran, 1992.
24. Tall & Thomas, 1991, p. 127
25. Küchemann, 1981
26. Weinberg, et al., 2004
27. Weinberg, et al., 2004, p. 7
28. Weinberg, et al., 2004, p. 8

29 Christou, Vosniadou, & Vamvakoussi, 2007
30 Stacey & MacGregor, 1994 describe this conjoining with fractions, music, and Chemistry. Chalouh & Herscovics, 1988, describe this conjoining with place value.
31 Chalou & Herscovics, 1983; Macgregor & Stacey, 1997
32 Chalou & Herscovics, p. 155
33 MacGregor & Stacey, 1997, p. 10
34 MacGregor & Stacey, 1997
35 Booth, 1984
36 MacGregor & Stacey, 1997 and Christou, Vosniadou, & Vamvakoussi, 2007
37 Chalouh & Herscovics, 1983
38 See discussion of 'concatenation' in Christou, Vosniadou, & Vamvakoussi, 2007, p. 287 and MacGregor & Stacey, 1997.
39 Booth, 1984, 1988
40 Matz, 1980 and Kieran, 1990, 1992
41 Christou & Vosniadou, 2012
42 Fujii, 1992, 2003 and Steinle, et al., 2009
43 Fujii, 2003 and Steinle, et al., 2009
44 Fujii, 2003, pp. 1–56.
45 Steinle, et al., 2009, 492
46 Booth, 1984; Chalou & Herscovics, 1983; Wagner & Parker, 1993; Warren, 1999
47 Booth, 1984 and Stephens, 2005
48 Steinle, et al., 2009, 497
49 Filloy & Rojano, 1984
50 Olivier, 1988, Booth, 1988, Chalouh & Herscovics, 1983, Wagner & Parker, 1993
51 Olivier, 1988, Booth, 1988, Chalouh & Herscovics, 1983, Wagner & Parker, 1993
52 Collis, 1974, 1978; Tall & Thomas, 1991, p. 126
53 Chalouh & Herscovics, 1988 and Kieran, 1981
54 Trigueros, Ursini, & Reyes, 1996, and Wagner, Rachlin, & Jensen, 1984
55 Sfard & Linchevski, 1994 and Tall & Thomas, 1991
56 Stacey & MacGregor, 1994, p. 5.
57 Stacey & MacGregor, 1994
58 Tall & Thomas, 1991
59 Tall & Thomas, 1991, p. 132.
60 Lannin, 2005; Sutherland & Rojano, 1993
61 Steinle, et. al., 2009
62 Küchemann, 1978, p. 25.
63 Küchemann, 1981
64 Warren, 1999
65 Steinle, 2009.
66 Hart & Kuchemann, 2005
67 MacGregor & Stacey, 1997, p. 16.
68 Küchemann, 1978, 1981. Other researchers agree that there are developmental considerations to learning algebra (Filloy & Rojano, 1989; Halford, 1978; Herscovics & Linchevski, 1994; Sfard & Linchevski, 1994)
69 Knuth et al., 2005, and MacGregor & Stacey, 1997
70 Carraher et al., 2001
71 MacGregor & Stacey, 1997
72 Booth, 1988; Galvin & Bell, 1977; MacGregor, 1986; Oliver, 1984; and Pimm, 1987
73 Tirosh, Even, & Robinson, 1998, p. 60.
74 McNeil et al. 2010, p. 629
75 McNeil et al. 2010
76 Warren, 1999

77 Weinberg, et al., 2004, p. 10
78 This entry is based on Booth, 1984; Fujii, 1993, 2003; Küchemann, 1978, 1981; Olivier, 1988; Stacey & MacGregor, 1997; and Warren, 1999
79 Christou, Vosniadou, & Vamvakoussi, 2007, based on Booth, 1984, Küchemann, 1981, & Wagner, 1981
80 Trigueros & Ursini, 2008; Ursini & Trigueros, 2001, 2004
81 Knuth et al., 2005
82 Trigueros & Ursini, 2003, p. 4
83 Knuth et al., 2005, p. 73
84 Knuth et al., 2005
85 Warren, 1999
86 Trigueros & Ursini, 2008; Ursini & Trigueros, 2001, 2004
87 Based on Küchemann, 1978, 1981, and Weinberg, Stephens, McNeil, Krill, Knuth, Alibli, 2004
88 Greer, 1994
89 Knuth et al., 2005
90 Trigueros & Ursini, 2008; Ursini & Trigueros, 2001, 2004
91 Trigueros & Ursini, 2008, pp. 4–338
92 Brown, Hodgen, & Küchemann, 2014
93 Brown, Hodgen, & Küchemann, 2014, p. 178
94 Sfard & Linchevski, 1994
95 Hewitt, 2012, p. 141
96 Stacey & MacGregor, 1994 describe this conjoining with fractions, music, and Chemistry. Chalouh & Herscovics, 1988, describe this conjoining with place value.
97 Mercer, 2000, p. 67
98 Kaput, 1992
99 Quoted from Kieran, 2007, p. 716 and based on MacGregor & Stacey, 1997.
100 Radford, 2010
101 Carraher et al., 2001
102 Tall & Thomas, 1991, p. 126
103 Tall & Thomas, 1991, p. 127
104 Hewitt, 2012
105 Based on Hewitt, 2012
106 Kaput, 1992
107 Arzarello, 1991; Arzarello et al., 1993, 1994; MacGregor and Stacey, 1993b; Mason, 1996; Lee, 1996; Rico et al., 1996
108 Booth, 1984, p. 20.
109 Franzblau & Warner, 2001
110 Franzblau & Warner, 2001, p. 192
111 Franzblau & Warner, 2001
112 Kieran, 2007, p. 716
113 Sfard & Linchevski, 1994
114 Cooper, et al., 1997
115 Warren, 1999
116 Stacey & MacGregor, 1997
117 Wilson, Ainley, & Bills, 2003
118 See Barrett, et al., 2006
119 Clement, Lochhead, & Monk, 1981
120 Lochhead, 1980
121 Rosnick & Clement, 1980, p. 3
122 Herscovics, 1989, Mestre, 1988, Mestre & Gerace, 1986; Spanos et al., 1988
123 Graf et al., 2004

124 Kirshner et al., 1991
125 Clement, 1982
126 Kinzel, 1999
127 Clement, 1982
128 MacGregor & Stacey, 1993a, p. 219. Also, Cohen & Kanim, 2005, and Palm, 2008.
129 Kieran, 1990
130 Fisher, Borchert, & Bassok, 2011
131 Kaput & Sims-Knight, 1983, and Fisher, 1988 found that not using initial letter variables helped students understand while Cooper, 1986, Fisher, 1988, and Clement, 1982 did not have success with that strategy.
132 Cooper, 1986
133 MacGregor & Stacey, 1993a
134 Fisher, Borchert, & Bassok, 2011
135 MacGregor & Price, 1999, p. 459
136 English & Warren, 1994
137 Kim et al., 2014
138 Clement, 1982 and with a computer program in Kim et al., 2014's study.

REFERENCES

Arzarello, F. (1991). Pre-algebraic problem solving. In J. P. Ponte, J. F. Matos, J. M. Matos, & D. Fernandes (Eds.), *Mathematical problem solving and new information technologies* (pp. 155–166). NATO ASI Series F, *89*. Berlin: Springer-Verlag.

Arzarello, F., Bazzini, L. & Chiappini, C. (1993). Cognitive processes in algebraic thinking: Towards a theoretical framework. In I. Hirabayashi, N. Nohda, K.Sheigematsu, & F-L. Lin (Eds.), *Proceedings of the 17th International Conference for the Psychology of Mathematics Education,* (Vol. 1, pp. 138–145). Ibaraki, Japan: University of Tsukuba, Tsukuba.

Arzarello, F., Bazzini, L., & Chiappini, G. (1994). *The process of naming in algebraic problem solving.* Paper presented at the 18th Annual Conference of the International Group for the Psychology of Mathematics Education (PME), Lisbon, Portugal.

Barrett, J. E., Clements, D. H., Klanderman, D., Pennisi, S.-J., & Polaki, M. V. (2006). Students' coordination of geometric reasoning and measuring strategies on a fixed perimeter task: Developing mathematical understanding of linear measurement. *Journal for Research in Mathematics Education, 37*(3), 187–221.

Bazzini, L. (1999). *On the construction and interpretation of symbolic expressions.* Paper presented at the Proceedings of the First Conference of the European Society for Research in Mathematics Education.

Booth, L. R. (1984). *Algebra: Children's strategies and errors.* Windsor, Berkshire: NFER-Nelson.

Booth, L. R. (1988). Children's difficulties in beginning algebra. In *The ideas of algebra, K–12, 1988 NCTM Yearbook* (pp. 20–32). Reston, VA: National Council of Teachers of Mathematics.

Brown, M., Hodgen, J., & Küchemann, D. (2014). Learning experiences designed to develop multiplicative reasoning: Using models to foster learners' understanding. In P. C. Toh, T. L. Toh, & B. Kaur (Eds.), *Learning experiences that promote mathematics learning* (pp. 187–208).Singapore: World Scientific.

Carraher, D., Schliemann, A. D., & Brizuela, B. M. (2001). Can young students operate on unknowns? In M. van den Heuvel-Panhuizen (Ed.), *Proceedings of the 25th conference of the International Group for the Psychology of Mathematics Education,* (Vol. 1, pp. 130–140). Utrecht, The Netherlands: PME.

Chalouh, L., & Herscovics, N. (1983). *The problem of concatenation.* Paper presented at the Fifth Annual Meeting of the North American Chapter of the International Group for the Psychology of Mathematics Education, Montreal.

Chalouh, L., & Herscovics, N. (1988). Teaching algebraic expressions in a meaningful way. In A. F. Coxford (Ed.), *The ideas of algebra, K–12* (1988 Yearbook, pp. 33–42). Reston, VA: National Council of Teachers of Mathematics.

Christou, K. P., & Vosniadou, S. (2012). What kinds of numbers do students assign to literal symbols? Aspects of the transition from arithmetic to algebra. *Mathematical Thinking & Learning, 14*(1), 1–27.

Christou, K., Vosniadou, S., & Vamvakoussi, X. (2007). Students' interpretations of literal symbols in algebra. In S. Vaosniadou, Baltas, A., & Vamvakoussi, X. (Eds.), *Reframing the conceptual change approach in learning and instruction.* New York, NY: Elsevier.

Clement, J. (1982). Algebra word problem solutions: Thought processes underlying a common misconception. *Journal for Research in Mathematics Education, 13*(1), 16–30.

Clement, J., Lochhead, J., & Monk, G. (1981). Translation difficulties in learning mathematics. *American Mathematical Monthly, 88,* 286–290.

Coady, C., & Pegg, J. (1993). *An exploration of student responses to the more demanding Küchemann test items.* Paper presented at the 16th Annual Mathematics Education Research Group of Australasia Conference.

Cohen, E., & Kanim, S. E. (2005). Factors influencing the algebra "reversal error." *American Journal of Physics, 73*(11), 1072–1078.

Collis, K. (1974). *What do we know abut K–14 students' learning of algebra?* Paper presented at the National Council of Teachers of Mathematics national symposium, The Nature and Role of Algebra in the K - 14 Curriculum, Washington, DC.

Collis, K. F. (1978). Operational thinking in elementary mathematics. In J. A. Keats, K. F. Collis & G. S. Halford (Eds.), *Cognitive development: Research based on a Neo-Piagetian approach* (pp. 221–248). Bath: The Pitman Press.

Cooper, M. (1986). The dependence of multiplicative reversal on equation format. *Journal of Mathematical Behavior, 5*(2), 115–120.

Cooper, T. J., Boulton-Lewis, G., Atweh, B., Pillay, H., Wilss, L., & Mutch, S. (1997). *The transition from arithmetic to algebra: Initial understanding of equals, operations, and variables.* Paper presented at the 21st Conference of the International Group for the Psychology of Mathematics Education, Lahti, Finland.

Davis, R. B. (1975). Cognitive processes involved in solving simple algebraic equations. *Journal of Children's Mathematical Behavior, 1*(3), 7–35.

English, L., & Warren, E. (1994). *The interaction between general reasoning processes and achievement in algebra and novel problem solving.* Paper presented at the 17th Annual Mathematics Education Research Group of Australasia Conference.

Falkner, K. P., Levi, L., & Carpenter, T. P. (1999). Children's understanding of equality: A foundation for algebra. *Teaching Children Mathematics, 6*(4), 232–236.

Filloy, E., & Rojano, T. (1984). *From an arithmetical to an algebraic thought: A clinical study with 12–13 years old.* Paper presented at the Sixth Annual Meeting of the

North American Chapter of the International Group for the Psychology of Mathematics Education, Madison: University of Wisconsin.

Filloy, B., & Rojano, T. (1989). Solving equations: The transition from arithmetic to algebra. for the learning of mathematics. *An International Journal of Mathematics Education, 9*(2), 19–25.

Fisher, K. M. (1988). The students-and-professors problem revisited. *Journal for Research in Mathematics Education, 19*, 260–262.

Fisher, K., Borchert, K., & Bassok, M. (2011). Following the standard form: Effects of equation format on algebraic modeling. *Memory & Cognition, 39*(3), 502–515.

Franzblau, D. S., & Warner, L. B. (2001). From Fibonacci numbers to fractals: Recursive patterns and subscript notation. *Yearbook* (National Council of Teachers of Mathematics), *2001*, 186–200.

Fujii, T. (1993). *A clinical interview on children's understanding and misconceptions in school mathematics.* Paper presented at the 17th PME International Conference.

Fujii, T. (2003). *Probing students' understanding of variables through cognitive conflict problems: Is the concept of variable so difficult for students to understand?* Paper presented at the 27th International Group for the Psychology of Mathematics Education Conference, Honolulu, HI.

Galvin, W., & Bell, A.W. (1977). *Aspects of difficulties in the solution of problems involving the formulation of equations.* Nottingham, UK: Shell Centre for Mathematical Education, University of Nottingham.

Graf, E. A., Bassok, M., Hunt, E., & Minstrell, J. (2004). A computer-based tutorial for algebraic representation: The effects of scaffolding on performance during the tutorial and on a transfer task. *Technology Instruction Cognition and Learning, 2*, 135–170.

Greer, B. (1994). Extending the meaning of multiplication and division. In G. Harel & J. Confrey (Eds.), *The development of multiplicative reasoning in the learning of mathematics* (pp. 61–85). Albany, NY: SUNY Press.

Halford, G. S. (1978). An approach to the definition of cognitive developmental stages in school mathematics. *British Journal of Educational Psychology, 48*, 298–314.

Hart, K., & Kuchemann, D. (2005). *Children's understanding of mathematics: 11–16.* Eastbourne, UK: Anthony Rowe, Ltd.

Herscovics, N. (1989). Cognitive obstacles encountered in the learning of algebra. In S. W. C. Kieran (Ed.), *Research issues in the learning and teaching of algebra* (Vol. 4, pp. 60–86). Reston, VA: National Council of Teachers of Mathematics.

Herscovics, N., & Linchevski, L. (1994). A cognitive gap between arithmetic and algebra. *Educational Studies in Mathematics, 27*(1), 59–78.

Hewitt, D. (2012). Young students learning formal algebraic notation and solving linear equations: Are commonly experienced difficulties avoidable? *Educational Studies in Mathematics, 81*(2), 139–159.

Hodgen, J., Brown, M., Coe, R., & Küchemann, D. (2012). Surveying lower secondary students' understandings of algebra and multiplicative reasoning: to what extent do particular errors and incorrect strategies indicate more sophisticated understandings? In J. C. Sung (Ed.), *Proceedings of the 12th International Congress on Mathematical Education* (ICME-12) (pp. 6572–6580). Seoul, Korea: International Mathematics Union.

Hodgen, J., Küchemann, D., Brown, M., & Coe, R. (2008). *Children's understandings of algebra 30 years on*. Paper presented at the British Society for Research into Learning Mathematics.

Kaput, J. (1992). Technology and mathematics education. In D. Grouws (Ed.), *Handbook of research on mathematics teaching and learning* (pp. 515–556). New York, NY: MacMillan Publishing.

Kaput, J. J., & Sims-Knight, J. (1983). Errors in translations to algebraic equations: Roots and implications. *Focus of Learning Problems in Mathematics, 5*(3), 63–78.

Kieran, C. (1981). Concepts associated with the equality symbol. *Educational Studies in Mathematics, 12,* 317–326

Kieran, C. (1990). Cognitive processes involved in learning school algebra. In P. N. J. Kilpatrick (Ed.), *Mathematics and cognition: A research synthesis by the international group for the psychology of mathematics education* (pp. 96–112). Cambridge, UK: Cambridge University Press.

Kieran, C. (1992). The learning and teaching of school algebra. In D. Grouws (Ed.), *Handbook of research on mathematics teaching and learning* (pp. 390–419). New York, NY: Free Press.

Kieran, C. (2007). Learning and teaching algebra at the middle school through college levels. In F. Lester (Ed.), *Second handbook of research on mathematics teaching and learning* (pp. 707–762). Charlotte, NC: Information Age Publishing.

Kim, S.-H., Phang, D., An, T., Yi, J. S., Kenney, R., & Uhan, N. A. (2014). POETIC: Interactive solutions to alleviate the reversal error in student–professor type problems. *International Journal of Human-Computer Studies, 72*(1), 12–22.

Kinzel, M. T. (1999). Understanding algebraic notation from the students' perspective. *Mathematics Teacher, 92*(5), 436–441.

Kirshner, D., Awtry, Y., McDonald, J., & Gray, E. (1991). *The cognitivist caricature of mathematical thinking*. Paper presented at the Thirteenth Annual Conference of the North American Chapter of the International Group of the Psychology of Mathematics Education, Blacksburg, VA.

Knuth, E., Alibali, M., McNeil, N., Weinberg, A., & Stephens, A. (2005). Middle school students' understanding of core algebraic concepts: Equivalence & variable. *ZDM, 37*(1), 68–76.

Küchemann, D. (1978). Children's understanding of numerical variables. *Mathematics in school, 7*(4), 23–25.

Küchemann, D. (1981). Algebra. In K. M. Hart (Ed.), *Children's understanding of mathematics* (pp. 102–119): London, UK: John Murray.

Kuchemann, D. (2005). Algebra. In K. M. Hart (Ed.), *Children's understanding of mathematics: 11-16* (11th revised ed.). United Kingdom: Antony Rowe Publishing Services.

Küchemann, D., Hodgen, J., & Brown, M. (2011). English school students' understanding of algebra, in the 1970s and now. *Der Mathematikunterricht, 57*(2), 41–54.

Laborde, C. (1990). Language and Mathematics. In P. N. J. Kilpatrick (Ed.), *Mathematics and cognition: a research synthesis by the international group for the psychology of mathematics education* (pp. 53–69). Cambridge, UK: Cambridge University Press.

Lannin, J. K. (2005). Generalization and justification: The challenge of introducing algebraic reasoning through patterning activities. *Mathematical Thinking & Learning, 7*(3), 231–258.

Lee, L. (1996). An initiation into algebraic culture through generalization activities. In N. Bednarz, C. Kieran, & L. Lee (Eds.), *Approaches to algebra* (pp. 87–106). Dordrecht/Boston/London: Kluwer.

Lochhead, J. (1980). Faculty interpretations of simple algebraic statements: The professor's side of the equation. *Journal of Mathematical Behavior, 3,* 29–37.

MacGregor, M. E. (1986). A fresh Look at fruit salad algebra. *The Australian Mathematics Teacher, 42*(3), 9–11.

MacGregor, M., & Price, E. (1999). An exploration of aspects of language proficiency and algebra learning. *Journal for Research in Mathematics Education, 30*(4), 449–467.

MacGregor, M., & Stacey, K. (1993a). Cognitive models underlying students' formulation of simple linear equations. *Journal for Research in Mathematics Education, 24*(3), 217–232.

MacGregor, M., & Stacey, K. (1993b). *Seeing a pattern and writing a rule.* Paper presented at the Seventeenth International Conference for the Psychology of Mathematics Education, University of Tsukuba, Japan.

MacGregor, M., & Stacey, K. (1993c). What is X? *Australian Association of Mathematics Teachers, 49*(4), 28–30.

MacGregor, M., & Stacey, K. (1997). Students' understanding of algebraic notation: 11–15. *Educational Studies in Mathematics, 33*(1), 1–19.

Mason, J. (1996). Expressing generality and roots of algebra. In C. K. N. Bednarz, & L. Lee (Eds.), *Approaches to algebra: Perspectives for research and teaching* (pp. 65–86). Dordrecht, Netherlands: Kluwer.

Matz, M. (1980). Building a metaphoric theory of mathematical thought. *Journal of Mathematical Behavior, 3(*1), 93–166.

May, K., & Van Engen, H. (1959). Relations and functions. In *The Growth of Mathematical Ideas, Grades K–12* (Twenty-fourth Yearbook of the National Council of Teachers of Mathematics, pp. 65–110). Washington, DC: NCTM.

McNeil, N. M., Weinberg, A., Hattikudur, S., Stephens, A. C., Asquith, P., Knuth, E. J., & Alibali, M. W. (2010). A is for apple: Mnemonic symbols hinder the interpretation of algebraic expressions. *Journal of Educational Psychology, 102*(3), 625–634.

Mercer, N. (2000). *Words and minds.* London: Routledge.

Mestre, J. (1988). The role of language comprehension in mathematics and problem solving. In R. Cocking & J. Mestre (Eds.), *Linguistic and cultural influences on learning mathematics* (pp. 201–220). Hillsdale, NJ: Erlbaum.

Mestre, J., & Gerace, W. (1986). The interplay of linguistic factors in mathematical translation tasks. *Focus on Learning Problems in Mathematics, 8*(1), 59–72.

Oliver, A. I. (1984). *Developing basic concepts in elementary algebra.* Paper presented at the Eighth International Conference for the Psychology of Mathematics Education, Sydney.

Olivier, A. I. (1988). *The construction of an algebraic concept through conflict.* Paper presented at the 12th International Conference on Psychology of Mathematics Education, Veszprém, Hungary.

Palm, T. (2008). Impact of authenticity on sense making in word problem solving. *Educational Studies in Mathematics, 67,* 37–58.

Pimm, D. (1987). *Speaking mathematically: Communication in mathematics classrooms.* London: Routledge & Kegan Paul

Radford, L. (2010). Elementary forms of algebraic thinking in young students. In M. M. Pinto & T. F. Kawasaki (Eds.), *Proceedings of the 34th conference of the International Group for the Psychology of Mathematics Education,* Vol. 4 (pp. 73–80). Belo Horizonte, Brazil: PME.

Rico, L., Castro, E., & Romero, I. (1996). The role of representation systems in the learning of numerical structures. In L. Puig & A. Gutierrez (Eds.), *Proceedings of the 20th conference of the International Group for the Psychology of Mathematics Education* (Vol. 1, pp. 87–102). Spain: University of Valencia.

Rosnick, P., & Clement, J. (1980). Learning without understanding: The effect of tutoring strategies on algebra misconceptions. *Journal of Mathematical Behavior, 3*(1), 3–27.

Schoenfeld, A. H., & Arcavi, A. (1988). On the meaning of variable. *Mathematics Teacher, 81*(6), 420–427.

Sfard, A. (1991). On the dual nature of mathematical conceptions: Reflections on processes and objects as different sides of the same coin. *Educational Studies in Mathematics, 22*(1), 1–36.

Sfard, A., & Linchevski, L. (1994). The gains and the pitfalls of reification—The case of algebra. *Educational Studies in Mathematics, 26,* 191–228.

Spanos, G., Rhodes, N., Dale, T., & Crandal, J. (1988). Linguistic features in mathematical problem solving. In R. Cocking & J. Mestre (Eds.), *Linguistic and cultural influences on learning mathematics* (pp. 221–240). Hillsdale, NJ: Erlbaum.

Stacey, K., & MacGregor, M. (1994). *Algebraic sums and products: Students' concepts and symbolism.* Paper presented at the 18th International Conference for the Psychology of Mathematics Education, Lisbon, Portugal.

Stacey, K., & MacGregor , M. E. (1997). Ideas about symbolism that students bring to algebra. *The Mathematics Teacher, 90*(2), 110–113.

Steinle, C., Gvozdenko, E., Price, B., Stacey, K., & Pierce, R. (2009). *Investigating students' numerical misconceptions in algebra.* Paper presented at the Crossing Divides: The 32nd Annual Conference of the Mathematics Education Research Group of Australasia, Palmerston North, New Zealand.

Stephens, A. C. (2005). Developing students' understandings of variable. *Mathematics Teaching in the Middle School, 11*(2), 96–100.

Sutherland, R., & Rojano, T. (1993). A spreadsheet approach to solving algebra problems. *Journal of Mathematical Behavior, 12,* 353–383.

Tall, D., & Thomas, M. (1991). Encouraging versatile thinking in algebra using the computer. *Educational Studies in Mathematics, 22*(2), 125–147.

Tirosh, D., Even, R., & Robinson, N. (1998). Simplifying algebraic expressions: Teacher awareness and teaching approaches. *Educational Studies in Mathematics, 35*(1), 51–64.

Trigueros, M. & Ursini, S. (1999). *Does the understanding of variable evolve through schooling?* Paper presented at the Conference of the International Group for the Psychology of Mathematics Education. Haifa, Israel.

Trigueros, M., & Ursini, S. (2003). First year undergraduates difficulties in working with the concept of variable. *CBMS, Research in Collegiate Mathematics Education, 12,* 1–29.

Trigueros, M., & Ursini, S. (2008). *Structure sense and the use of variable.* Paper presented at the 32nd Conference of the International Group for the Psychology of Mathematics Education.

Trigueros, M., Ursini, S., & Reyes, A. (1996). College students' conceptions of variable. *Proceedings of the XX PME International Conference* (pp. 315–22.). Valencia, Spain.

Ursini, S., & Trigueros, M. (2001). *A model for the uses of variable in elementary algebra.* Paper presented at the 25th Conference of The International Group for the Psychology of Mathematics Education., Utrecht, Netherlands.

Ursini, S., & Trigueros, M. (2004). *How do high school students interpret parameters in algebra?* Paper presented at the 28th Conference of the International Group for the Psychology of Mathematics Education, Bergen, Norway.

Usiskin, Z. (1988). Conceptions of school algebra and uses of variables. In A. Coxford & A. Schulte (Eds.), *The ideas of algebra, K–12* (pp. 8–19). Reston, VA: National Council of Teachers of Mathematics.

Wagner, S. (1981). Conservation of equation and function of variable. *Journal for Research in Mathematics Education, 12,* 107–118.

Wagner, S., & Parker, S. (1993). Advancing algebra. In P. S. Wilson (Ed.), *Research ideas for the classroom: High school mathematics* (pp. 120–139). New York, NY: Macmillan.

Wagner S., Rachlin S. L., & Jensen R. J. (1984). *Algebra learning project—Final report.* Department of Mathematics Education, University of Georgia.

Warren, E. (1999). *The concept of variable: Gauging students' understanding.* Paper presented at the 23rd Conference of the International Group for the Psychology of Mathematics Education, Haifa, Israel.

Weinberg, A. D., Stephens, A. C., McNeil, N. M., Krill, D. E., Knuth, E. J., & Alibli, M. W. (2004). *Students' initial and developing conceptions of variable.* Paper presented at the Annual Meeting of the American Educational Research Association, San Diego, CA.

Wilson, K., Ainley, J., & Bills, L. (2005). *Naming a column on a spreadsheet: Is it more algebraic?* Paper presented at the Sixth British Congress on Mathematics Education, University of Warwick.

CHAPTER 3

ALGEBRAIC RELATIONS

> The notion of equivalence relation
> is arguably one of the most fundamental ideas of mathematics
> —*Asghari, 2009*

INTRODUCTION

One of the fundamental components of an Algebra course is learning to solve equations and inequalities. "Students' understanding of core algebraic concepts of variable and equivalence influences their success in solving problems, the strategies they use, and the justification they give for their solutions."[1] They acquire a new level of mathematical power when they comprehend the relationships and processes of equation solving. Solving for x, finding x, and solving for the unknown are all ways that we characterize the use of algebraic relations for a purpose. However, 60% of 16-to-18-year-old students in one study believed that $k = 5$ was not an equation because there was no operation, 65% thought that $a = a$ was not an equation, and others thought that $3w = 7w - 4w$ was not an equation because it was in the wrong order.[2] A solid understanding of algebraic relations begins with a strong understanding of what an equation is and the role of the equal sign. In addition to these characterizations, the literature on the teaching and learning of algebra tells us we must also consider the fundamental arithmetic skills that students use when manipulating the terms in an expression and the structure of the task itself. In the end, students need to become fluent enough with these skills and concepts that they can apply solution strategies successfully across a variety of tasks.

One- and Two-Step Equations*[1]** represent a class of algebraic relations that are often the easiest for students to solve. These equations often take the form of $ax + b = c$ where a, b, and c are constants. Students may visualize what has been "done to x" and "undoing what has been done" as an informal way of approaching the task that is often successful. For example, in the equation $x - 6 = 7$, students can verbalize the task and say, "x is a number such that when I take away 6, I get 7." From there, the transition to knowing that I can find x if I have 6 and 7 is a small cognitive leap. The literature demonstrates that because of this, ***One- and Two-Step Equations are a good entry point for learning to work with algebraic relations.[3]

UNDERSTANDING THE ELEMENTS IN AN ALGEBRAIC RELATION

There are various elements of algebraic equations and inequalities (such as rational coefficients, negative numbers, and solutions involving zero) that can be sources of potential student errors. Students may struggle when applying skills learned in number operations to the basic algorithms of solving equations. Errors made in these lower-level concepts lead to difficulties for students trying to perform the basic operations necessary to solve an equation or inequality.[4]

Rational Numbers are commonly seen in standard algebra tasks. However, many students struggle to accurately apply operations on rational numbers when given a task such as the following:[5]

$\frac{1}{2}$ of the freshman class takes band during first period. $\frac{2}{5}$ of the freshman class takes physical education. What fraction of the class takes something else?

In this task, the student must operate on rational numbers with unlike denominators. Even if the student can accurately translate the task into an equation (see ***Modeling: Translating Word Problems into Algebraic Sentences***), those unable to perform the necessary addition or subtraction will not be able to find the answer. The literature indicates that students need to review fundamental operations on rational numbers as they move into algebra. This review should begin with whole numbers and continue through fractions, as students who have difficulties with operations on fractions will struggle even more with operations on rational expressions.

---

[1] ***Bold and italicized*** print indicates a reference to a section in the book

In the same way that **Rational Numbers** are a challenge, students will often struggle with **Equations Involving Negative Numbers**. These challenges take two forms. First, students might not recognize that a negative number is in the expression as they focus their attention on other features like the presence of variables. Second, once the negative expression is recognized, the student might not have the necessary skills to operate on it. In the following series of tasks, we can see these challenges illustrated:[6]

1. $12 - x = 7$
2. $4 - x = 5$
3. $-4 - x = 10$
4. $-x = 7$

The first task seems to be the easiest for students as it can be solved through reasoning (12 minus what is seven?). The remaining tasks are more complex as their solutions are not as transparent. For example, what can I subtract from four to get five? It is now in the student's best interest to perform operations on this equation to find the unknown value. But how do you "undo" subtraction of x? What do I get when I take five from four? In these situations, the teacher might try to come up with a representation of the equation that is fixed in the real world. However, many of the models for looking at **Equations Involving Negative Numbers** are problematic as well. Some of the issues involved with equations have their roots in students' conceptions of what variables can represent, particularly when negatives are involved (see **Variables & Expressions: Representation**).

In some situations, students can perform operations accurately, but then fail to understand the meaning of the simplified relation. Some algebraic relations have elements that lead to solutions like $0x = 7$, $0 = 7$, or $x = 0$ once simplified. Students will see these algebraic relations with **Solutions Involving Zero** and misunderstand what can be done with the zero. Teachers will need to generate adequate representations, either graphically or in words, that can demonstrate how the students should interpret the resulting answer.

TYPES OF ALGEBRAIC RELATIONS

There are also features of algebraic relations—such as the location of the variable, the number of variables, and the presence of an inequality—that bring additional challenges. The procedures used for solving a task are often dependent on the type of algebraic relation presented. The steps necessary to solve an equation with x terms (or another variable) on both sides of the equal sign follow naturally from the steps necessary to solve tasks in which the variable appears only once, but students often fail to see this progression. The presence of an inequality further complicates the process. The operations necessary to solve an inequality are very similar to those used to solve equations, but why do we "flip the sign" when we divide by a negative, and how are we supposed to remember to do that?

As students transition to more complex tasks, error patterns appear. When students begin **Solving Equations with Variables on Both Sides**, they are prone to arithmetic errors. Because these tasks are more complex, arithmetic errors can go unnoticed by students: is difficult to reason the solution in words as they did in **One- and Two-Step Equations**. Furthermore, when given an equation like $6x - 2 = 3x + 4$, we often use informal language to characterize the procedures necessary to simplify the equation. Language like "move the $3x$ to the other side" can lead students to ignore the sign of the coefficient and make the following error:

$$6x - 2 = 3x + 4 \qquad 6x + 3x - 2 = 4$$

The $3x$ term has "moved," but the resulting equation is inconsistent with the original. Similar difficulties can occur when students misunderstand the meaning of the equal sign.[7] Given an equation like $3x - 2x + 7 = 4x + 4$, students might respond by subtracting $4x$ from every x-term in the equation. The result is $(3x - 4x) - (2x - 4x) + 7 = (4x - 4x) + 4$. The teacher's challenge is to help students understand the equal sign as the balance point rather than as an operator that implies a flow from left to right to complete an answer.

Once students have progressed through **One- and Two-Step Equations** and **Solving Equations with Variables on Both Sides**, it is common for algebra curricula to move to solving **Inequalities**. The procedures used for operating on inequalities are very similar on the surface to what students have experienced with operating on equations. If the students focus too much on solving inequalities algebraically, and neglect numeric and graphical solutions, the distinction of this class of algebraic relations can be unclear. How is it different when the equal sign is replaced by a "greater than" symbol? The procedural knowledge of how to handle division by negative numbers is insufficient to prevent errors; students need additional strategies to effectively operate on order relations. Once again, the root of students' struggles with inequalities may lie in their ability to understand what a variable can represent. It is a big leap to comprehend how a variable can be a range of possible solutions rather than a single answer (see **Variables & Expressions: Letter Used as a Functional Relationship**).

We then discuss **Factoring** as a means of solving quadratic equations. Back in 1989, NCTM advocated for a "decreased emphasis on factoring and an increased emphasis on using computer utilities and graphing calculators to solve problems that involve factoring."[8] Some researchers argue that factoring should not be taught at all; it has very little applicability because it requires integer coefficients between 10 and -10. "Rote mechanical methods like FOIL (First Outer, Inner, Last for factoring trinomials) . . . contribute little to student understanding."[9] Instead, students should rely on completing the square and the quadratic formula because those can be applied more universally.[10] Yet most algebra classes still spend quite a bit of time teaching students to factor. With the advent of phone apps like Wolfram Alpha that will factor anything and show you the steps, solutions, and representations, it is inevitable that factoring will lose its importance. That said, we focus here on students' struggles with understanding the mechanics of factoring, but even more so on how they understand the meaning of solutions and the structures involved in the process.

SOLUTION STRATEGIES

By understanding the different elements of algebraic relations and recognizing the nuances of different types of algebraic relations, the student can develop a toolkit of successful solution strategies without just seeing the toolkit as an unrelated list of procedures to memorize. It should be our goal for students to have a variety of strategies that they can use as needed. The strategies should be both formal (graphing, algebraic algorithms, successive approximation) and informal (guess-and-check, reasoning, intuition).

Formal procedures are a reliable way to solve algebraic relations. When employed properly, these procedures work every time, regardless of the complexity of the task. However, students will never truly understand a procedure until they have internalized it, realized the impact of operations on the structure and not the solution (i.e., the answer doesn't change when you add 3 to each side), and made it their own. Watching repeated demonstrations of formal procedures is insufficient. This does not mean that the teacher has no role in the development of student understanding of procedures; rather, students need guidance when developing new ideas. Such guidance can come in the form of a rich variety of examples that offer the opportunity for students to see patterns and common characteristics across a variety of scenarios. The development of student understanding means that we nurture ***Student Intuition and Informal Procedures*** as well, while exploring the role of formal procedures in their work. Informal procedures that are mathematically correct but less efficient might make more intuitive sense to students. Students can be confused if teachers imply the methods are incorrect, rather than less efficient. Both formal and informal procedures enhance different aspects of student understanding, and one should not be sacrificed for the other.

As students develop their toolkits for solving equations, it is critical that they understand how these different strategies are related. They must also know how and when to select the proper strategy, given a particular context. This is a demonstration of ***Flexible Use of Solution Strategies***. Knowing how to solve an equation by graphing it is an element of procedural knowledge. Knowing when it is best to solve equations graphically is conceptual knowledge.[11] For example, when given an equation like $3x + 4 = 2 + 3x + 2$, will the student employ algebraic manipulation to solve it? Should we take the time to enter the equations into a calculator? While both strategies can generate a correct answer, it would be most efficient to examine the task and see that the two sides of the equation are the same. We can immediately jump to the identity as the solution, and no further action is needed. A quick check of a variety of values of x will confirm our idea. In contrast, if asked to solve $3(x + 5) = 20$, a graphical or algebraic solution might be more efficient. Thinking through this task, or trying to plug in values for x, might eventually give us a correct solution, but it would be laborious. Having a variety of strategies, and knowing when to use each, is the hallmark of true conceptual understanding.

• • • • •

Student thinking about algebraic relations is complex. True conceptual understanding of solution strategies is dependent on a solid arithmetic foundation and an understanding of the features of various tasks. Tasks can include rational numbers and negative numbers and solutions involving zero. Each of these scenarios comes with its respective considerations. As students move from simple one-step and two-step equations to equations with variables on both sides, and to inequalities, it is critical that they understand how these different levels of complexity impact the use of the procedures they have developed. As they progress through these procedures, they will demonstrate conceptual understanding. We will take each of these ideas and illustrate how they appear in tasks, student work, and student thinking. We will share the specific mathematical issues under consideration, and potential teaching strategies.

ONE- AND TWO-STEP EQUATIONS: X ON ONE SIDE ONLY

Common Core Standards: 6.EE.5, 6, 7, 7.EE.4a, 8.EE.7, A-CED.1

An equation expresses an equivalent relationship between two expressions. Solving equations requires an understanding of the use of variables and the relationship between operations. Linear equations are the simplest kind of equations to solve because they are built by a sequence of invertible arithmetic operations with which students are familiar. Instruction in solving these types of equations often will focus on "undoing what has been done to x." However, this can be problematic for cases like:

$$-5x = 10$$

If students do not understand the symbols, they become confused about whether to add 5 or divide by -5. In this case, teachers can ask students to compare and contrast $5x$, $5+x$ and $5*x$. In each case, what has been "done to x"?

Four metaphors are often used to help students understand the ideas of an equation:[12]

1. A story about a number
2. A recipe
3. A balance
4. A function machine

In regard to *a story about a number*, consider the problem $[4(x - 1) - 2] \div 2 = 9$. This can be interpreted as a story: "There was a number x. You take away 1, then multiply by 4, then subtract 2, then divide by 2. The answer is 9. The *recipe* metaphor considers that problem similarly as the process of getting 9. Instead of seeing the equation as a sequence of procedures, the *balance* model conveys the sense of equilibrium between the sides (see Figure 3.1). The *function machine* metaphor is discussed in detail in **Patterns & Functions: Function Machines.** Research indicates that teachers may assume students automatically transfer the metaphors to the new situation of an equation. However, students may remember the procedures for solving, while not understanding equivalence between the left and right sides of an equation. Each metaphor may help with certain aspects of the concept of equation, but teachers should be aware of each metaphor's limitations.

Linear equations in which the variable only appears on one side of the equal sign are a good starting point for students learning to solve equations because manipulation of variables is minimal. In this class of equations, the manipulations involve explicit numbers. Previous misconceptions about number properties can confound the solving of these types of equations, though. For example, in the problem in Table 3.1, we see that the student does not understand when the Commutative Property of Addition can be applied. Or, when given an equation

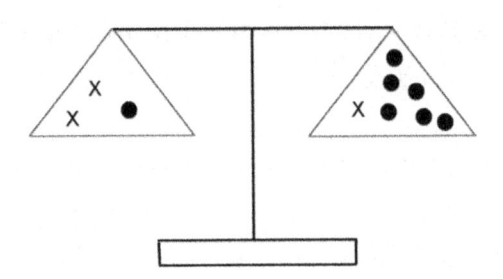

FIGURE 3.1. Balance model for 2x + 1 = x + 6.

such as $5x - \frac{3}{2} = 1$, misconceptions about rational numbers can lead to incorrect solutions.

To develop complete understanding of linear equations, students need to see equations as a whole, recognizing them as an entire structure rather than seeing them as an accumulation of numbers, unknowns, and operations. The main idea is to transform the original equation into a simpler one with the solution $x = a$, based on relationships between operations. Even when students understand the purpose and procedures for solving, they do not necessarily understand the concepts.

In general, students who rely on computational thinking have more difficulty solving equations than students who use relational thinking. Relational thinking means that the students are able to take advantage of relationships between numbers to rewrite expressions in ways that make the computation easier. For example, a student thinking relationally might rewrite 47 + 38 as 50 + 35, which is somewhat simpler to compute. Knowing the properties of numbers and operations is an advantage when moving from concrete numbers to abstract manipulation of variables. Students struggle with the relations when they only spend time with the procedures. Consider Problem 2 in Table 3.2, in which students are asked to think conceptually about the impact of procedures on equations.[13] The problem is a bit more abstract and demonstrates the challenges that students face when they don't understand the relationships in equations. See *Variables & Expressions: Acceptance of Lack of Closure* and *Letter Evaluated* for more insight regarding this problem.

TABLE 3.1. Problem 1: Variable on One Side of the Equation

5 − 2x = -15
2x − 5 = -15
2x = -10
x = -5

TABLE 3.2. Problem 2: Relationships in Equations

Problem	Response
If $e + f = 8$, what is $e + f + g$?	Correct: $8 + g$ (41%)[a]
	$8g$ (4%)
	12 (13%)
	9 (2%)
	some other number (10%)

[a]Percent responses from Kuchemann's (1981, 2,820 13- to 15-year-old students) study. Correct response percent similar in Hodgen, et al.'s (2008, 1,810 12- to 14-year-old students) study.

When introducing solution strategies for equations, teachers often begin with simple equations where the answer is apparent and the algebraic procedures are not necessary. However, giving students easier equations and insisting they use algebraic methods to solve them does not encourage algebraic skill development or conceptual comprehension of equivalence (see **Student Intuition and Informal Procedures**). Instead, we should challenge students right away with equations whose procedures require the use of algebraic methods, such as when the solutions are not whole numbers.

Some teachers depend on models for use in introductions as well. It is important to remember that what may seem like "real-life" situations may not be real for all students. For example, the balance model (see Figure 3.1) may be confusing for students who have never seen a balance. One must know the students and their experiences when selecting a model that will make sense to them. Students should be encouraged to verify that their solutions solve the original equation: checking the validity of their solutions leads to improved confidence. However, care should be taken when asking the students to justify their steps. If forced to explain their steps, students may only do the steps that they can explain. It is critical that teachers probe thinking to uncover misconceptions, rather than just looking at student work.

One of the areas that has been well researched is students' difficulties in understanding the equal sign.[14] Researchers found that few preservice teachers understand that many students "hold misconceptions about the equal sign."[15] Because elementary students typically handle only problems that flow from left to right, they often develop an "operational conception" of the equal sign. They see it as an indication to "do something," or as the end of the problem ("and the answer is..."), instead of a symbol that signifies the equivalence of two expressions.[16] Researchers have found that math textbooks encourage these misconceptions by "rarely present(ing) equal signs in contexts most likely to elicit a relational interpretation," such as having operations on both sides.[17] As a result, students may believe equations should be read left to right and therefore perceive an equation such as $5 = x + 2$ as backward. Students with these conceptions of the equal sign ultimately have difficulty solving equations. Recent efforts in elementary mathematics classrooms encourage teachers to help students develop a "relational conception" of equations and also help them consider how parts of an equation fit together. Researchers have found that elementary students as young as first grade are capable of understanding the relational meaning of the equal sign.[18]

Sometimes teachers try to help students with equation solving by introducing phrases that indicate short cuts or easy ways to remember or do problems. These can have unintended consequences, such as with the phrase "Change sides, change signs" for solving equations by addition or subtraction (see Figure 3.2). Researchers find that rote strategies such as "change sides, change signs" can have harmful effects:[19] such an approach to solving equations may confuse students by giving them the impression that the equal sign is a magic change-maker. This type of strategy encourages students to implement procedures without understanding the relational nature of the equal sign. Furthermore, the approach fails for a problem such as $5x = 10$.

Another commonly used phrase is "whatever you do to one side, you have to do to the other." Researchers found that the vagueness of the phrase resulted in students' different interpretations. For instance, some students interpreted that phrase to mean "use only subtraction to simplify an equation," or "get rid of" nuisance variables to make simplifying the problem easier.[20] Therefore, it is critical that teachers take responsibility for developing that sense of relationship and balance instead of relying on catchy phrases that may help students through a narrow range of problems and have negative long-term consequences. "A sophisticated and flexible understanding of the equals sign (=) is important for arithmetic competence and for learning further mathematics, particularly algebra."[21]

Also, students sometimes adopt informal procedures in order to solve equations. For instance, in solving the equation $2x - 3 = 5$, a student might write $2x - 3 = 5 + 3 = 8 = 4$. In this case, even though the equal sign is incorrectly used, students can follow their reasoning and arrive at a correct answer.[22] One researcher refers to this tendency as "notational abbreviation."[23] This approach may be successful with simple equations, but for more complex equations this procedure can cause significant problems with understanding. It is important for teachers to help students see this strategy as understandable, but not algebraically correct. The goal is for students to comprehend the balance perspective of the equal sign.

$$x + 5 = 9$$
$$5 \rightarrow -5$$
$$x = 9 - 5$$

FIGURE 3.2. "Change Sides, Change Signs."

Algebraic Relations • 109

Teachers

- Equations with variables on one side are a good introduction to solving equations.
- Encourage students to imagine the entire equation as a unit in balance. Models can be helpful, but informal reasoning for simpler equations should also be encouraged.
- Forcing students to apply procedures to simple tasks can frustrate them and hinder their mathematical reasoning. Challenge students with difficult tasks where procedures are tools that help.
- "Generalization can only develop from a broad range of experiences. So students need to encounter, early on, equations which have other than small positive integer coefficients, or solutions."[24]
- Analyzing the nature of students' errors when solving equations may be insightful for making instructional decisions.[25]
- The Center for Algebraic Thinking's *Algebra Equation Builder* app[26] (see Figure 3.3) gives students practice in maneuvering numbers, letters, and operations in order to create a true algebraic equation with a variable on one side of the equation.

FIGURE 3.3. Algebra Equation Builder App: Variable on One Side.

RATIONAL NUMBERS

Common Core Standards: 6.NS.6, 6.EE.7, 7.NS.1, 2, 7.EE.1,3

Researchers find that students have many misconceptions when it comes to fractions.[27] One of the reasons for difficulties with fractions is that students are often encouraged to memorize algorithms without developing an understanding of them. Understanding rational numbers and being fluent with operations on rational numbers are important building materials for algebra. In order to be fluent with rational numbers, students need to understand five concepts:

1. Part–whole relationships,
2. Ratios,
3. Quotients,
4. Measures, and
5. Operators.[28]

Developing the ratio meaning of fraction is important to students' ability to develop proportional reasoning. Developing the rate meaning of fraction is important for students to develop an understanding of slope. Understanding of operations on fractions is the basis for operations with rational numbers in algebraic equations and rational expressions. Teachers will need to review the meaning of fractions and use of common denominators during algebra courses.

To determine the level of instruction needed, there are certain types of pre-assessment tasks that can provide insight into student understanding.[29] For example, when given a task such as $\frac{5}{12} + \frac{3}{8} = ?$, a common incorrect solution is to add the numerator and denominator to get $\frac{8}{20}$. In this case, manipulatives and other visual models can help students comprehend the mistake.

This type of error is compounded when rational expressions contain variables. Some students will evaluate $\frac{a}{b} + \frac{a}{c}$ as $\frac{2a}{b+c}$ because they feel that common numerators can be operated in the same way as common denominators.[30] To help students overcome this misconception, the teacher could ask students to pick different numbers for a, b, and c to see that their answer would only be correct (in real numbers) when $a = 0$.

The tasks discussed in this section may serve to assess students' abilities to simplify expressions using learned algorithms. In addition, much can be gained by asking students to complete non-routine tasks using rational expressions. For instance, see Problem 3 in Table 3.3.

This type of task does not have an algorithm that can be easily applied, so it depends on students' inductive reasoning and rational number sense.

TABLE 3.3.

Problem 3: Rational Numbers as *n* Gets Very Large, What Happens to 1/*n*?

a) It gets close to 1
b) It gets close to 0
c) It gets very large

Source: Brown & Quinn, 2006.

Informal estimation is a valuable precursor to formal operations and allows students to confirm how reasonable an answer might be, as in Problem 4 in Table 3.4. In this case, students should be able to reason that the expression is "almost 1 – almost $\frac{1}{2}$." So the answer should be *b*. Even if students do not know exact values for rational numbers, they should be able to reason about the relative sizes of the numerator and denominator. Even if students don't know the exact value of terms such as $\frac{a}{5}$, they can use their understanding of the relative value of a term with a variable divided by 5. Informal procedures and student intuition are valuable components of mathematical understanding and should be cultivated (see **Student Intuition and Informal Procedures**).

Rational expressions in algebra lend themselves to unique forms of errors and misconceptions as well. For example, understanding how to simplify fractions in Problem 5 of Table 3.5 can lead to the incorrect algebraic simplifications in Problems 6 and 7. In Problem 7, only 22.8% of students got the final correct answer of $\frac{x+5}{x+4}$. Of the others, 38.9% struggled to make any progress, 24.4% canceled the

TABLE 3.4. Problem 4: Estimation

Problem	Response[a]
$\frac{12}{13} - \frac{3}{7}$ is about how much?	
a) 1	15.4% chose a) or c)
b) $\frac{1}{2}$	32.2% said they didn't know
c) 0	
d) I don't know.	

[a] Percent of 143 students aged 13–15 years in Brown & Quinn (2006)

TABLE 3.5. Problems with Rational Expressions

Problem 5
$\dfrac{8}{12} = \dfrac{4*2}{4*3} = \dfrac{2}{3}$

Problem 6
$\dfrac{x+y}{x^2} = \dfrac{1+y^2}{x}$ or $\dfrac{x}{x+y^2} = \dfrac{1}{1+y^2}$

Problem 7
$\dfrac{x^2+3x-10}{x^2+2x-8} = \dfrac{(x+5)(x-2)}{(x+4)(x-2)} = \dfrac{(x+5)}{(x+4)} = \dfrac{5}{4}$

Source: Hall (2002); Rossi (2008)

x^2 term and stopped, and 11.1% factored the numerator and denominator correctly but did not cancel the $(x-2)$ term.[31] Generally, students get the idea that they can cancel terms from the top and bottom but misunderstand when that is appropriate.

Teachers

- Researchers suggest the following strategies for helping students make the transition to algebra with a solid understanding of rational numbers:[32]
- Fraction operations should be developed as a generalization of whole number operations, and students should be given the chance to construct their own algorithms.
- The development of the formal definitions of fraction operations should progress through students' experiences and algorithms, and also prepare students for the abstraction necessary in algebra.
- When teaching about fractions, start with manipulation of concrete objects and the use of pictorial representations such as unit rectangles and number lines.
- Develop fraction notation, but avoid introducing formal procedures. Students need tasks that require reasoning above formal steps.
- Focus on building a broad base of experience that will be the foundation for a progressively more formal approach to learning fractions.

Some of the manipulatives that could be used to develop the meaning of fraction and operations with fractions are Cuisenaire Rods, pattern blocks, Geoboards, number lines, and fraction bars. These manipulatives are commonly available tools found in many schools and are also freely available as online tools through Shodor, NCTM, and the National Library of Virtual Manipulatives. Another tool

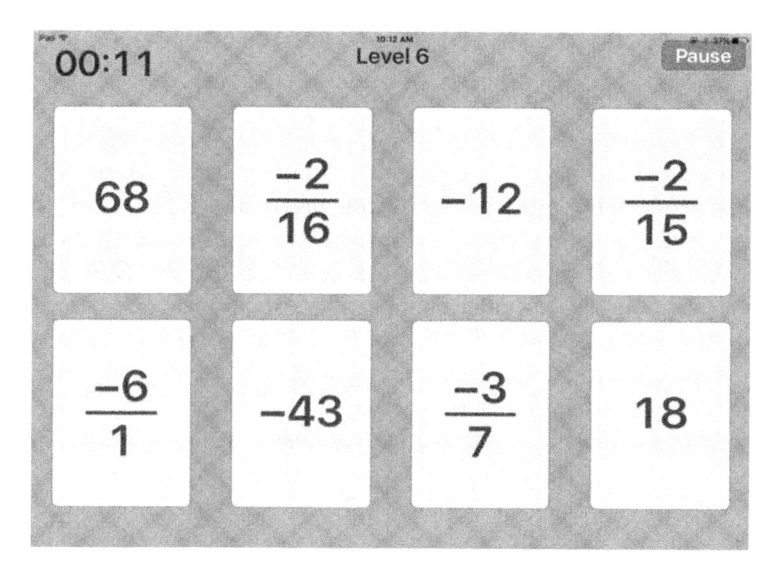

FIGURE 3.4. Algebra Card Clutter App.

is the Center for Algebraic Thinking app, *Algebra Card Clutter* (see Figure 3.4), which can help students develop understanding of the relative sizes of fractions by having students organize cards from lowest to highest.

EQUATIONS INVOLVING NEGATIVE NUMBERS

Common Core Standards: 6.EE.2b, 5, 6, 7, 7.EE.4a, 8.EE.7, A-CED.1

There are foundational concepts that students must understand in order to solve equations and simplify expressions. When numbers in the expression or equation are negative, additional consideration must be given. Solving equations often involves subtracting the same quantity from both sides to simplify the equation. However, this can lead to subtracting a negative from a negative, which often confuses students. Likewise, multiplication by a negative can lead to similar confusion. Some research indicates that working with negative numbers actually has some benefits for students trying to understand algebra. Where "real world" models of negative numbers such as elevators or money have limitations, students can make full use of operations with signed numbers in algebraic equations.[33] The study of functions is more complete and meaningful when you can integrate negative numbers. "Signed number tasks can help students move beyond the conception that equations display 'an action on one side and its result on the other' to an 'equivalence relation of transformations.'"[34]

Negative numbers have traditionally been challenging for students to understand. Rote learning of operations with negative numbers has been shown to limit "depth and clarity of the concept beyond what was taught."[35] Researchers suggest that the development of students' understanding about negative numbers progresses through a number of ways of thinking and levels of acceptance:[36]

- Problems in which negative numbers are not encountered; where a is always greater than b in $a - b$.
- The idea of opposite quantities is encountered and a sense of symmetry is developed. Students have a sense of a number—e.g., 5—and its opposite, so that $5 - 5 = 0$.
- The result of an operation or solution to a problem or equation is an isolated negative number, e.g., $7 - 9 = -2$. This extends the operations of addition and subtraction to the left on a number line when a larger number is subtracted from a smaller number.
- A formal sense of the integers as a set of numbers that contains positive and negative whole numbers. Understanding the number line and that numbers can extend left and right. Bigger numbers are further to the right.
- An understanding of the set of rational numbers and then the set of real numbers, including fractional parts of negative numbers and a sense of infinity between negative integers (e.g., -5.678 . . . with as many numbers after the decimal as you want).
- Understanding the dual meanings of negative numbers. Numbers to the left are smaller than numbers to the right. However, there is also a sense of

magnitude of negativity. That is, -11 is greater in debt or deeper in a hole than -3.

Students begin learning about integers with the assumption from early arithmetic that addition increases quantities and subtraction decreases.[37] This can be a difficult, naive conception for students to let go of and may impact their understanding of negatives when variables are involved.

Researchers have found that students' difficulties with negative numbers fall into three categories of understanding:[38]

1. The number system and direction and magnitude of the number:[39]
    1. understanding meaning of negative number,
    2. representing negative numbers on a number line, and
    3. ordering negative numbers on a number line.
2. Arithmetic operations:[40]
    1. show operation on a number line,
    2. conduct operations,
    3. write operation appropriate to context, and
    4. write a problem with a given operation.
3. The meaning of the minus sign:[41]
    1. interpret and explain.

At the heart of this issue is students' typical difficulty with basic arithmetic with negative numbers. Teachers will see consistent errors when students lack a strong knowledge base for operations on negatives, reflecting these categories of errors. Operationally, these errors can be classified into five types:[42]

1. Adding a number to its additive inverse (the student has ignored the negative sign):

$$-3 + 3 = 6$$

2. Adding a positive number to a negative number (the student thinks addition should result in a positive number or a number looking bigger than the addends):

$$-8 + 3 = 5 \text{ or } 11 \text{ or } -11$$

3. Subtracting a positive number from a negative number (the student thinks that the smaller number should be subtracted from the larger number, or the student has ignored the sign):

$$-3 - 8 = -5, 5, \text{ or } 11$$

4. Subtracting a positive number from a smaller positive number (the student has ignored the order of terms):

$$3 - 8 = 5$$

5. Subtracting a positive number from 0 (the student has ignored the zero, assuming it does nothing):

$$0 - 8 = 8 \text{ or } 0$$

There are other difficulties that arise with negatives in equations, separate from those arithmetic difficulties. Equations are full of other symbols. For instance, it is easy for a student's eye to skip over minus signs while trying to focus on "more important" things like variables. It is not always clear to a student what way is appropriate to eliminate an unwanted term or factor from an equation when that term or factor includes a minus sign.

Arithmetic issues are compounded when considering negative variables, as many students have significant difficulty interpreting the value of a variable when it is negative (see ***Variables & Expressions: Representation***). When presented with a task like $13 - x = 5$, students can often reason through the task without formal operations in the same way elementary school students might deduce what goes in the box for a task like $13 - \square = 5$. However, researchers find that students can lose sight of this intuitive approach to problems and instead struggle to get x by itself procedurally (see ***Students' Intuition and Informal Procedures***). In $13 - x = 5$, the problem is simple intuitively but complex procedurally (having to deal with the negative of a variable). The negative in this problem does not pose as much of a cognitive challenge for students, perhaps because it is simple subtraction of a smaller positive number from a larger; so, in a sense, there are no "negative" numbers involved. This changes, though, as the tasks become more grounded in integer computations. Examples like $5 - x = 7$ or $-5 - x = 11$ pose more of a cognitive challenge.

As students begin solving multi-step algebraic relations (see ***Solving Equations with Variables on Both Sides***), the negative terms continue to cause problems. In the following task, a student has detached the negative preceding the $3x$ and this produces the error.

$$2 - 3x + 6 = 2x + 18$$

$$-2x \qquad\qquad -2x$$

$$2 - 1x + 6 = 18$$

It is helpful to know and recognize common student errors. However, nothing can compare to a direct teacher-student conversation to find out what a student is thinking. Incorrect answers can result from either misconceptions or simple mistakes, and it is important to determine the difference. In the conversation in Table 3.6, an interviewer poses a task and the student presents and defends an incorrect response.

Initially, this error appears to be an error of the third type described above. If we only look at the answer (-5), then we cannot know if the student thinks that the smaller number should be subtracted from the larger number or if the student has ignored the sign. Based on the verbal response, it is still unclear which of the errors has occurred (or if this is a new type of error). In this situation, the teacher needs to consider how to frame questions to access the student's way of thinking in order to have a clear understanding of the mathematical issues at play. Perhaps the teacher could ask, "How is this different from 8 – 3 or -8 + 3?"

We know that it is easier for students to learn to work with positive numbers because they can map their understanding of operations and properties to physical materials. Developing a similar way of understanding negative numbers is harder because negative amounts do not appear physically in their world. A comprehensive understanding of "negative" includes:[43]

- An understanding of subtraction/operations,
- An understanding of direction,
- An understanding of value or magnitude, and
- A symmetrical understanding of opposites.

Teachers can help students develop their understanding of negative numbers by using phrases to remind them of strategies, such as "like signs positive" and "unlike signs negative" when learning multiplication with negatives. However, these "rules" are often vaguely remembered by students and can be misinterpreted and misapplied—for example, with addition.[44]

To help students understand negatives, teachers do use physical representations such as manipulatives (like integer chips, see below for an example), number lines, and several different metaphors (e.g., money/debt, temperature, time, elevators, or football). These ideas will help students bridge the gap between the meaning of negative numbers and the everyday world. However, manipulatives, number lines, and metaphors should be about mathematical ideas to be developed, not the procedures to be followed to use them; and, the mathematical symbols should always be linked.

Research indicates that students' understanding of negative numbers improves when they are aware of the limitations of the models (manipulatives, number lines, and metaphors).[45] For example, a money/debt model could be used to develop the meaning of a negative number. A debt of $7 can be represented by -7. But teachers must make sure that students distinguish the '-' in front of the 7 from the subtraction operation. For example, if you owe $2 to your father and $3 to your mother, then your total debt could be found by computing (-2) + (-3) = -5 which means you owe $5. It is possible that teachers represent this situation by

-2–3, which is arithmetically correct but creates confusion between the negative sign and the sign for the subtraction operation. This shortcut could confuse students when they try to develop the meaning of the negative sign as distinct from the subtraction sign. The money/debt metaphor can confuse students in problems like $(-8) - (-2)$. Some students tend to translate this expression incorrectly as "You owe \$8 and then you owe another \$2." This ignores the fact that the subtraction sign in the middle indicates that you are taking away from your debt, which does not necessarily make sense in the students' world, although it may in the real world.

Having a mental number line is also often considered a component of number sense that helps to build an understanding of negative numbers.[46] Internalization of negative numbers is the stage when a person becomes skillful in performing subtractions. Understanding the numerical system and relative size of the numbers, including the number zero, is an important part of having a mental number line. Absolute value is also an important and very powerful concept to comprehend in this aspect.

The problem $-x = 7$ is one of the most difficult equations for students to understand. They consider $-x$ statically just like -4 and treat them as the same. Students can struggle with the idea that $-x = 7$ could give $x = -7$ because to them, the right-hand side of the original equation would have had to be -7. A problem such as $6x = 7$ is easier for students to work than $-5x = 10$. They may not recognize that -5 and x are being multiplied. The negative sign persuades students that there is a subtraction of 5 that needs to be reversed with addition rather than multiplication by -5.

It is essential that students are grounded in variables and notation before solving equations. Students need to distinguish between the x and the minus sign as being two distinct pieces of information. Being precise with language is important when working with negative numbers. For example, in a problem such as "6 − -3," saying "6 minus minus 3" rather than "6 subtract negative 3" can hinder students' understanding.[47] Thinking of the negative sign as signaling the opposite of x could help with relational reasoning (see ***Variables & Expressions: Representation*** for further discussion on students' thinking with negative variables).

Teaching Strategies

Once teachers can recognize potential student errors and have an understanding of the mathematical issues involved, the focus moves to teaching strategies that are effective for addressing these misconceptions. Using different models for negative numbers and knowing the pros and cons of each model will help students to understand negative numbers and their operations.[48] Teachers can then move to a parallel discussion using negative coefficients of variables. Researchers have found that reasoning about negative numbers using metaphors can be helpful to some for understanding their use in that narrow context, but confusing to others when they don't already understand the concept.[49] In fact, "integers do not appear as tools that are necessary for solving problems in the world, but as an artificial, purely mathematical lens that we can apply to real-world contexts when that is the game to be played."[50] The "real-world" problems we typically use in school do not need negative numbers to be solved. In any case, teachers should alternate among models to compensate for the imperfection of each model; however, such alternations should be done in concert with pointing out the issues in the model so students can appreciate how negative numbers can be used.

Some commonly used models are:

- *Debt or owing:* We need to be careful as we use this model because not all operations can be modeled naturally in this way. For example, subtracting a negative number from a negative number can become problematic for some students. Instead of the debt or owing model for negative numbers and operations, we can use the story of a letter carrier delivering checks or bills to an address and the person at the address keeps track of the bills and checks as they are delivered. In this story line, subtracting a negative number from a negative number can be conveyed by delivering a bill to the wrong address and the postman comes back to take away the wrong bill.
- *Two-color integer chips* (sometimes called Equilibrium): One color represents positive numbers, and the other color represents the negative numbers. With this model, most of the operations are modeled naturally. However, for subtracting negatives, pairs of negative and positive chips need to be generated, and this is somewhat artificial. For example, for 4 – (-3) four black chips are displayed then 3 pairs of black and white chips are formed to display zeros, then three white chips are taken away to display the value of the operation, 7, with seven black chips (see Figure 3.5).
- *Elevators:* Numbers are used to represent both position (the third floor) and the action (going up three floors). It might be hard for students to grasp these two meanings of the number; however, this is a model that aligns well with the number line.
- *Time:* A scale from B.C. to A.D. could be used to model numbers. However, time is already a hard concept for students to grasp, and adding and subtracting dates does not make much sense.[51]
- *Temperature:* This model may be beneficial to use if students already have a conceptual understanding of temperature, particularly what it means to be "below zero." However, if they don't grasp the temperature idea it might be hard for them to understand what it means to add or subtract temperatures. This is another model that aligns well with the number line, though, for those students who do understand temperature.
- *Number line model for operating with negative numbers:* In this model, addition and subtraction operations indicate which direction you face: for addition, you face the positive numbers and for subtraction, you face the negative numbers. The sign of the number indicates how you move on the number line: if you have a positive number, then you move forward

Step 1

● ● ● ●

Step 2

● ● ● ●
● ● ●
○ ○ ○ } zero

Step 3 (taking away -3)

● ● ● ●
● ● ●

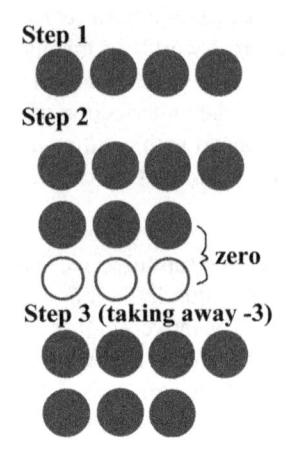

FIGURE 3.5. Integer Chips Modeling 4 – (-3) = 7.

in the direction you are facing; if you have a negative number, you move backward in the direction you are facing. For example, for (-7) + (-2) you start at -7 on the number line and since you are adding, you face positive numbers. Since you have -2, you move backward 2 units facing positive numbers and land on -9. This model helps students to operate with negative numbers systematically. However, it is possible for students to memorize the steps to produce the answers without really understanding the meaning behind the operation. Thus, providing an alternative model to develop the meaning of negative numbers is highly suggested. The *Hop the Number Line* app (see Figure 3.6) can be a useful tool for helping students practice their thinking with addition and subtraction of integers on the number line.

- *Computer animations:* Animations, such as a seagull flying above an ocean and then diving into the sea after fish, have been found helpful for students trying to understand the dynamic between negative numbers.[52]

Understanding the benefits and drawbacks of these models is helpful when designing lessons or addressing individual student needs. As students move beyond arithmetic operations, researchers find that there are additional teaching strategies when working with equations that contain negative terms:

- Relational thinking ("the same as") can help students solve questions like $5 - x = 7$ or $-5 - x = 11$. Teachers should emphasize that $5 - x$ is the same as 7. Simpler tasks could be solved using reasoning instead of procedural operations that introduce the potential for errors. This is also an effective strategy for **Solving Equations with Variables on Both Sides**.

FIGURE 3.6. Hop the Number Line App.

- Asking students to justify (not just describe) their steps in writing forces them to do only steps that they can explain. Ignoring signs becomes difficult under such types of questioning, and defending the validity of their solutions may also improve students' confidence.
- Common errors should be discussed at the appropriate point, both in lessons and in textbooks (especially at the introductory level), because it prevents development of bad habits and misconceptions on the part of the student.
- Interviewing students rather than just looking at written work will help uncover misconceptions. This is especially true for students who are English Language Learners, as they may have differing conceptions of various terms.
- Algebra tiles extend the strategy of two-color integer chips to variables. Positive and negative x are represented by different colors and are different sizes than positive and negative numbers (see Figure 3.7).

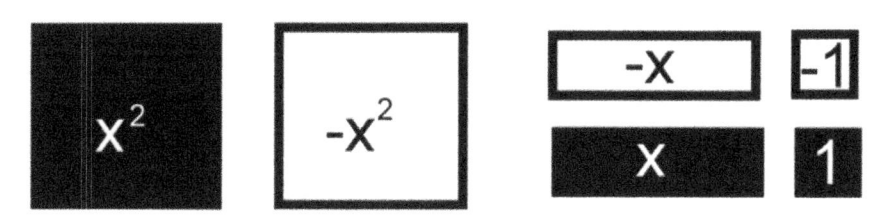

FIGURE 3.7. Algebra Tiles with Negatives.

Teachers

- Students need to have strong arithmetic skills in order to solve equations in algebra. Review of operations on negative numbers is critical even in algebra.
- Take the opportunity to interview students as well as analyze their individual work. This can reveal the actual mathematical issues underlying misconceptions and faulty logical reasoning.
- Use models for reviewing operations on negative numbers. Researchers found metaphors as generally useful depictions of positive and negative numbers.
- The Center for Algebraic Thinking offers two apps to give students practice with Integers. The first, *Hop the Number Line* (see previous Figure 3.6), asks students to place the rabbit on the correct spot on the number line in order to combine the integers accurately. The second, *Algebra Card Clutter* (see previous Figure 3.4), has students organize cards from lowest to highest. It has levels for students to organize cards that include fractions, decimals, exponents, absolute value, and square roots.

INEQUALITIES

Common Core Standards: 6.NS.7a, 6.EE.5, 8, 7.EE.4b

For many students, inequalities represent a difficult next step after equations. The Trends in International Mathematics and Science Study (TIMSS) asked students to solve the inequality: $9x - 6 < 4x + 4$. Only 17% of eighth-grade students internationally could solve that problem accurately.[53] In the United States, 21% correctly solved the inequality. The procedures for solving inequalities are deceptively similar to those for solving equations, but there are significant differences that students often overlook. For example, division by a negative number requires a change in the sense of the inequality. Why does it change? Sometimes we ask students to "break apart" the inequality and develop two simultaneous solutions. Sometimes it is best to simply reason about the inequality as a comparison of two expressions. Sometimes students need to simplify the inequality down to a single appearance of the variable. When should that happen? Why?

Teachers can help students recognize these differences between solving equations and solving inequalities. When teachers introduce inequalities, students will try to connect what they already know about how to solve equations. In word problems, the context of a problem can help students recognize the differences between situations of inequality and equality.[54] Students must realize that with an inequality, you are not seeking a precise answer: the inequality has many potential solutions—an interval of values. Students often depend on algebraic methods for solving equations and inequalities, and use strategies interchangeably (and sometimes incorrectly). This misconception is understandable because equations and inequalities "look" similar. Students struggle with solving linear and quadratic inequalities, especially if they are provided with only one method of solving them. Typical student difficulties when solving inequalities are:[55]

- When multiplying equations on both sides by a negative number, students forget to change the direction of the inequality sign.
- Students may treat an inequality as an equation or even change the inequality to an equation.
- Students have difficulties translating words such as "at least" or "not more than" into mathematical symbols using inequalities (see ***Modeling: Translating Word Problems into Algebraic Sentences***).
- Students have a hard time accepting and understanding solutions to inequalities:
  - If the solution results in a single answer, the empty set, or the set of all real numbers, students are often confused.
  - In early algebra, students have difficulty thinking about solutions to inequalities as numbers other than positive integers.

o Working with solutions to quadratic equations, students struggle to understand the solutions. For instance, for the problem $x^2 > 25$, students often respond $x > \pm 5$ instead of $x < (-5)$ or $x > 5$. When asked to explain this solution, students resorted to the procedures they were used to applying when solving equations.

In order to address these potential struggles, the literature points to a number of effective strategies. First, students can be asked questions that demand more than just a procedure to identify an answer. For example, try to answer the following questions:[56]

1. Is there a value for x that will make $(2x - 6)(x - 3) < 0$ true?
2. Let p and q be odd integers between 20 and 50. For these values, is $5p - q > 2p + 15$ always true, sometimes true, or never true?
3. Is there a value for x that will make the following statement true?

$$(6x - 8 - 15x) + 12 > (6x - 8 - 15x) + 6?$$

Each of these problems does not require students to solve for x or simplify. Researchers suggest that this approach may be effective at "eliciting a greater variety of anticipatory behaviors"[57] because it can encourage students to predict the answer, rather than perform any actions. In Question 1, the factored form of the inequality is $2(x - 3)^2$. Students need to recognize that no values can make that inequality true because you will only get positives with the squared term. In Question 2, the task is not looking for a single solution, but rather whether it can be true and when. In Question 3, students should be able to recognize the structure of the problem—both sides contain $6x - 8 - 15x$ and $12 > 6$—and deduce that any value would work. It might be helpful for students to begin with guess-and-check to get a sense of the question, and then see where graphical and algebraic solutions take them.

Question 3 taps into the similar ideas that are mentioned in ***Algebraic Relations: Student Intuition and Informal Procedures.*** That is, instead of automatically beginning to use a set of procedures, it is good for students to look at the structure of the inequality and recognize the commonalities that lead to a conclusion about the relationship. Consider two students' responses to Question 3 in Table 3.7. Both students combined like terms to simplify the equation. However, Student 2 began by noticing that the expressions in the parentheses were the same. Teachers should encourage students to develop and work with their intuition to seek understanding first, and then what procedures might be appropriate.

Second, when introducing inequalities, it can be helpful to explicitly distinguish between different classes of equations and expressions. In each of the following, students can have difficulties understanding them as equations or inequalities:[58]

- Identities: $\cos^2 x + \sin^2 x = 1$

TABLE 3.7. Interviews on Inequality Question 3

Student 1	I was taught to combine like terms. I was taught this (>) is actually an equal sign. To solve it I would solve an equation [she obtained -9x + 6 = -9x and then wrote 6 > 0]. Umm, that doesn't [seem] right, because x has canceled out. What did I do wrong? ... OK. Is there a value for x that will make the following statement true? Maybe there isn't.
Student 2	The stuffs in the parentheses are the same. Umm, OK, first I guess I would combine all like terms ... [he got -9x + 4 > -9x – 2]. Umm, now it's asking is there a value for x that will make the following statement true. Umm, let me see, I think 4 and -2, so you have a common term (i.e., -9x). OK, so it's, you have a -9, so anything [positive] that you multiply will [make it] a negative number, and this (+4) is positive. Let's see, yes, there is a value because... this, this [left] side will be greater. I guess, if it (-9x) was positive then, so is this side (-9x). So any negative number would make the statement true. ... Umm, I think all numbers would make the statement true.

Source: Lim (2006)

- Non algebraic: $\int f(x)dx = x^2 + C$
- Equations/Inequalities with more than one unknown: $3x + 4y = a$ or $3x + 4y \geq a$
- Trivial equations/inequalities: $x = 2, x \geq 2$,
- Functions: $f(x) = 2x + 1$
- Inequalities and Expressions: $3x+2 > 4x +5$

It is important to give students the opportunity to notice the subtle differences between the types of equations and inequalities and how they appear graphically and symbolically. Developing meaning for symbolic representations of equations or inequalities (including linear and second and third degree) helps students to understand what the solutions represent or even judge if an answer makes sense. Designing lessons where equations and inequalities are compared and contrasted can have the additional benefit of promoting students' better understanding of the equal sign.

Third, teachers are encouraged to provide more than one solution method for solving inequalities. Illustrating multiple solution strategies helps students to become more flexible in solving problems in different topics.[59] Students who are introduced to more than one method for solving inequalities, for example, cope better when they are stuck because they can try another method. Different strategies for working with inequalities include:[60]

- Algebraic manipulation (using operations on the inequality to simplify),
- Using the number line (graphing the solution set on a single line),
- The coordinate system graphical representation method (i.e., graph the equation and determine on which side of the graph the desired values are located),

TABLE 3.8. The Sign-Chart Method

$(3x - 3)(5x + 15) > 0$	$-\infty$	-3	1	∞
Sign of $3x - 3$		negative	negative	positive
Sign of $5x + 15$		negative	positive	positive
Sign of $(3x - 3)(5x + 15)$		positive	negative	positive

Based on Dobbs & Peterson (1991); McLaurin (1985); Tsamir & Reshef (2006)

- The sign-chart method (see Table 3.8), and
- the logical connectives method (i.e., creating a system of equations that are connected to each other by "or" or "and").

In particular, researchers have found that students struggle with logical connectors and knowing whether a response requires "or" or "and."[61] For instance, with the inequality $\frac{x-5}{x+2} < 0$, students wrote the answer $x - 5 > 0$, $x + 2 < 0$ without any logical connection. This is also true of inequalities with absolute value signs, such as $|x| < 3$ and $|x - 2| < 1$, which 65% and 77.1% of high school students, respectively, were unable to solve with the correct logical connection.[62]

Research indicates that students have more success with graphical approaches than algebraic approaches because the "visual characteristics can help students to understand the differences between equalities and inequalities and the meanings of the different signs used to express these relationships."[63] Graphing each side of an inequality as a separate graph and doing a logical comparison can help promote an understanding of the result. As noted above, students generally have quite a bit of difficulty with logical connectives. They have a difficult time understanding when "or" or "and" is appropriate for a solution, as is evident for the problem in Table 3.9 and the interview in Table 3.10.[64]

If students are already familiar with graphing functions, then solving inequalities with graphic methods can help them to understand the solution better. The

TABLE 3.9. Problem 8: Rational Inequality

Problem	Response
$\frac{x-5}{x+2} < 0$	$x + 2 < 0\ x - 5 > 0$
$x < -2$	$x > 5$

Source: Tsamir, Almog, & Tirosh, 1998; Tsamir & Almog, 2001

TABLE 3.10. Interview on Rational Inequality and Logical Connectives

Interviewer:	What is one supposed to do with the two results?
Student:	I think that the solution is "or."
Interviewer:	How did you get $x > 5$?
Student:	By imposing a positive numerator.
Interviewer:	And how did you get $x < -2$.
Student:	By imposing a negative denominator.
Interviewer:	So, what should the connection between the two be?
Student:	"and" sorry, ah "or" [thinks] I would like it to be "and" but it is impossible, thus it is probably "or."

Source: Tsamir, Almog, & Tirosh, 1998, p. 133

graphical method for solving inequalities also helps these students to answer questions related to the functions (solutions, etc.). Furthermore, by focusing on graphical methods of solving equations and inequalities, students see representations of equations and inequalities that look different, they are less likely to incorrectly apply strategies from one type to the other, and the visuals help them interpret the results[65] (also see ***Algebraic Relations: Flexible Use of Solution Strategies***).

Teachers

- Choose tasks that demand more than just a procedure to identify an answer. Tasks should highlight the ways that inequality solutions differ from solutions to equations.
- Take time to analyze the features of different tasks in algebra (inequalities, different classes of equations, expressions, etc.). Students need to know the meaning behind the form of each task.
- Provide more than one method to solve inequalities and help students to choose a method they understand well.
- Help students develop the habit of thinking about what the inequality question is asking first, before starting to manipulate the variables and numbers.
- Research suggests that having students compare inequalities with the equal sign "facilitated a relational understanding of the equal sign more than a lesson in which students learned about the equal sign alone in the same amount of time."[66]
- The Center for Algebraic Thinking's *Inequality Kickoff* app (see Figure 3.8) gives students practice in thinking about inequalities with variables.

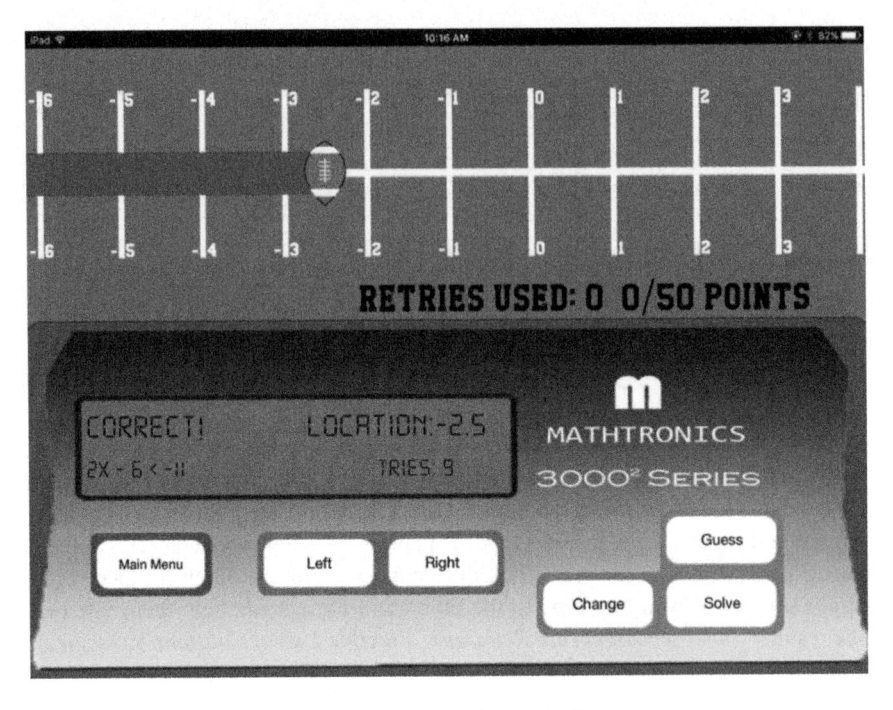

FIGURE 3.8.   Inequality Kickoff App.

SOLVING EQUATIONS WITH VARIABLES ON BOTH SIDES

Common Core Standards: 6.EE.2, 3, 7, 7.RP.2a, 2b, 8.EE.7a, 7b, A-REI.2, 3

There are many errors students can make when solving equations, including simplification errors, sign errors, "unbalanced" errors (adding different quantities to each side of the equation), and arithmetic errors. A solid foundation of arithmetic can help students understand the rationale behind the symbolic manipulations that go into solving an equation. Students' selection of a solution strategy depends extensively on the strategies employed in lessons by their teachers.

As a student solves an equation with variables on both sides of the equal sign, there are a number of opportunities for simplification errors to occur. Given a task such as $2x + 3 = 6x - 5$, imagine that the student begins by getting all of the variables "onto the left side:"

1. Subtract $6x$ from both sides $-4x + 3 = -5$,
2. Subtract 3 from both sides $-4x = -8$, then
3. Divide both sides by -4 $x = 2$.

In step 1, some students may ignore the sign of the variable and just "move the x to one side." In this case, we would see $8x + 3 = -5$. In step 3, it is common for students to make one of two errors. Some students divide a negative by a negative and get a negative answer. Others "divide backward" and incorrectly get the error of $x =$. These errors can be extended when there are multiple like terms on one side. When working on tasks like $5x - 3x + 1 = 4x - 3$, it can be common for students to add $3x$ to every x term regardless of the equal sign. Simply teaching students to combine like terms prior to adding or subtracting terms from both sides of the equation may miss addressing this apparent lack of understanding of the equal sign's role. Another common error for students is the "deletion error," in which they eliminate numbers and variables incorrectly.[67] For instance, $2yz - 2y$ may be equated to $2z$ (because students may misapply the fact that $2y - 2y = 0$), or $3x + 5 = y + 3$ is simplified to $x + 5 = y$ (because students believe $3x - 3 = x$).

Growing research indicates the power of worked examples for students learning to solve equations.[68] In particular, comparison of correct and incorrect solutions can help students focus on certain features of a problem or typical mistakes.[69] For instance, in Table 3.11 Alex and Morgan each simplified the given expression differently and solved an equation differently. Researchers found that students who compared and contrasted alternative solution methods increased their procedural proficiency more than students who simply reflected on the same solution methods one at a time.[70] An important component in the success of these comparison tasks is teachers' ability to bring out the important mathematics in the discussions of the problems.

TABLE 3.11. Comparing Algebraic Strategies

Alex's Way	Morgan's Way
$3x(5x + 2) + 4(5x + 2)$	$3x(5x + 2) + 4(5x + 2)$
$15x^2 + 6x + 20x + 8$	$(3x + 4)(5x + 2)$
$15x^2 + 26x + 8$	$15x^2 + 6x + 20x + 8$
	$15x^2 + 26x + 8$
$45y + 90 = 60y$	
$135y = 60y$	$45y + 90 = 60y$
$75y = 0$	$90 = 15y$
$y = 0$	$6 = y$

During discussions, students should be required to justify (not just report) their steps. Asking the students to justify their steps forces them to only do steps that they can explain. This is difficult when solving equations is taught as a system of procedures, rather than using algebraic properties. Students should explicitly use the reflexive, symmetric and transitive properties of equality to improve their understanding of the use of equal sign and teachers should reinforce the idea of inverses.[71] For example, instead of going from $x + 6 = 3x + 1$ directly to $2x + 1 = 6$, teachers could show the in-between steps by indicating the properties being used: "$x + 6 = 3x + 1$ subtracting x from both sides, we get $6 = 2x + 1$ and by the symmetric property of equality, we arrive at $2x + 1 = 6$."

During instruction, solving single unknown equations by introducing a second variable can translate an algebraic problem to a graphical one. For example, $2x + 3 = 4x + 1$ can be rewritten as $y = 2x + 3$ and $y = 4x + 1$. These two functions can then be compared graphically to find the common solution. Common errors should be discussed and analyzed at the appropriate point, especially at the introductory level, taking care to not create confusion. Such discussion helps students recognize faulty reasoning, prevent formation of students' bad habits and development of inaccurate constructions, and see errors as opportunities for learning.[72] As students can have selective attention, there is the danger that they will simply remember "the teacher said something about..." and store the error in memory. It is important that when teachers are proactive in discussing common errors that they formatively assess the impact of that instruction to ensure that the message intended was the message received.

There are different challenges when there are multiple appearances of a variable in an equation than with a single appearance of a variable. The complexity increases when a second variable is introduced, such as in the following problem:[73]

The perimeter of a rectangle is five times its width. Its length is twelve meters. What is its width?

The problem results in the equations $P = 5a$, $P = 24 + 2a$. Researchers find that "algebraic competences that deal with handling a single unknown are not spontaneously extended to two-unknown cases."[74] Just because students understand how

TABLE 3.12. Construct Map of Equivalence

Level	Description
1. Rigid Operational	View the symbol = as an operator. Consider only equations with operations on the left to be properly formed. For example, 3 + 4 = ?.
2. Flexible Operational	View = as an operator. Accept as properly formed equations those that contain operations on the left *or* right. For example, ? = 3 + 4.
3. Basic Relational	Implicitly view = as a relation signaling that the same value is on each side, but unable to define it as such. Accept a wide range of arithmetic equations as properly formed, including those with expressions on both sides. For example, 3 + ? = 4 + 5.
4. Comparative Relational	Explicitly view = as a relation signaling that the same value is on each side and able to define it as such. Accept a wide range of arithmetic equation types with operations on both sides with multidigit numbers or multiple instances of a variable. Draw on arithmetic principles (commutativity, associativity, and inversion) in order to evaluate and solve equations in terms of their structural properties. For example, recognize that 3 + 5 = 5 + 3 and 6 + 9 = 7 + 8 are true by drawing on the commutative and associative properties of addition, respectively.
5. Structural Relational	View = as a structural tool through the transitive property. Relate equations in a context to each other. For example, if $x = 5 + y$ and $x = 3 + 4y$ then $5 + y = 3 + 4y$.

Based on Jones, Inglis, Gilmore, & Dowens (2012, p. 167) and Rittle-Johnson, Matthews, Taylor, & McEldoon (2011, p. 87)

to solve equations such as $2x + 5 = 9$ does not mean they will understand the dynamics when there are two different variables involved. One of the abilities needed to solve the perimeter problem above is based on a more complex understanding of the equal sign. Besides the understanding of the equals sign as signaling the same value on both sides of the equation, "the notion of *substitution* is also an important part of a sophisticated understanding of mathematical equivalence."[75] The idea of substitution, in this case, is based on the transitive property: if $a = b$, and $b = c$, then $a = b$; c can be substituted in for b in the first equation. For the above problem, P can be replaced in the second equation with $5w$ from the first equation to get $5a = 24 + 2a$. The sense of "sameness" for equivalence must extend not only within an equation, but across equations.[76] This structural sense of equality is the next step in advancing students' mathematical power. Consider the construct map of the development of the idea of equivalence in Table 3.12.

Teachers

- Require student to justify solutions using algebraic properties, rather than just reporting steps.
- Introduce graphical solutions as a way to understand the meaning of a linear equation with variables on both sides. Consider graphing each side of the equation separately prior to transformation.

- Certain procedural habits (initially combining like terms, working vertically, etc.) can prevent common errors. Reinforce these habits during instruction and assessment.
- Be sure to discuss the potential for all three types of solutions: one solution, no solution, and infinitely many solutions.
- It is helpful for students to align their work vertically rather than horizontally to show the development of their ideas toward a solution.
- Students should be expected to check the validity of their solutions both for verification and for growth in their confidence.
- The Center for Algebraic Thinking's *Algebra Equation Builder* app (see previous Figure 3.3), can also help students build equations with variables on both sides. Students must consider the operations, numbers, and variables as they build a true equation.

FACTORING (FACTORIZATION)

Common Core Standards: A-REI.4.b, F-IF.8.a

Textbooks describe factoring as the reversal of multiplying polynomials or changing an expression from a sum of terms to a product.[77] Students have varying beliefs about what factoring is:[78]

Reversal: The process of undoing or reversing a multiplication by applying the distributive property or multiplication of binomials or both.
Deconstructive: The process of breaking down or simplifying an expression.
Evaluative: A process that you can check by multiplying out your answer.
Formal: A factored expression that is in the form of a product.
Numeric: Factoring has to do with decomposing numbers into products of primes.

Each of these beliefs includes some truth about factoring. Students may hold more than one of these beliefs simultaneously, but if they only hold one in isolation, they are likely to struggle. For instance, a student that only sees factoring as *reversal*, or 'unFOILing' as some students put it, might factor $(3ax^2 - 4ax)$ as $(3x - 4)(ax + 0)$ and be unable to see how to factor $(3x + 6 - ax - 2a)$. A student who has the *deconstructive* belief might factor $(x^2 + 5x - 24)$ as $x(x + 5 - 3*8)$, changing each term into its factors. A student with the *numeric* belief may focus on the numbers, such as factoring $(x^2 + 5x - 24)$ as $(x^2 + 5x + 3*2*2*2)$ or $(3ax^2 - 4ax)$ as $(3axx - 2*2ax)$. It is important for teachers to understand what students believe factoring is, as that will guide how students approach problems. Knowing their belief systems, teachers can work to expand any limited definitions.

When working with quadratic equations, students are often confused by the various forms that quadratics take:[79]

Factored form: $y = a(x - r)(x - s)$
Standard form: $y = ax^2 + bx + c$
Vertex form: $y = a(x - h)^2 + k$

Students need to recognize when a particular format is the most appropriate for a context and how the structure of each form lends itself to different kinds of information about the function. They need the procedural skills to be able to transform an equation into one of the forms or from one form into another easily.

There are the basic methods for solving quadratic equations:

- Factoring
- Factoring by grouping
- Applying the quadratic formula
- Completing the square

In regard to *factoring*, there are two primary methods to handling an expression such as $3x^2 + 7x + 2$.[80]

1. The *inspection method.* $3x^2$ can be factored with $3x$ and x. 2 can be factored as 2 and 1 or -1 and -2. Based on the middle term of $7x$ the factored form is $(3x + 1)(x + 2)$.
2. The *decomposition method.* Multiply $3x^2$ and 2 to get $6x^2$. Decompose $7x$ into the sum of two terms whose product is $6x^2$ or $7x = 6x + x$. Factor $3x^2 + (6x + x) + 2$ by grouping: $3x^2 + 6x + x + 2 = (3x^2 + 6x) + (x + 2) = 3x(x + 2) + 1(x + 2) = (3x + 1)(x + 2)$.

Strategies such as the diamond method may help students organize their thinking and figure out the factors for trinomials (See Figure 3.12 for factoring the equation $x^2 - 8x + 15 = 0$. See the Center for Algebraic Thinking's *Diamond Factor* app (Figure 3.9) for practice on that skill).

Understanding the dynamic and patterns between numbers in multiplication and division is at the heart of factoring equations. Many secondary mathematics students struggle with basic multiplication table facts, negative factors, and multiple factors. This may "make factoring simple quadratics ($ax^2 + bx + c, a = 1$) a

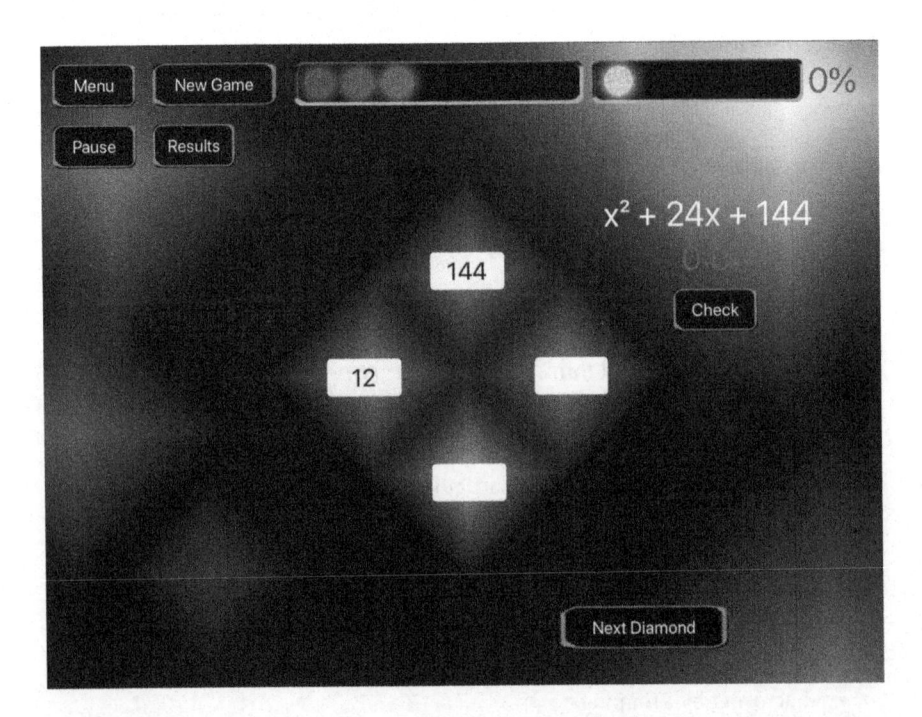

FIGURE 3.9. Two Numbers That Multiply to the Top Number and Add to the Bottom Number from the *Diamond Factor* App.

considerable challenge, while non-simple quadratics ($ax^2 + bx + c, a \neq 1$) become almost impossible. In both cases, students need to rely on procedural knowledge (e.g., multiplication facts) and conceptual understanding (e.g., the relationship between a, b and c)."[81] Students' errors in solving quadratic equation can also be attributed to "their weaknesses in mastering topics such as fractions, negative numbers and algebraic expansions."[82] Another possible source of error with factoring is students' difficulties with the idea of exponents. Some believe that t^2 is equal to $2t$ because it is $t*t$, which is the same as $2t$, so $t^2 - 2t = 0$.[83] Others struggle with how to handle subtraction involving terms with exponents. For instance, when solving $t^2 - 2t = 0$, rather than factoring, one student wrote $1t^2 - 2t^1 = 0$, subtracted, and got the result $-1t = 0$, because "you have to subtract powers as well" (subtracting the constants 2 from 1 and the exponent 1 from 2).[84]

Researchers found that when asked to solve an equation such as $(x - 3)(x - 5) = 0$, secondary and university students and teachers multiplied the binomials to get a trinomial first, then factored it back before solving.[85] Students without "structure sense" may not realize that the quadratic trinomial and its factorized equivalent are "different interpretations of the same structure."[86]

Structure sense, as it applies to high school algebra, can be described as a collection of abilities. These abilities include the ability to:

- See an algebraic expression or sentence as an entity,
- Recognize it as a previously met structure,
- Divide an entity into sub-structures,
- Recognize mutual connections between structures,
- Recognize which manipulations it is possible to perform, and
- Recognize which manipulations it is useful to perform.[87]

Researchers found that students without structure sense and with only procedural understanding tend to remove any parentheses or brackets first.[88] In interviews about strategies for solving quadratic equations, 11th-grade students said: "I get rid of the brackets. The fewer brackets the better" and "First I always open the brackets."[89] In one study, over 80% of errors solving the problem $(2x - 3)(x + 2) = 0$ involved expansion errors.[90] Table 3.13 shows some of the common errors found. As mentioned earlier, an important part of learning about factoring is developing structure sense—not just focusing on getting solutions, but knowing the role of parentheses and having a sense of purpose for manipulations.

Students also tend to make the following error:

TABLE 3.13. Errors Solving Factored Equations

$2x - 3)(x + 2) = 0$	$(2x - 3)(x + 2)$	$(2x - 3)(x + 2) = 0$
$5 + 2$	$x + 5$	$2x + x - 3 + 2 = 0$
$= 7$	$= 5x$	$3x - 1 = 0$

$$x^2 - 10x + 21 = 12$$
$$\Rightarrow (x-7)(x-3) = 12$$
so, either $x - 7 = 12$ or $x - 3 = 12$.

They do not understand the null-factor law or remember why or when you want to set each binomial equal to a number (zero). Students may follow procedures blindly, leading to generalization of those procedures inappropriately, as in this case. Sometimes students may also choose the constants in the binomial expressions as the solutions rather than setting each factor equal to zero to solve. For example, students might say that the solutions of $(x + 3)(x + 5) = 0$ are 3 and 5. It is important to reinforce students' understanding that they are looking for what value of x would make each binomial factor equal zero ($x + 3 = 0$ $x = -3$ or $x + 5 = 0$ $x = -5$).[91] In Table 3.14 the interview demonstrates that some students might see the two solutions of a quadratic equation as the two numbers to input simultaneously, rather than each being a single potential solution. That is, if $x = -3$, $((-3) + 3)((-3) + 5) = 0$ because $(0)(2) = 0$.

Since factoring of quadratics is the writing of polynomials as a product of polynomials, students need to have both a strong conceptual understanding of multiplication of polynomials and the procedural knowledge to retrieve basic multiplication facts effectively. "It is useful for students to have conceptual knowledge of how products of terms relate to one another (i.e., exponent laws, addition of like terms, etc.). With this understanding, students can do the necessary procedural steps in factorization but also step back and ask themselves if the results make sense."[92]

Manipulatives such as algebra tiles have been shown to be effective for some students in learning the structural relationships of polynomials and conducting factoring.[93] Concrete models can help students visualize abstract ideas. Algebra Tiles and Algebra Discs can be used to represent variables and constants, as in Figure 3.10; they are used to represent polynomials for factoring or expansion. The ability to physically represent each of the terms and manipulate the tiles or discs into polynomials can support students' efforts to develop understanding of factoring. Algebra Tiles appear to be a stronger representation because they

TABLE 3.14. Variables as Two Different Values in Same Equation

Interviewer:	What number will you substitute?
Student:	The first x is three and the second x is five.
Interviewer:	Why do you use that method?
Student:	I substituted numbers and got three minus three equals zero, five minus five equals zero. Zero multiplied by zero equals zero. It is a true sentence.
Interviewer:	Can you check your answers?
Student:	$(3 - 3)(5 - 5) = 0$ It equals zero. It is a true sentence.

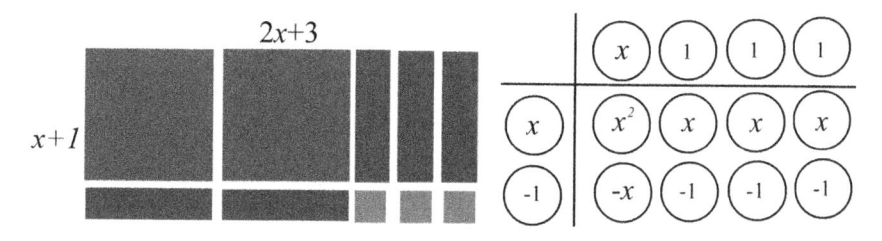

FIGURE 3.10. Algebra Tiles for factoring $x^2 + 5x + 3$ (left). Algebra Discs for factoring $x^2 - 2x - 3$ (right). Based on Hoong, et al. (2010)

demonstrate "'factorisation as forming rectangle and finding length/breadth given area' using the underlying idea of area conservation more conspicuously, thus strengthening its sense-making potential."[94]

A word of caution, however, when using manipulatives. In her article "Magical Hopes; Manipulatives and the Reform of Math Education," Ball discusses the importance of not relying on the tools to teach.[95] Instead, she and others encourage the important role teachers have in facilitating meaning making through the use of manipulatives.[96] Good questioning practices can help students bridge the gap between the physical representations and the abstract ideas. Writing out the symbolic representation of the physical manipulatives is an important part of the process, as students will not necessarily make that connection on their own. "Students who successfully make connections between physical representations and mathematical representations have created meaning of mathematical ideas."[97]

Factoring by grouping is based on a strong facility with the distributive property. Consider the following problem:

$$5y + 6x + 10y^2 + 3$$
$$10y^2 + 5y + 6x + 3$$
$$5y(2y+1) + 3(2y+1)$$
$$(5y+3)(2y+1)$$

In this problem, students must understand that they need to group the first line into terms that have something in common, such as $5y$, and then use the distributive property twice to come up with the final factored form. Researchers find that common errors students make with the distributive property are based on what the factors are. For instance, for the expression $mv-mu$ students may factor it as $m^2(vu)$ or $m^2(v-u)$.[98] One possible source of this type of error is students' difficulties with the idea of exponents. Some believe that t^2 is equal to $2t$ because it is $t * t$, which is the same as $2t$, so $t^2 - 2t = 0$.

Again, visual representations might help students comprehend the structural dynamics when grouping (see Figure 3.11). Consider the expression: $xy + 2y +$

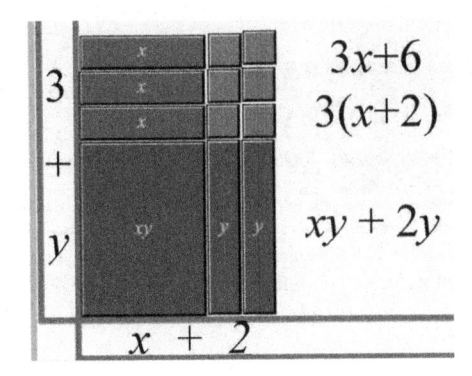

FIGURE 3.11. Algebra Tiles for Factoring by Grouping the Expression $xy + 2y + 3x + 6$.

$3x + 6$ and the use of algebra tiles to demonstrate the organization, structure, and dynamic between terms and expressions when grouping.

When working on real-world problems, it is rare that an equation will be factorable, yet most exercises in school textbooks are factorable, except when working with the *quadratic formula*.[99] Accordingly, some mathematicians argue that completing the square and the quadratic formula are the most usable strategies. Research on how students think about and understand the quadratic formula is unfortunately limited. Generally, however, students struggle with understanding the derivation of the formula, leading to mindlessly plugging in values for variables to acquire solutions.[100] Students often make procedural errors, particularly with the negative signs (i.e., finding $-b$ when b is negative).[101] They also tend to have conceptual difficulty with the two different roots, due to the $\pm$ sign. They may have difficulty in determining the values for a, b, and c, sometimes incorrectly identifying those values before writing the equation in the form $ax^2 + bx + c = 0$.[102] Students may believe that an equation must have three terms to be a quadratic equation. They may need to be reminded that b and c can equal 0, but a cannot be 0, as that would make the equation linear and the fraction undefined.

It is generally accepted that visual, geometric approaches are most effective in helping students learn how to *complete the square*.[103] For instance, for the equation $x(x + 10) = 39$, a student might divide the shaded rectangle in Figure 3.12 into two parts (with size $5x$) and rearrange them into the second figure. A new square is created in the lower right corner with area 5 x 5 or 25. Adding that area "completes the square." Procedurally, $x^2 + 10x + 25 = 39 + 25$ or $(x + 5)^2 = 64$. Taking the square root of both sides leads to a solution of $x = 3$. The solution of -13 does not apply in this case because a negative solution cannot be a measurement of a side of the rectangle. The inability to get a negative solution can therefore be a limitation of the geometric model.

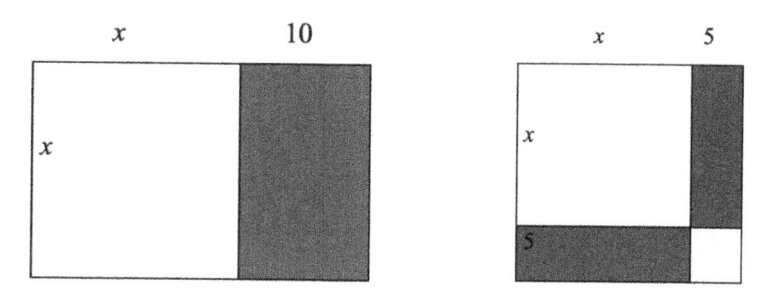

FIGURE 3.12. Geometric Model for Completing the Square.

Finally, researchers have found that students can often factor and solve quadratic equations, but are confused about the concept of a variable (see *Variables & Expressions*) and of a "solution" or "root" to a quadratic equation.[104] In the section *Variables & Expressions: Conservation of Variables*, we noted that some students responded to a problem such as $x + x + x = 12$ by stating that (2, 5, 5) or (10, 1, 1) would be acceptable solutions, not understanding the consistency of a variable within an equation. Many students do not realize that if a variable appears twice in an equation—for example, with $x^2 - 8x + 15 = 0$, or $(x - 3)(x - 5) = 0$—then it has the same value in the different places in which it appears.[105] For example, in the equation $(x - 3)(x - 5) = 0$, even if students obtained the correct solutions, $x = 3$ and $x = 5$, some students thought that the two xs in the equation stood for different variables, indicating they lacked relational understanding and relied on rote procedures.[106] Some students do not understand why there would be two solutions. "Part of the problem is that students who generate solutions to an equation often do not understand that the only number(s) which 'make the equation true' are the solutions."[107] Research indicates that many students do not understand what quadratic equations are, from a mathematical point of view, so they struggle with the meaning of the solutions.[108] Just because students can factor, complete the square, or use the quadratic formula does not mean that they understand quadratic equations. Teachers should work with students to understand the structures and conceptual meaning of manipulations (relational understanding) of equations and expressions, along with developing their procedural skills (instrumental understanding).[109] "We do not want meaningless symbol manipulation; if students use symbolic expressions, we want them to use the symbols with understanding."[110]

Teachers

- Boosting students' confidence in their multiplication facts may be a first step in preparing them for success with factoring.

- Discuss the meaning of solutions with students often, rather than emphasizing the procedural skill alone.
- In one study, Algebra I students who were learning to factor used graphing calculators to investigate graphs of families of quadratics, to explore real-world applications of factoring, and to apply the data collected to concepts of factoring. As a result, their basic skills in factoring were the same as a control group while being able to engage in higher-order thinking.[111]
- When plugging in values for the quadratic formula, encourage students to use parentheses to help them manage negative values. For example, $b = -4$, $-b = -(-4)$.
- Use parentheses/brackets as an intermediate tool to help students see and understand structure. For instance, $xy + 2y + 3x + 6 = (xy + 2y) + (3x + 6) = y(x + 2) + 3(x + 2)$.

FLEXIBLE USE OF SOLUTION STRATEGIES

Common Core Standards: 6.EE.2, 3, 7, 7.RP.2a, 2b, 8.EE.7a, 7b, A-REI.2, 3

Procedural or instrumental knowledge of solving algebraic relations has two parts: knowledge of the format and syntax of algebraic relations, and knowledge of rules and algorithms useful for solving them. It is knowing the rules without necessarily understanding the reasoning.[112] Conceptual or relational knowledge refers to knowledge of the underlying structure of mathematics.[113] It is knowing why you do what you do. In contrast to procedural knowledge, conceptual knowledge is "knowledge rich in relationships and includes the understanding of mathematical concepts, definitions, and fact knowledge. Both procedural and conceptual knowledge are considered as necessary aspects of mathematical understanding."[114] Some researchers argue that the distinction between procedural knowledge and conceptual knowledge is a false one—that they are so interwoven that it is really impossible to distinguish between them.[115] They are mutually reinforcing, rather than antagonistic.[116]

Students need to know a variety of procedures for accurately and efficiently solving equations. Equally important, but often overlooked, is the need to understand how these solution strategies are related and the ability to select the proper strategy for a given context. Students need to be "prepared to notice and able to perceive" features of the structure of an equation in order to determine a potential solution strategy.[117] These strategies are characteristic of conceptual knowledge for solving algebraic relations.

When students are confronted with non-routine tasks, their overdependence on procedures for solving equations can be problematic. For example, when asked to solve the equation $2(3x + 4) = 6x - 5$, 57% of high school students used symbolic manipulation as their first strategy, rather than recognize that $6x$ was on both sides of the equation, so there would be no solution. When pressed to explain the relationship between graphs of each side of the equation, only 59% could answer correctly. Almost one third of students in another study needed to compute the solutions to pairs of equations such as $x + 2 = 5$ and $x + 2 - 2 = 5 - 2$ in order to determine whether they were equivalent.[118] Their orientation was not to look holistically at the problem first before conducting procedures. Researchers interviewed pairs of students about their solutions to tasks like these and the responses (see Table 3.15) were typical.[119] As mentioned earlier, teachers need to help students learn structure sense in order to help them develop flexible use of strategies on problems such as these.

What is intellectually needed to solve a problem is different from the set of math skills needed to solve a problem. Students may know the properties, like the Distributive Property, but not think to use them to solve a problem. Flexible use of solution strategies seems to have two main components: *versatility* of the available interpretations and the *adaptability* of the perspective.[120] Versatility involves

TABLE 3.15. Interview on Flexible Use of Solution Strategies

Interviewer:	Could you have solved these in another way?
Student 1:	Guess-and-check
Student 2:	I mean you could try to get . . . all the x values . . . on one side and then move all the others on the one side, pl—plug it into a graph.

knowing multiple ways of approaching or interpreting problems. Adaptability involves being able to apply an appropriate approach in a specific context, based on the character of the task. When solving equations, students can use guess-and-check, graphical methods, algebraic relations, or table strategies—or they can examine the structure of the problem to find the answer. However, when confronted with a task, students tend to respond in the following ways:

- Overuse of guess-and-check. Initial equations can often be solved easily using this strategy, so students begin to rely on it.
- Random use of alternating methods. Students will change strategies, correctly or incorrectly, whenever the task changes.
- Overuse of a single strategy. Students will choose one procedure and apply it to every equation, regardless of the task.

When a rule-based approach is used in algebra, it can cause confusion for students because they may memorize rules but not understand the appropriate use of—or concepts behind—the rules. The reinforcement of a strategy by solving a large set of exercises may help students to pass tests; however, this does not mean that they are learning algebra with understanding. When teaching students to solve equations, it can be helpful to have them compare multiple methods side by side. This process strengthens conceptual and procedural knowledge because:

- It draws their attention to important features of the solution, and
- It helps them to allow for the possibility of multiple solution strategies.

When comparing methods, it is critical for the teacher to draw students' attention to the important features—this makes them actively compare the methods. "Show-and-tell" is not enough. Developing flexibility in solving equations requires students to know which strategies are more efficient than others under particular conditions. Instead of focusing on the difficulty of a question that students can solve, teachers can assess the sophistication level of the solution strategy. One potential guide for how to assess sophistication of these strategies is in Table 3.16.

In one study of an algebra classroom in China, researchers identified the strategy of *teaching with variation* as a key reason for the success of East Asian students in comparison with their Western counterparts on international tests.[121] Teaching with variation involves "illustrating the essential features by using different forms

TABLE 3.16. Assessing Use of Solution Strategies

0		Unable to answer question
1	a	Known basic facts
	b	Counting techniques
	c	Inverse operation
2		Guess-and-check
3	a	Cover up
	b	Working backward
	c	Working backward, then known facts (1a)
	d	Working backward, then known guess-and-check (2)
4		Formal operations/equation as object
5		Using a diagram

Based on Linsell (2010)

of visual materials and sometimes highlighting the essence of a concept by changing the non-essential features" (see Table 3.17 for an example).[122]

Teachers need to understand that students who can solve one-step equations correctly may not grasp inverse operations strategies sufficiently to move to two-step equations. No matter the structure of the equation (one-step versus two-step, one unknown versus two), students have difficulty solving equations with division in them. Teachers need to be aware that even if students can solve two-step equations correctly, they might not understand the strategy they use, or alternative strategies.

TABLE 3.17. Examples of Teaching with Variation

Teacher: Look at the following equations and answer this question: How many types can they be divided into?

$30 - 3x = 0$ $90 - 4x = 8$ $10 + 5x = 12.5$ $8.5 - 4x = 0.1$
$8x - 5x = 6$ $20 + 30 = 50$ $50 > 30$

Teacher: Point out which of the following are equations and which are not.

$2x = 1$ $3x + 4 = 7$ $4y - 3 = 5$ $3x + 4y = 12$ $x^2 - 1 = 0$
$x^2 + y^2 = 1$ $2x - 3 > 2$ $2 + x = 8$ $1 + 8$
$6 = 8$

Teacher: Do equations with one unknown and such different formats as below share a common solving process?

$5x - 2 = 8$ $\frac{1}{4}x = -\frac{1}{2}x + 3$ $4(x + 0.5) + x = 17$

$\frac{1}{7}(x+14) = \frac{1}{4}(x+20)$ $\frac{1}{5}(x+15) = \frac{1}{2} - \frac{1}{3}(x-7)$

Teachers

- Provide tasks that can be solved using a variety of methods (guess-and-check, table, algebraic, graphical). Students need to be comfortable with all methods.
- When comparing methods, it is critical for the teacher to draw students' attention to the important features of the task that make different methods more efficient.
- De-emphasize the difficulty of the task and instead focus on the sophistication of the solution strategy. Introduce concepts of elegance and efficiency in regard to solution strategies.

SOLUTIONS INVOLVING ZERO

Common Core Standards: 6.EE.5, 6, 7, 7.EE.4a, 8.EE.7, A-CED.1

Piaget (1960) said that negative numbers and the number zero were among the greatest discoveries in mathematics history. This can be a difficult concept for students to understand. To see there are "no apples" on the table is different than seeing "zero apples." "Not having any" is an easier concept than "having zero."[123] The number zero can have many different interpretations in students' minds and can cause them difficulty when thinking about negatives (see **Equations with Negative Numbers**) and equations. Researchers suggest multiple concepts of zero among students:[124]

- *Nil zero:* Zero has no value and students act as if it wasn't there.
- *Place value zero:* Zero is used as a placeholder in a large number when there is none of that place value (over half of students up to the eighth grade could not write a number such as "two hundred thousand forty three").[125]
- *Implicit zero:* **The zero** does not appear in writing, but is used in solving a task. A student might solve a problem by thinking about obtaining a zero in the process. For example, $5 - 17 = 5 - 5 - 12 = 0 - 12 = -12$ might be the thought process while $5 - 17 = -12$ is the only thing written.
- *Total zero:* The combination of number opposites. For example, $34 + (-34) = 0$.
- *Arithmetic zero:* The result of an arithmetic operation.
- *Algebraic zero:* The result of an algebraic operation or the solution of an equation.

Students are often confused by the appearance of 0 in an equation and frequently interpret it to mean that the solution of the equation is also zero. For example, when solving the equation: $2x - 3 = 2x + 4$, if they arrive at $0x = 7$ they might conclude that $x = 0$. There are cases like $3x + 3 = 2x + 3$ that produce a legitimate solution of $x = 0$, and students need exposure to these as well. The equation $4x + 8 = 4x + 8$ is an identity, which means that while students might end up with $0 = 0$ as a solution, the solution set is the set of all real numbers. This range of equations emphasizes the need for reasoning (see **Student Intuition and Informal Procedures**) and checking the solution to see if their solution is correct.

It is important for students to understand the results in which a single solution is not evident. Consider the following problem of an inequality with no solution:[126]

Mr. A solved the inequality $1 - 2x < 2(6 - x)$ as follows:

$$1 - 2x < 2(6 - x)$$
$$1 - 2x < 12 - 2x$$
$$-2x + 2x < 12 - 1$$
$$0 < 11$$

Here Mr. A got into difficulty.

1. Write down your opinion about Mr. A's solution.
2. Write down your way of solving this inequality $1 - 2x < 2(6 - 2x)$ and your reasons.

This problem was written in order to provoke cognitive conflict with students with the following key features: the disappearance of x, the expression "Here Mr. A got into difficulty," and the answer $0 < 11$. Fourteen percent of students responded with a correct answer. Of those who answered incorrectly, 34% responded with an inequality including x (e.g., $x > \frac{18}{11}$) and 18% did not have an x (e.g., $0 > 11$).[127] Many students struggled with their perception that they must get an answer in a particular form, such as $x > a$, claiming that a final answer without x was not possible. Other students followed strict procedures without significant understanding of the meaning of the solutions, treating x as merely an object in transforming an expression. In a study of high school students, most did not know how to interpret the results of their manipulation of symbols when the result was an identity or no solution.[128]

To overcome these challenges, teachers can apply a number of strategies that develop students' understanding. Connecting the solutions to real-life situations such as sharing cookies can help with understanding. If students get $0x = 7$ (for example) and conclude that $x = \frac{7}{0} = 0$, teachers could ask them whether they can they take 7 cookies and share them among 0 people as a way of explaining why division by zero is undefined. In general, asking the students to justify their steps forces them to do only steps that they can explain. Furthermore, students should be expected to check the validity of their solutions for confidence improvement. The checking process becomes important in contextual problems because the meaning that is associated with the check will confirm the correct interpretation of the solution. One final strategy is to prevent students from aligning work horizontally (e.g., $8x + 4 = 28$, $8x = 24$, $x = 3$). To show the progression of ideas leading to a solution more clearly, they should align their work vertically (see **Equal Sign**).

Teachers

- Give students an opportunity to work with equations involving zero, but have explicit discussions about the solutions and what they mean.
- Provide examples that allow for contrasting special case solutions like $x = 0$, $x = x$, and $x =$ no solution.
- Require students to check validity of solutions and to justify this validity.

STUDENTS' INTUITION AND INFORMAL PROCEDURES

Common Core Standards: 6.EE.2, 3

We never really grasp a concept until we internalize it and make it our own. Watching someone else work through an example will not help us internalize an idea nearly so well as working through several problems ourselves. At the same time, we usually need at least a little guidance for new ideas. It is important to strike a balance between these needs and to provide a rich enough variety of examples for students to work through that they can see patterns emerge and recognize common principles at play in seemingly different situations. Formal procedures, while useful and important, do not generally enhance understanding or intuition, and development of intuition should not be sacrificed for the sake of formal procedures.

Students learn from constructing meaning from multiple examples. They will not learn a procedure by seeing one clear example (as is done in most textbooks). They need to try things over and over to test their procedures and see what works. For example, consider the task

$$20x + 5 = 5x + 65.$$

Algebra texts often funnel students into thinking they should transform the equation into $15x = 60$, and then divide by 15 to find x. However, other legitimate solution strategies exist. For example, students could also:

- Do the computation without recording any steps and indicate that $x=4$;
- Graph both sides of the equation as functions and indicate that the x value of the intersection is 4;
- Bring all the terms to the left, factor, and then write down the value that will produce an output of 0; or
- Divide all the coefficients by 5 and then solve.

"Transmitting algebraic rules by the teacher, memorizing them by students and practicing them in a mechanical way is worthless."[129] Students' misconceptions and incorrect ways of solving equations are documented in multiple ways. The underlying student difficulties indicate that students are unaware of the reasons for the manipulations they perform, and they do not grasp the meaning of an equal sign.

When students evaluate expressions, their procedures may fall into one of the categories below:

- Automatization: Students do it, but cannot explain how or why they did what they did ("I don't know why, it's just what my teacher said to do").
- Formulas: Students apply a memorized formula (like difference of 2 squares).

TABLE 3.18. Students' Procedures

Interviewer:	Can you solve $\frac{1}{3}(x+4) = 2$ for x?
Student:	Yes, I can multiply both sides by 3 over 1, then subtract 4 from both sides and get $x = 2$.
Interviewer:	Can you solve for x without writing anything on your paper? Can you do this one in your head?
Student:	That is a lot to keep track of in my head. I would probably make a mistake because I'm not good at mental math. I could guess and check but that always takes forever. I thin k it would be hard but I could probably do it.

- Guessing-substituting: Students guess what the simplified version is, and then plug in a number like 2 or 3 to check if they are correct.
- Preparatory modification of the expression: Students change something in the expression before evaluation (like changing subtracting a negative to addition or division by an integer into multiplying by a fraction).
- Concretization: Students convert variables into a concrete idea (like $3x + 2x$ is 3 apples + 2 apples = 5 apples).
- Rules: Students identify a memorized rule (like when adding $3x$ and $2x$ you add the coefficients and leave the variable alone).
- Quasi-rules: Students apply rules, but do it inconsistently (sometimes raising numbers to an exponent, sometimes not).
- Invented procedures: Students will use their own invented procedures (correct or incorrect) rather than use what they have been taught in school.

In Table 3.18, the student solved in a procedural way. Instead, the student could have applied an informal procedure and solved the equation by thinking it through. "One third of something ($x + 4$) is 2, so that 'something' must be 6. Something (x) plus 4 is 6, so that 'something' is 2. Then, $x = 2$." This method is often described as the *cover up* method. Students cover up part of the equation ($x + 4$ in this example) in order to intuitively think about the problem. Researchers find that students can solve complex problems intuitively that they could not solve procedurally, such as the one in Figure 3.13. The app *Cover Up* by the Center for Algebraic Thinking provides practice in using this skill.

Presenting mathematical concepts through standalone examples and repetitious practice does not foster understanding. We appear to teach algorithms too soon and assume that once taught, they are remembered. This approach does not help students overcome misconceptions. Instead, we should provide examples of problems that help students see the benefit of using informal procedures, as in the example above. Other examples help them to see the necessity for formal procedures, as in a task like: $\frac{6}{a} = 15$. When students encounter tasks like this, they can

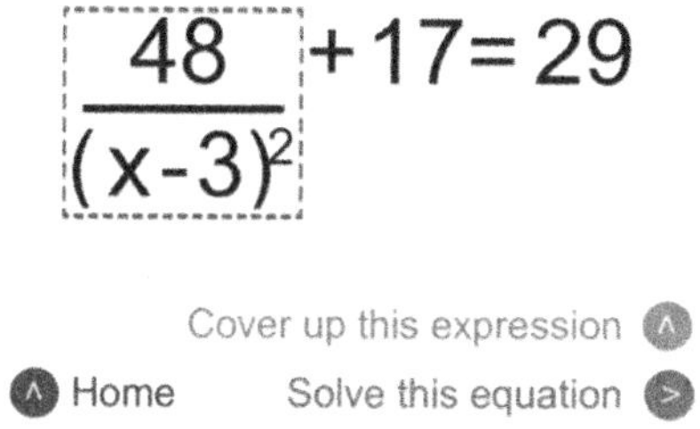

$$\frac{48}{(x-3)^2} + 17 = 29$$

Cover up this expression

Home   Solve this equation

FIGURE 3.13.   The Cover Up iOS App.

see the value of procedures. It is difficult to reason "6 divided by something is 15" but it is fairly straightforward to multiply both sides by a, and then divide by 15.

Equations can be thought of as using metaphors as well. Four types are:

1. A story of something happening to x: For $4(x + 2) = 4$, "There is a number. You add 2 to the number, and then multiply by 4 and the number becomes 4." This helps with single-variable equations with no variable on the other side. However, in this particular case, it can reinforce the misconception that you do a procedure to what is on the left to get the answer on the right of the equal sign.
2. The function machine: Students are given a set of inputs and outputs and need to find the rule. This works well for data provided in a table format. The app *Function Mystery Machine* (see Figure 3.14) by the Center for Algebraic Thinking provides practice in using this skill.
3. The recipe: For $y = 2a + b$, "To make y, you need to double a and add it to b." This connects known formulas like $A = L \times W$ to algebraic equations. It also shows students that procedures can happen on the right side of an equation.
4. The balance model: Two expressions are seen to be equal and if you operate on one, you must operate equally on the other (this should be used with caution with students who are not at the formal-operational mode according to Piaget).

In order to enhance students' successful use of their intuition and informal strategies, research suggests that teachers should help them develop their orientation toward interpreting the problem and predicting and foreseeing an approach.[130] Often, when students are presented with a problem, they tend to rush to action without analyzing the problem overall. Table 3.19 identifies the different ways students might think in approaching a problem situation, and their level of productive anticipation. Anticipation is the application of a way of thinking to a sit-

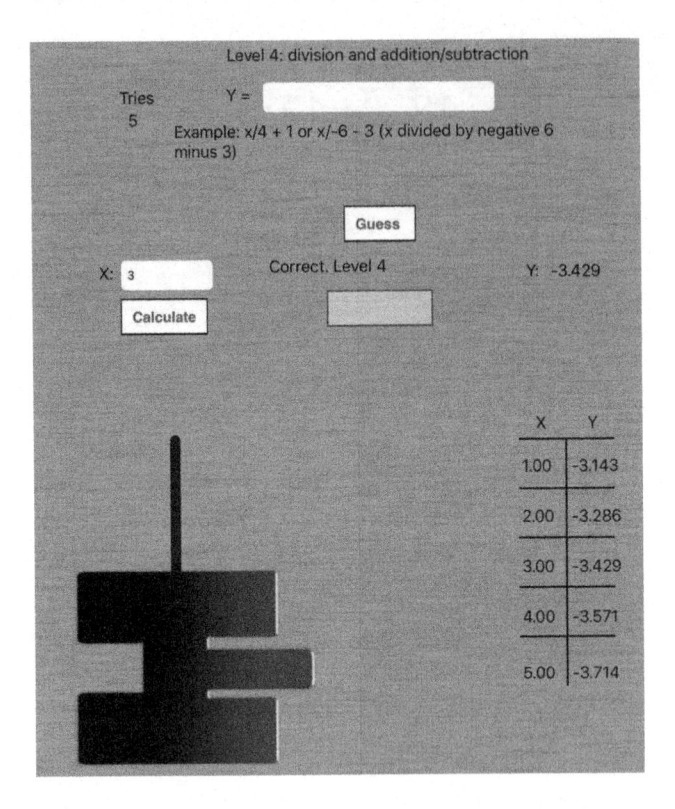

FIGURE 3.14. The Function Mystery Machine iOS app.

uation—a foreseeing of what is to come, based on some evidence or experience. These Ways of Thinking provide insight into students' anticipation and prediction, but are not listed in a hierarchal manner. The table may help teachers consider the different ways of thinking students might employ as they use their intuition or informal strategies. Understanding what students take into account as they engage with a problem may also help teachers know how to increase the sophistication of students' anticipation or intuitive approaches.

A student's errors are actually natural steps to understanding—they can provide the teacher insight into the pupil's thinking. Students need to confront the conflict between their misconceptions and the principles they have learned, or else the connections may not be made. Research with 11th-grade students resulted in a suggestion for considering "Ways of Thinking" (Table 3.19) when it comes to encouraging students' use of intuition and informal methods to predict a result and foreseeing an action during problem solving. One problem used in the study was mentioned earlier in the discussion of ***Inequalities:***

TABLE 3.19. Ways of Thinking

Way of thinking	Description
Impulsive anticipation	Spontaneously proceeds with an action that comes to mind without analyzing the problem situation and without considering the relevance of the anticipated action to the problem situation.
Tenacious anticipation	Maintains and does not re-evaluate one's way of understanding (prediction, problem-solving approach, claim, or conclusion) of the problem situation in light of new information.
Explorative anticipation	Explores an idea to gain a better understanding of the problem situation.
Analytic Anticipation	Analyzes the problem situation and establishes a goal or a criterion to guide one's actions.
Interiorized anticipation	Spontaneously proceeds with an action without having to analyze the problem situation because one has interiorized the relevance of the anticipated action to the situation at hand. *Interiorized* means the student has not only internalized a particular way of understanding (i.e., gained the ability to autonomously and spontaneously apply one's way of understanding to another similar situation) but has also reorganized and abstracted the way of understanding to a higher level of understanding.[131]
Association-based prediction	Predicts by associating two ideas without establishing the basis for making such an association.
Comparison-based prediction	Predicts by comparing two elements or situations in a static manner.
Coordination-based prediction	Predicts by coordinating quantities or attending to relationships among quantities.

Source: Lim (2006, pp. 2–106)

Is there a value for x that will make the following statement true?

$$(6x - 8 - 15x) + 12 > (6x - 8 - 15x) + 6?$$

A student who had *impulsive anticipation*, for instance, might immediately begin to combine like terms and isolate a variable. A student who thought with *analytic anticipation* might notice that there were xs on each side of the inequality and that there was an opportunity to simplify what was inside the parentheses. The student might ultimately combine like terms and isolate the variable, as the student with *impulsive anticipation* might, but he or she would not act "without thinking." The student who had *coordination-based prediction* might recognize that there were two quantities in parentheses that had identical structures. Then he or she might reason that the effect on the inequality would be equivalent, leaving the $12 > 6$. Using these descriptions as a guide, teachers might identify students' ways of thinking and look for opportunities to develop students' sophistication in anticipation, reasoning, and analysis of problems prior to and during action.

Teachers

- Avoid introducing procedures with linear equations. This appears to eliminate student intuition regarding the meaning of equations. Encourage students to develop hypotheses of how to approach problems and discuss the merits and disadvantages of different approaches.
- Consider working toward strategies based on patterns. For instance, showing students $5*x = 10$, students should be able to intuitively come up with an answer, rather than perform procedures. Extend those examples and help students articulate a strategy for handling problems with that structure.
- Provide examples of problems that help students see the benefit of using informal and non-standard procedures.
- Contrast intuition-invoking examples with problems that help students see the benefits of procedures.
- Use metaphors (stories, recipes, function machine, balance model, etc.) when describing equations.

ENDNOTES

1. Tahir & Cavanagh, 2010.
2. Godfrey & Thomas, 2008.
3. Linsell, 2008.
4. Ding & Li, 2010.
5. Brown & Quinn, 2006, 2007a, 2007b.
6. Vlassis, 2002.
7. Knuth, Stephens, McNeil, & Alibali, 2006.
8. Kennedy, 1991, p. 118.
9. Kennedy, 1991, p. 118 based on Usiskin, 1988.
10. Bossé & Nandakumar, 2005.
11. Attorps, 2003.
12. MacGregor, 1999.
13. Kuchemann, 1981; Hodgen, Kuchemann, Brown, & Coe, 2008.
14. de Lima & Tall, 2006; Kieran, 1981; Knuth, Stephens, McNeil, & Alibali, 2006.
15. Stephens, 2006, p. 249.
16. McNeil & Alibali, 2005; Molina & Ambrose, 2008.
17. McNeil, Grandau, Knuth, Alibali, Stephens, Hattikudur, & Krill, 2006, p. 367.
18. Carpenter et al., 2003; Koehler, 2004; Saenz-Ludlow & Walgamuth, 1998.
19. Bodin & Capponi, 1996; de Lima & Tall, 2006; Freitas, 2002; Vaiyavutjamai, 2004a, 2004b.
20. O'Rode, 2011, p. 25.
21. Jones, Inglis, Gilmore, & Dowens, 2012, p. 166.
22. Barcellos, 2005; Sakpakornkan & Harries, 2003.
23. Barcellos, 2005.
24. Falle, 2005.
25. Hall, 2002.
26. All Center for Algebraic Thinking apps are available at no cost for iOS devices, except *Math Flyer* and *Help Me Choose!* which are $0.99.
27. Behr, Harel, Post, & Lesh, 1992; Brown & Quinn, 2007a; Brown & Quinn, 2007b; Lamon, 2011.
28. Kieran, 1980, p. 134.
29. Brown & Quinn, 2006.
30. Rossi, 2008.
31. Based on 180 9th- to 11th-grade students in Hall, 2002.
32. Brown & Quinn, 2006.
33. Peled & Carraher, 2008.
34. Peled & Carraher, 2008, p. 325.
35. Gates, 1995, p. 283.
36. Based on Gallardo, 2002; Gallardo, 2003; Peled, 1991.
37. Hefendahl-Hebeker, 1991; Kilhamn, 2011.
38. Adapted from Atiparmak, 2010 and Kilhamn, 2009.
39. Ball, 1993; Schwarz, Kohn, & Resnick, 1993.
40. Schwarz, Kohn, & Resnick, 1993; Vlassis, 2004.
41. Gallardo, 1995.
42. Based on Hayes, 1994; Hall, 2002; Hayes, 1996; and Seng, 2010.
43. Vlassis, 2004.
44. Hayes, 1994, 1996.
45. Kilhamn, 2008, 2009; Schwarz, Kohn, & Resnick, 1993.
46. Fischer & Rottman, 2005.
47. Hayes, 1994.
48. Peled, Mukhopadhyay, & Resnick, 1989 and Schwarz, Kohn, & Resnick, 1993.

49 Peled & Carraher, 2008.
50 Whitacre, Bishop, Lamb, Philipp, Schappelle, & Lewis, 2011, p. 916.
51 Gallardo, 2003.
52 Atiparmak & Ozdogan, 2010.
53 Mullis, Martin, Gonzalez, Gregory, Garden, O'Connor, Chrostowski, & Smith, 2000.
54 Verikios & Farmaki, 2010.
55 Based on Tsamir & Almog, 2001 and Kroll, 1986.
56 Lim, 2006.
57 Lim, 2006, p. 104.
58 Based on Attorps, 2003.
59 Tsamir & Reshef, 2006.
60 Tsamir, Almog, & Tirosh, 1998; Tsamir & Reshef, 2006.
61 Tsamir & Almog, 2001; Tsamir, Almog, & Tirosh, 1998.
62 Almog & Ilani, 2012.
63 De Souza, Nogueira de Lima, & Campos, 2015. Based on Kieran, 2004 and Radford, 2004 and also validated by Tsamir, Almog, & Tirosh, 1998.
64 Dreyfus & Eisenberg, 1982; Parish, 1992; Tsamir, 1998, 2001.
65 Dreyfus & Eisenberg, 1982; Tsamir, Almog, & Tirosh, 1998
66 Hattikudur & Alibali, 2010, p. 29.
67 Carry, Lewis, and Bernard, 1980; Hall, 2002, Matz, 1982.
68 Hu, Ginns, Bobis, 2015; Kirschner, Sweller, Clark, 2006; Renkl, 2005; Sweller & Cooper, 1985; Star, Pollack, Durkin, Rittle-Johnson, Lynch, Newton, & Gogolen, 2015.
69 Star, Pollack, Durkin, Rittle-Johnson, Lynch, Newton, & Gogolen, 2015.
70 Rittle-Johnson & Star, 2007; Rittle-Johnson & Star, 2009.
71 Hall, 2002.
72 Hall, 2002.
73 Filloy, Rojano, & Solares, 2010.
74 Filloy, Rojano, & Solares, 2010.
75 Jones, Inglis, Gilmore, & Dowens, 2012, p. 166.
76 Jones & Pratt, 2012.
77 McKeague, 1990; Pulsinelli & Hooper, 1991; Nanney & Cable, 1991; Aufman & Barker, 1991; Nustad & Wesner, 1991.
78 Rauf, 1994.
79 Kotsopoulous, 2007; Parent, 2015.
80 Hoffman, 1976 from Sonnerhed, 2009.
81 Ayres, 2000; Kotsopoulos, 2007.
82 Zakaria & Maat, 2010.
83 de Lima & Tall, 2006.
84 de Lima & Tall, 2006, pp. 4–238.
85 Lim, 2000; Sarwadi & Shahrill, 2014; Vaiyavutjamai, Ellerton, & Clements, 2005.
86 Hoch & Dreyfus, 2004, p. 51. Also based on Novotna & Hoch, 2008.
87 Hoch & Dreyfus, 2004, p. 51.
88 de Lima & Tall, 2006; Hoch & Dreyfus, 2004.
89 Based on interviews with 92 11th-grade students in Hoch & Dreyfus, 2004.
90 Sarwadi & Shahrill, 2014.
91 Kotsopoulos, 2007.
92 Kotsopoulos, 2007, p. 22.
93 Bossé & Nandakumar, 2005; Leitze & Kitt, 2000; Leong et al., 2010; Sharp, 1995; Sobol, 1998; Sonnerhed, 2011; Vaiyavutjamai and Clements 2006.
94 Hoong, et al., 2010, p. 21.
95 Ball, 1992.
96 Ball, 1992; Wood, Cobb, & Yackel, 1995.

97   Sharp, 1995, p. 1, based on Kaput, 1989.
98   Yahya & Shahrill, 2015.
99   Bossé & Nandakumar, 2005.
100 Lim, 2000.
101 Yahya & Shahrill, 2015.
102 Burger, et al., 2008; Yahya & Shahrill, 2015.
103 Allaire & Bradley, 2001; Vinogradova, 2007.
104 Vaiyavutjamai & Clements, 2006; Zakaria & Maat, 2010.
105 Vaiyajavutjamai & Clements, 2006.
106 Kotsopoulos, 2007; Law & Shahrill, 2013; Pungut & Shahrill, 2014; Sarwadi & Shahrill, 2014; Vaiyavutjamai, 2004; Vaiyavutjamai, Ellerton & Clements, 2005; Vaiyavutjamai & Clements, 2006.
107 Vaiyajavutjamai & Clements, 2006, p. 51.
108 Sonnerhed, 2011.
109 Didiş, M. G., Baş, S., & Erbaş, 2011; Skemp, 1976; Sonnerhed, 2009.
110 Sherin, 2001, p. 479.
111 Wilkins, 1995.
112 Skemp, 1976.
113 From Attorps, 2003 based on Hiebert & Lefevre, 1986.
114 Hiebert & Lefevre, 1986, as quoted in Attorps, 2003, p. 2.
115 Star, 2005; Baroody, Feil and Johnson, 2007; Star, 2007; Kieran, 2013; Star & Stylianides, 2013; Foster, 2014.
116 Foster, 2014.
117 Sfard & Linchevski, 1994.
118 Steinberg, Sleeman, & Ktorza, 1990.
119 Huntley, Marcus, Kahan, & Miller, 2007.
120 Sfard & Linchevski, 1994, p. 204–5.
121 Gu, 1994; Gu, Huang, & Marton, 2004.
122 Li, Peng, & Song, 2011.
123 Anthony & Walshaw, 2004, based on Sheffield and Cruikshank, 2001.
124 Anthony & Walshaw, 2004; Gallardo, 2005; Gallardo, 2006.
125 From Anthony & Walshaw, 2004, p. 38, citing Crooks and Flockton, 2002.
126 Fujii, 2003.
127 Based on 123 lower secondary students in Fujii, 2003.
128 Huntley, Marcus, Kahan, & Miller, 2007.
129 Demby, 1997, p. 68.
130 Lim, 2006.
131 Lim, 2006, p. 2–106.

REFERENCES

Allaire, P. R., & Bradley, R. E. (2001). Geometric approaches to quadratic equations from other times and places. *The Mathematics Teacher, 94*(4), 308–313.

Almog, N., & Ilany, B. (2012). Absolute value inequalities: High school students' solutions and misconceptions. *Educational Studies in Mathematics, 81*(3), 347–365.

Altiparmak, K., & Ozdogan, E. (2010). A study on the teaching of the concept of negative numbers. *International Journal of Mathematical Education in Science and Technology, 41*(1), 31–47.

Anthony, G., & Walshaw, A. (2004). Zero: a "none" number? *Teaching Children Mathematics, 11*(1), 38.

Asghari, A. (2009). Experiencing equivalence but organizing order. *Educational Studies in Mathematics, 71*(3), 219–234.

Attorps, I. (2003). *Teachers' images of the 'equation' concept.* Paper presented at the Third Conference of the European society for Research in Mathematics Education, Bellaria, Italy.

Aufman, R. N., & Barker, V. C. (1991). *Intermediate algebra.* Boston: Houghton Mifflin.

Ayres, P. L. (2000). *An analysis of bracket expansion errors.* Paper presented at the 24th Annual Conference of the International Group for the Psychology of Mathematics Education Hiroshima, Japan.

Ball, D. (1992). Magical hopes: Manipulatives and the reform of math education. *American Educator: The Professional Journal of the American Federation of Teachers, 16*(2), 14–18, 46–47.

Ball, D. (1993). With an eye on the mathematical horizon: Dilemmas of teaching elementary school mathematics, *Elementary School Journal, 93*, 379–397.

Barcellos, A. (2005). *Mathematical misconceptions of college-age algebra students.* Unpublished dissertation. University of California, Davis.

Baroody, A. J., Feil, Y., & Johnson, A. R. (2007). An alternative reconceptualization of procedural and conceptual knowledge. *Journal for Research in Mathematics Education, 38*, 115–131.

Behr, M. J., Harel, G., Post, T. R., & Lesh, R. (1992). Rational number, ratio, and proportion. In D. Grouws (Ed.), *Handbook of research on mathematics teaching and learning.* New York, NY: Macmillan Publishing.

Bodin, A., & Capponi, B. (1996). Junior secondary school practices. In A. J. Bishop, M. A. Clements, C. Keitel, J. Kilpatrick, & C. Laborde (Eds.), *International handbook of mathematics education* (pp. 565–614). Dordrecht, The Netherlands: Kluwer Academic Publishers.

Bossé, M. J., & Nandakumar, N. R. (2005). The factorability of quadratics: Motivation for more techniques. *Teaching Mathematics and its Applications: An International Journal of the IMA, 24*(4), 143–153.

Brown, G., & Quinn, R. J. (2006). Algebra students' difficulty with fractions: An error analysis. *Australian Mathematics Teacher, 62*(4), 28–40.

Brown, G., & Quinn, R. J. (2007a). Fraction proficiency and success in Algebra: What does research say? *Australian Mathematics Teacher, 63*(3), 23–30.

Brown, G., & Quinn, R. J. (2007b). Investigating the relationship between fraction proficiency and success in Algebra. *Australian Mathematics Teacher, 63*(4), 8–15.

Burger, E., Chard, D., Hall, E., Kennedy, P., Leinwand, S., Renfro, F., Roby, T., Seymour, D., & Waits, B. (2008). *Holt California Algebra 1, Student Edition.* New York, NY: Holt, Rinehart, & Winston.

Carpenter, T. P., Franke, M. L., & Levi, L. (2003). *Thinking mathematically: Integrating arithmetic and algebra in elementary school.* Portsmouth, NH: Heinemann.

Carry, L. R., Lewis, C., & Bernard, J. (1980) *Psychology of equation solving; An information processing study*, Austin, TX: University of Texas at Austin, Department of Curriculum and Instruction.

Crooks, T., & Flockton, L. (2002). *Mathematics assessment results 2001. National education monitoring.* Dunedin, New Zealand: Educational Assessment Research Unit.

de Lima, R., & Tall, D. (2006). *The concept of equations: What have students met before?* Paper presented at the 30th Conference of the International Group for the Psychology of Mathematics Education, Prague, Czech Republic.

De Souza, V., Nogueira de Lima, R., & Campos, T. (2015). A functional graphic approach to inequations. *Revista Lanioamericana de Investigacion en Matematica Educativa, 18*(1), 109–125.

Demby, A. (1997). Algebraic procedures used by 13-to-15-year-olds. *Educational Studies in Mathematics, 33*(1), 45–70.

Didiş, M. G., Baş, S., & Erbaş, A. (2011). *Students' reasoning in quadratic equations with one unknown.* Paper presented at the 7th Congress of the European Society for Research in Mathematics Education, Rzeszów, Poland.

Ding, M., & Li, X. (2010). *An analysis of the distributive property in U.S. and Chinese elementary mathematics texts.* Paper presented at the Annual Meeting of the American Educational Research Association, Denver, CO.

Dobbs, D., & Peterson, J. (1991). The sign-chart method for solving inequalities. *Mathematics Teacher, 84*, 657–664.

Dreyfus, T., & Eisenberg, T. (1982). Intuitive functional concepts: A baseline study on intuitions *Journal for Research in Mathematics Education, 13*, 360–380.

Falle, J. (2005). *From arithmetic to Algebra: Novice students' strategies for solving equations.* Paper presented at the 28th annual conference of the Mathematics Education Research Group of Australasia.

Filloy, E., Rojano, T., & Solares, A. (2010). Problems dealing with unknown quantities and two different levels of representing unknowns. *Journal for Research in Mathematics Education, 41*(1), 52–80.

Fischer, M. H., & Rottman, J. (2005). Do negative numbers have a place on the mental number line? *Psychology Science, 47*(1), 22–32.

Foster, C. (2014). *'Can't you just tell us the rule?' Teaching procedures relationally.* Paper presented at the 8th British Congress of Mathematics Education, University of Nottingham.

Freitas, M. A. de. (2002). *Equação do primeiro grau: métodos de resolução e análise de erros no ensino médio. [Equations in the first degree: Methods of resolution and analysis of errors in high school].* Master's Dissertation. São Paulo: PUC-SP.

Fujii, T. (2003). *Probing students' understanding of variables through cognitive conflict problems: Is the concept of variable so difficult for students to understand?* Paper presented at the 27th International Group for the Psychology of Mathematics Education Conference, Honolulu, HI.

Gallardo, A. (1995). Negative numbers in the teaching of arithmetic. In D. Owens, M. Reeds, & G. Millsaps, (Eds.), *Proceedings of the 17th Annual Meeting for the Psychology of Mathematics Education*, (North America Chapter, vol. 1, pp. 158–163.) ERIC Clearinghouse for Science, Mathematics and Environmental Education, Ohio, USA.

Gallardo, A. (2002). The extension of the natural-number domain to the integers in the transition from arithmetic to algebra. *Educational Studies in Mathematics, 49,* 171–192.

Gallardo, A. (2003). *It is possible to die before being born? Negative integers subtraction: A case study.* Paper presented at the 27th International Group for the Psychology of Mathematics Education Conference, Honolulu, HI.

Gallardo, A. (2005). *The duality of zero in the transition from arithmetic to algebra.* Paper presented at the 29th Conference of the International Group for the Psychology of Mathematics Education, Melbourne.

Gallardo, A., & Hernandez, A. (2006). *The zero and negativity among secondary school students.* Paper presented at the 30th Conference of the International Group for the Psychology of Mathematics Education (3: 153–160). Prague, Czech Republic.

Gates, L. (1995). *Product of two negative numbers: An example of how rote learning a strategy is synonymous with learning the concept.* Paper presented at the 18th Annual Mathematics Education Research Group of Australasia Conference. Salma Tahir Macquarie University.

Godfrey, D., & Thomas, M. O. J. (2008). Student perspectives on equation: The transition from school to university. *Mathematics Education Research Journal, 20*(2), 71–92.

Gu, L. (1994). *Theory of teaching experiment: The methodology and teaching principle of Qinpu.* Beijing: Educational Science Press.

Gu, L., Huang, R., & Marton, F. (2004). Teaching with variation: A Chinese way of promoting effective mathematics learning. In L. Fan, N. Wong, J. Cai, & S. Li (Eds.), *How Chinese learn mathematics: Perspectives from insiders.* Singapore: World Scientific.

Hall, R. (2002). An analysis of thought processes during simplification of an algebraic expression. *Philosophy of Mathematics Education, 15.*

Hattikudur, S., & Alibali, M. W. (2010). Learning about the equal sign: Does comparing with inequality symbols help? *Journal of Experimental Child Psychology, 107,* 15–30.

Hayes, B. (1994). *Becoming more positive with negatives.* Paper presented at the 17th Annual Mathematics Education Research Group of Australasia Conference.

Hayes, B. (1996). *Investigating the teaching and learning of negative number concepts and operations.* Paper presented at the 19th Annual Mathematics Education Research Group of Australasia Conference: Melbourne, Australia.

Hefendehl-Hebeker, L. (1991). Negative numbers: Obstacles in their evolution from intuitive to intellectual constructs. *For the Learning of Mathematics, 11*(1), 26–32.

Hiebert, J., & Lefevre, P. (1986). Conceptual and procedural knowledge in mathematics: An introductory analysis. In J. Hiebert (Ed.), *Conceptual and procedural knowledge: The case of mathematics* (pp. 1–28). Hillsdale, NJ: Lawrence Erlbaum.

Hoch, M., & Dreyfus, T. (2004). *Structure sense in high school Algebra: The effect of brackets.* Paper presented at the 28th Annual Meeting of the International Group for the Psychology of Mathematics Education, Bergen, Norway.

Hodgen, J., Kuchemann, D., Brown, M., & Coe, R. (2008). Children's understandings of algebra 30 years on. *Proceedings of the British Society for Research into Learning Mathematics, 28*(3), 36–41.

Hoffman, N. (1976). Factorisation of quadratics. *Mathematics teaching, 76,* 54–55.

Hoong, L. Y., Fwe, Y. S., Yvonne, T. M. L., Subramaniam, T., Zaini, I. K. B. M., Chiew, Q. E., & Karen, T. K. L. (2010). Concretising factorisation of quadratic expressions. *Australian Mathematics Teacher, 66*(3), 19–24.

Hu, F., Ginns, P., & Bobis, J. (2015). Getting the point: Tracing worked examples enhances learning. *Learning and Instruction, 35,* 85–93.

Huntley, M. A., Marcus, R., Kahan, J., & Miller, J. L. (2007). Investigating high-school students' reasoning strategies when they solve linear equations. *The Journal of Mathematical Behavior, 26*(2), 115–139.

Jones, I., Inglis, M., Gilmore, C., & Dowens, M. (2012). Substitution and sameness: Two components of a relational conception of the equals sign. *Journal of Experimental Child Psychology, 113*(1), 166–176.

Jones, I., & Pratt, D. (2012). A substituting meaning for the equals sign in arithmetic notating tasks. *Journal for Research in Mathematics Education, 43*, 2–33.

Kaput, J. (1989). Linking representations in the symbol systems of algebra. In S. W. C. Kieran (Ed.), *Research issues in the learning and teaching of algebra* (Vol. 4 of *Research agenda for mathematics education*) (pp. 167–194). Reston, VA: National Council of Teachers of Mathematics.

Kennedy, P. A. (1991). Factoring by grouping: Making the connection. *Mathematics and Computer Education, 25*(2), 118–123.

Kieran, C. (1980). *The interpretation of the equal sign: Symbol for an equivalence relation vs. an operator symbol.* Paper presented at the Psychology of Mathematics Education Conference, Berkeley, CA.

Kieran, C. (1981). Concepts associated with the equality symbol. *Educational Studies in Mathematics, 12*(3), 317–326.

Kieran, C. (2004). The equation/inequality connection in constructing meaning for inequality situations. In M. J. Hoines & A. B. Fuglestad (Eds.), *Proceedings of the 28th Conference of the International Group for the Psychology of Mathematics Education, 1*, 143–147.

Kieran, C. (2013). The false dichotomy in mathematics education between conceptual understanding and procedural skills: An example from algebra. In R. K. Leatham (Ed.), *Vital directions for mathematics education research* (pp. 153–171). New York, NY: Springer New York.

Kieren, T. E. (1980). The rational number construct: Its elements and mechanisms. In T. E. Kieren (Ed.), *Recent research on number learning* (pp. 125–149). Columbus, OH: ERIC Clearinghouse for Science, Math, and Environmental Education.

Kilhamn, C. (2008). *Making sense of negative numbers through metaphorical reasoning.* Paper presented at the 6th Swedish Mathematics Education Research Seminar, Stockholm, Sweden.

Kilhamn, C. (2009). *The notion of number sense in relation to negative numbers.* Paper presented at the 33rd conference of the International Group for the Psychology of Mathematics Education, Tessaloniki, Greece.

Kilhamn, C. (2011). *Making sense of negative numbers.* (Doctoral Thesis), University of Gothenburg.

Kirschner, P. A., Sweller, J., & Clark, R. E. (2006). Why minimal guidance during instruction does not work: An analysis of the failure of constructivist, discovery, problem-based, experiential, and inquiry-based teaching. *Educational Psychologist, 41*(2), 75–86.

Knuth, E. J., Stephens, A. C., McNeil, N. M., & Alibali, M. W. (2006). Does understanding the equal sign matter? Evidence from solving equations. *Journal for Research in Mathematics Education, 36*(4), 297–312.

Koehler, J. L. (2004). *Learning to think relationally: Thinking relationally to learn.* Unpublished dissertation research proposal, University of Wisconsin-Madison.

Kotsopoulos, D. (2007). Unravelling student challenges with quadratics: A cognitive approach. *Australian Mathematics Teacher, 63*(2), 19–24.

Kroll, R. (1986). *Metacognitive analysis of the difficulties caused by intervening factors in the solution of inequalities*, Doctoral Dissertation. Georgia State University, Atlanta, Georgia.

Kuchemann, D. (1981). Algebra. In K. M. Hart (Ed.), *Children's understanding of mathematics* (pp. 102–119): London, UK: John Murray.

Lamon, S. J. (2011). *Teaching fractions and ratios for understanding: Essential content knowledge and instructional strategies for teachers*. New York, NY: Routledge.

Law, F. F., & Shahrill, M. (2013). *Investigating students' conceptual knowledge and procedural skills in trigonometry*. Paper presented at the Annual International Conference, Australian Association for Research in Education (AARE 2013), Adelaide, Australia.

Leitze, A. R., & Kitt, N. A. (2000). Using homemade algebra tiles to develop algebra and prealgebra concepts. *The Mathematics Teacher, 93*(6), 462–466.

Leong, Y. H., Yap, S. F., Yvoone, T. M., Mohd Zaini, I. K. B., Chiew, Q. E., Tan, K. L. K., & Subramaniam, T. (2010). Concretising factorisation of quadratic expressions. *The Australian Mathematics Teacher, 66*(3), 19–24.

Li, J., Peng, A., & Song, N. (2011). Teaching algebraic equations with variation in Chinese classroom. In J. Cai & E. Knuth (Eds.), *Early algebraization, Advances in mathematics education* (pp. 529–556). Berlin: Springer-Verlag.

Lim, T. H. (2000). *The teaching and learning of algebraic equations and factorisation in 'O' Level mathematics: A case study*. Unpublished M.Ed. dissertation, Universiti Brunei Darussalam, Brunei Darussalam.

Lim, K. H. (2006). *Characterizing students' thinking: Algebraic inequalities and equations*. Paper presented at the Annual meeting of the Psychology of Mathematics Education in North America.

Linsell, C. (2008). Solving equations: Students' algebraic thinking. In Ministry of Education (Ed.), *Findings from the New Zealand Secondary Numeracy Project 2007* (pp. 39–44). Wellington, New Zealand: Learning Media.

Linsell, C. (2010). A hierarchy of strategies for solving linear equations. In R. Hunter, B. Bicknell, & T. Burgess (Eds.), *Crossing divides: Proceedings of the 32nd annual conference of the Mathematics Education Research Group of Australasia* (Vol. 1). Palmerston North, NZ: MERGA.

MacGregor, M. E. (1999). How students interpret equations: Intuition versus taught procedures. In H. Steinbring, M. G. Bartolini-Bussi, & A. Sierpinska (Eds.), *Language and communication in the mathematics classroom* (pp. 262–270). Reston, VA: NCTM.

Matz, M. (1982). Towards a process model for high school algebra errors. In D. S. J. S. Brown (Ed.), *Intelligent tutoring systems* (pp. 25–50). New York, NY: Academic Press.

McKeague, C.P. (1990). *Intermediate algebra*. San Diego, CA: Harcourt Brace Jovanivich.

McLaurin, S.C. (1985). A unified way to teach the solution of inequalities. *Mathematics Teacher 78*, 91–95.

McNeil, N. M., & Alibali, M. W. (2005). Why won't you change your mind? Knowledge of operation patterns hinders learning and performance on equations. *Child Development, 76*(4), 883–899.

McNeil, N. M., Grandau, L., Knuth, E. J., Alibali, M. W., Stephens, A. C., Hattikudur, S., & Krill, D. E. (2006). Middle-school students' understanding of the equal sign: The books they read can't help. *Cognition and Instruction, 24*(3), 367–385.

Molina, M., & Ambrose, R. (2008). From an operational to a relational conception of the equal sign: Third graders' developing algebraic thinking. *Focus on Learning Problems in Mathematics, 30*(1), 61–80.

Mullis, I. V. S., Martin, M. O., Gonzalez, E. J., Gregory, K. D., Garden, R. A., O'Connor, K. M., Christowski, S. J., & Smith, T. A. (2000). *TIMSS 1999 International Mathematics Report: Findings from IES's Repeat of the Third International Mathematics and Science Study at the Eighth Grade.* Chestnut Hill, MA: Boston College.

Nanney, J. L., & Cable, J. L. (1991). *Intermediate algebra.* Dubuque, IA: Wm. C. Brown.

Novotna, J., & Hoch, M. (2008). How structure sense for algebraic expressions or equations is related to structure sense for abstract algebra. *Mathematics Education Research Journal, 20*(2), 93–104.

Nustad, H. L. & Wesner, T. H. (1991). *Principals of intermediate algebra with applications.* Dubuque: Wm. C. Brown.

O'Rode, N. (2011). Latino/a students understanding of equivalence. In K. Téllez, J. Moschkovich, & M. Civil (Eds.), *Latinos/as and mathematics education: Research on learning and teaching in classrooms and communities* (pp. 19–36). Information Age Publishing.

Parent, J. S. (2015). *Students' understanding of quadratic functions: Learning from students' voices.* University of Vermont, Graduate College Dissertations and Theses. Paper 376.

Parrish, C. (1992). Inequalities, absolute value, and logical connectives. *The Mathematics Teacher, 85*(9), 756–757.

Peled, I. (1991). Levels of knowledge about negative numbers: Effects of age and ability. In F. Furinghetti (Ed.), *Proceedings of the 15th international conference for the Psychology in Mathematics Education* (Vol. 3, pp. 145–152). Assisi, Italy: Conference Committee.

Peled, I., & Carraher, D. (2008). Signed numbers and algebraic thinking. In D. W. Carraher & M. L. Blanton (Eds.), *Algebra in the early grades* (pp. 303–328). New York, NY: NCTM.

Peled, I., Mukhopadhyay, S., & Resnick, L. B. (1989). *Formal and informal sources of mental models for negative numbers.* Paper presented at the 13th International Conference for the Psychology of Mathematics Education, Paris, France.

Piaget, J. (1960). *Introducción a la espistemología genética. I. El pensamiento matemático. Biblioteca de Psicología Evolutiva* [*Introduction to genetic epistemology: Mathematical thinking. Library of Evolutionary Psychology*]. Paidós, Buenos Aires, Argentina.

Pulsinelli, L., & Hooper, P. (1991). *Intermediate algebra: An interactive approach.* New York, NY: Macmillan.

Pungut, M. H. A., & Shahrill, M. (2014). Students' English language abilities in solving mathematics word problems. *Mathematics Education Trends and Research, 2014,* 1–11.

Radford, L. (2004). *Syntax and meaning.* Paper presented at the 28th International Conference of the International Group for the Psychology of Mathematics Education, Bergen, Norway.

Rauff, J. V. (1994). Constructivism, factoring, and beliefs. *School Science and Mathematics, 94*(8), 421.

Renkl, A. (2005). The worked-out-example principle in multimedia learning. In R. Mayer (Ed.), *The Cambridge handbook of multimedia learning* (pp. 229–245). Cambridge, MA: Cambridge University Press.

Rittle-Johnson, B., & Star, J. R. (2007). Does comparing solution methods facilitate conceptual and procedural knowledge? An experimental study on learning to solve equations. *Journal of Educational Psychology, 99*(3), 561–574.

Rittle-Johnson, B., & Star, J. R. (2009). Compared with what? The effects of different comparisons on conceptual knowledge and procedural flexibility for equation solving. *Journal of Educational Psychology, 101*(3), 529–544.

Rittle-Johnson, B., & Star, J. R. (2011). The power of comparison in learning and instruction: Learning outcomes supported by different types of comparisons. In B. Ross & J. Mestre (Eds.), *Psychology of learning and motivation: Cognition in education* (Vol. 55, pp. 199–226). San Diego, CA: Elsevier.

Rossi, P. S. (2008). An uncommon approach to a common algebraic error. *PRIMUS, 18*(6), 554–558.

Saenz-Ludlow, A., & Walgamuth, C. (1998). Third graders' interpretations of equality and the equal symbol. *Educational Studies in Mathematics, 35*(2), 153–187.

Sakpakornkan, N., & Harries, T. (2003). *Pupils' processes of thinking: Learning to solve algebraic problems in England and Thailand*. Paper presented at the British society for research into learning mathematics, Oxford, United Kingdom.

Sarwadi, H. R. H., & Shahrill, M. (2014). Understanding students' mathematical errors and misconceptions: The case of year 11 repeating students. *Mathematics Education Trends and Research*, 1–10.

Schwarz, B. B., Kohn, A. S., & Resnick, L. B. (1993). Positives about negatives: A case study of an intermediate model for signed numbers. *The Journal of the Learning Sciences, 3*(1), 37–92.

Seng, L. K. (2010). An error analysis of form 2 (Grade 7) students in simplifying algebraic expressions: A descriptive study. *Electronic Journal of Research in Educational Psychology, 8*(1), 139–162.

Sfard, A., & Linchevski, L. (1994). The gains and the pitfalls of reification—The case of algebra. *Educational Studies in Mathematics, 26*, 191–228.

Sharp, J. (1995). *Results of using algebra tiles as meaningful representations of algebra concepts*. Paper presented at the Annual Meeting of the Mid-Western Education Research Association, Chicago, IL.

Sheffield, L., & Cruikshank, D. (2001). *Teaching and learning elementary and middle school mathematics*. New York, NY: John Wiley & Sons.

Sherin, B. L. (2001). How students understand physics equations. *Cognition and Instruction, 19*(4), 479–541.

Skemp, R. R. (1976). Relational understanding and instrumental understanding. *Mathematics Teaching, 77*, 20–26.

Sobol, A. J. (1998). *A formative and summative evaluation study of classroom interactions and student/teacher effects when implementing algebra tile manipulatives with junior high school students* (Order No. 9830812). Unpublished Doctoral dissertation, St. John's University, New York. Retrieved from http://proxy.lib.pacificu.edu:2048/login?url=http://search.proquest.com/docview/304449341?accountid=13047.

Sönnerhed, W. W. (2009). *Alternative approaches of solving quadratic equations in mathematics teaching: An empirical study of mathematics textbooks and teaching material or Swedish Upper-secondary school.* Retrieved from http://www.ipd.gu.se/digitalAssets/1272/1272539_plansem_wei.pdf.

Sönnerhed, W. W. (2011). *Mathematics textbooks for teaching: An analysis of content knowledge and pedagogical content knowledge concerning algebra in mathematics textbooks in Swedish upper secondary education.* Dissertation, Gothenburg University. Retrieved from https://gupea.ub.gu.se/bitstream/2077/27935/1/gupea_2077_27935_1.pdf.

Star, J. R., Pollack, C., Durkin, K., Rittle-Johnson, B., Lynch, K., Newton, K., & Gogolen, C. (2015). Learning from comparison in algebra. *Contemporary Educational Psychology, 40,* 41–54.

Star, J. R. (2005). Reconceptualizing procedural knowledge. *Journal for Research in Mathematics Education, 36*(5), 404–411.

Star, J. R. (2007) Foregrounding procedural knowledge. *Journal for Research in Mathematics Education, 38*(2), 132–135.

Star, J. R., & Stylianides, G. J. (2013) Procedural and conceptual knowledge: exploring the gap between knowledge type and knowledge quality. *Canadian Journal of Science, Mathematics and Technology Education, 13*(2), 169–181.

Steinberg, R. M., Sleeman, D. H., & Ktorza, D. (1990). Algebra students' knowledge of equivalence of equations. *Journal for Research in Mathematics Education, 22*(2), 112–121.

Stephens, A. C. (2006). Equivalence and relational thinking: Preservice elementary teachers' awareness of opportunities and misconceptions. *Journal of Mathematics Teacher Education, 9*(3), 249–278.

Sweller, J., & Cooper, G. A. (1985). The use of worked examples as a substitute for problem solving in learning Algebra. *Cognition and Instruction, 2*(1), 59–89.

Tahir, S., & Cavanagh, M. (2010). *The multifaceted variable approach: Selection of method in solving simple linear equations.* Paper presented at the 33rd annual conference of the Mathematics Education Research Group of Australasia. Freemantle, Australia.

Tsamir, P., Almog, N., & Tirosh, D. (1998). *Students' solutions of inequalities.* Paper presented at the Conference of the International Group for the Psychology of Mathematics Education, Stellenbosch, South Africa.

Tsamir, P., & Almog, N. (2001). Students' strategies and difficulties: the case of algebraic inequalities. *International Journal of Mathematical Education in Science & Technology, 32*(4), 513–524.

Tsamir, P., & Reshef, M. (2006). Students' preferences when solving quadratic inequalities. *Focus on Learning Problems in Mathematics, 28*(1),37.

Tsamir, P., Almog, N., & Tirosh, D. (1998). Students' solution of inequalities. In A. Olivier & K. Newstead (Eds.), *Proceedings of the 22nd Conference of the International Group for the Psychology of Mathematics Education, 4,* 129–136. Stellenbosch, South Africa.

Vaiyavutjamai, P. (2004a). *Factors influencing the teaching and learning of equations and inequations in two government secondary schools in Thailand.* Unpublished Ph.D. dissertation, Universiti Brunei Darussalam, Brunei Darussalam.

Vaiyavutjamai, P. (2004b). Influences on student learning of teaching methods used by secondary mathematics teachers. *Journal of Applied Research in Education, 8,* 37–52.

Vaiyavutjamai, P., & Clements, M. A. (2006). Effects of classroom instruction on students' understanding of quadratic equations. *Mathematics Education Research Journal, 18*(1), 47–77.

Vaiyavutjamai, P., Ellerton, N. F., & Clements, M. A. (2005). Influences on student learning of quadratic equations. In P. Clarkson, A. Downton, D. Gronn, M. Horne, A. McDonough, R. Pierce, & A. Roche (Eds.), *Building connections: Research, theory and practice* (Vol. 2, pp. 735–742). Sydney: MERGA.

Verikios, P., & Farmski, V. (2010). From equation to inequality using a function-based approach. *International Journal of Mathematical Education in Science and Technology. 41*(4), 515–530.

Vinogradova, N. (2007). Solving quadratic equations by completing squares. *Mathematics Teaching in the Middle School, 12*(7), 403–405.

Vlassis, J. (2002). The balance model: Hindrance or support for the solving of linear equations with one unknown. *Educational Studies in Mathematics, 49*(3), 341–359.

Vlassis, J. (2004). Making sense of the minus sign or becoming flexible in "negativity." *Learning and Instruction, 14*(5), 469–484.

Whitacre, I., Bishop, J.P., Lamb, L., Philipp, R., Schappelle, B. & Lewis, M. (2011). *Integers: History, textbook approaches, and children's productive mathematical intuitions.* Paper presented at the 33rd Annual Meeting of the North American Chapter of the International Group for the Psychology of Mathematics Education, University of Nevada, Reno.

Wilkins, C. W. (1995). The effect of the graphing calculator on student achievement in factoring quadratic equations. *Dissertation Abstracts International, 56*, 2159A.

Wood, T., Cobb, P., & Yackel, E. (1985). From alternative epistemologies to practice in education: Rethinking what it means to teach and learn. In Stefe, L. & Gale, J. (Eds.) *Constructivism in education* (pp. 401–42). London: LEA.

Yahya, N., & Shahrill, M. (2015). The strategies used in solving algebra by secondary school repeating students. *Procedia—Social and Behavioral Sciences, 186*, 1192–1200.

Zakaria, E., & Maat, S. M. (2010). Analysis of students' error in learning of quadratic equations. *International Education Studies, 3*(3), 105–110.

CHAPTER 4

ANALYSIS OF CHANGE (GRAPHING)

> Understanding change is fundamental to understanding functions
> and to understanding many ideas presented in the news.
> —*Principles and Standards of School Mathematics, 2000*

INTRODUCTION

One of the most interesting graphs ever created is Minard's image depicting Napoleon's march on Moscow in June of 1812 (see Figure 4.1). The image is a physical graph imposed on a map of the region. The graph represents the size of Napoleon's army (initially 400,000) as he progressed from the Niemen River in Belarus on the left toward the upper right corner and Moscow. A couple of regiments broke off to scout at the initial top part of the graph. As the French army chased the Russian army east, the Russians burned everything so the French army would have no resources. Napoleon reached Moscow in October. The dark graph then shows his return march in concert with the temperature graph at the bottom, which is read right to left, starting at 0 degree Celsius. The temperatures steadily drop down to -30 degrees Celsius by the 6th of December, as Napoleon's return

How Students Think When Doing Algebra, pages 165–235.
Copyright © 2019 by Information Age Publishing
All rights of reproduction in any form reserved.

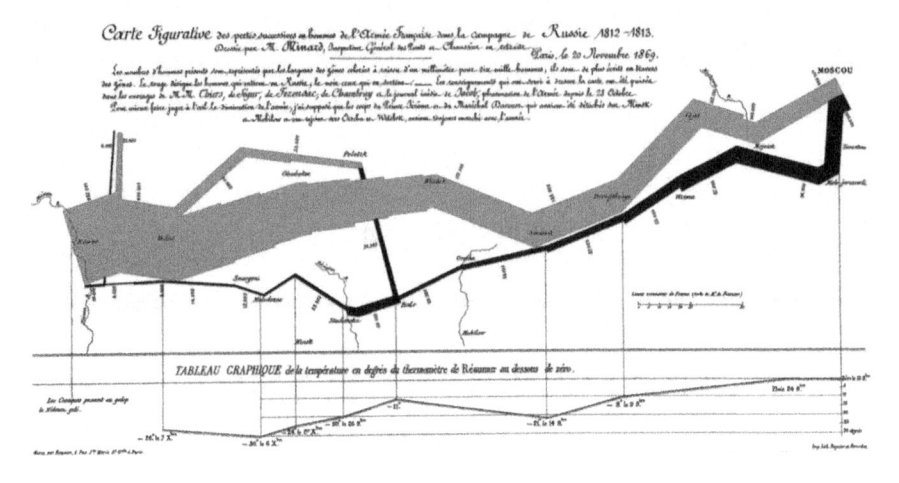

FIGURE 4.1. Minard's graphical map of Napoleon's march to Moscow. Tufte (2001)

was during the bitterly cold Russian winter. Because of the destruction on the way to Moscow, the French army needed to return by a different route. Due to the battles and starvation, only 10,000 troops were left when they returned to the Niemer River. What a powerful visual description of history!

This graph conveys change over time. It is complex and engaging. Can students understand the dynamic? Change is one of the fundamental features of our world that is represented mathematically in algebra. Analyzing change in various contexts is one of the four fundamental concepts of algebra described in the *Principles and Standards for School Mathematics*[1] and is embedded throughout the Common Core State Standards. One of the generative questions of mathematics is: How much has a given quantity changed?

Representing and understanding change requires multiple types of actions:

1. Construction (building a graph based on data or an equation),
2. Scaling (attending to marking axes),
3. Translation (considering multiple representations),
4. Classification (identifying a function that is represented in a graph),
5. Interpretation (making sense from a graph), and
6. Prediction (making conjectures about the patterns in a graph).[2]

In order to have a rich understanding of change, students need to develop those skills rel ated to the representation of change. Research indicates that students have considerable difficulty making the link between representations.[3]

In this chapter we begin with a discussion of **Ratio and Proportional Reasoning**, as this is the basis for moving on to **Understanding Rate of Change**. An examination of change starts with constant rates of change based on constant ratios in linear functions and then moves on to non-constant rates of change. While rate of change can be seen in many forms, a major component of an Algebra course is **Constructing Graphs from Real-World Data**, which represents change visually. Plotting ordered pairs of numbers on a coordinate system leads to students' observing patterns in the graphical representation of those ordered pairs. An essential aspect of graphing is knowing how choices made for **Scaling** the axes influence the nature of the visual representation. Teachers must help students with **Connecting Graphs to Algebraic Relationships** in order to understand how the symbolic representation is reflected in the visual representation and vice versa. This includes **Understanding Slope** and how the steepness or direction of the line communicates information about the change occurring in the problem context.

For example, in the beginning of this book we discussed the discouraging increase in the rates of K–12 students failing algebra in the United States. In writing this book, we hope to contribute to efforts to change those rates, which is to say to change the slope of that graph, so that fewer and fewer students fail algebra (or more and more students pass algebra) as a result of myriad efforts. If, for example, by the year 2020 we see that students are failing algebra at a lower rate than in 2014, that might be evidence of a correlation between our work in this text and the learning of algebra in schools in the United States. Described graphically, the slope of the line that portrays the failure rate of algebra students in the United States should decrease or become less steeply slanted upward. We hope it would "flatten out" or even decline, implying an alteration in the rate of change. When teaching about mathematical descriptions of change (of which graphing is one example) it is more effective to embed student understanding of graphing in real-world examples.

Recently, National Public Radio reported a rise in the number of stay-at-home parents in Oregon:

> The number of stay-at-home parents in Oregon is on the rise. A new report released Wednesday finds that one in five mothers is staying home to care for family while with fathers, it's one in a hundred. Oregon's Office of Economic Analysis focused on the state's working population, those between the ages of twenty-five and fifty-four. The report finds that, while mothers account for the vast majority of Oregon's stay-at-home parents, the share of fathers staying at home has doubled in the past decade.[4]

The story stated that one in five mothers was staying home to care for family. For fathers, the rate was one in one hundred. What is interesting in this significant discrepancy between the two genders is that the 1% rate for fathers staying home represents a doubling of the rate in the past decade. Overall, across both genders the number of stay-at-home parents is on the rise. Can students understand the dynamic of change in this article? Figure 4.2 was included in the report. Can students interpret this graph and the changes over time? What was happening in 2001? How is that different than 2010?

The Common Core State Standards for Mathematics note that "Mathematically proficient students can explain correspondences between equations, verbal

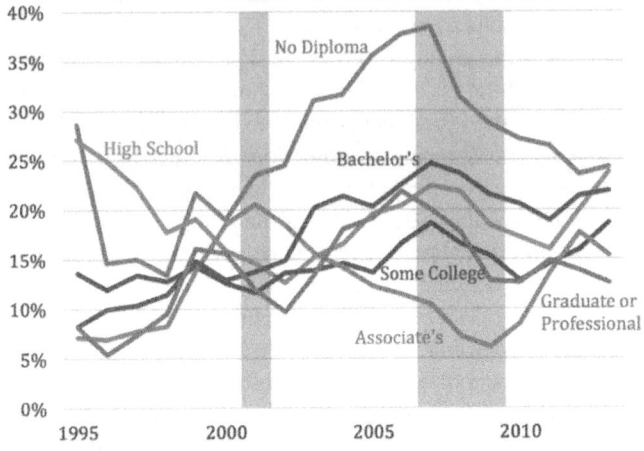

Stay-At-Home Moms by Educational Attainment
Share of Women in each group with children and are Not in the Labor Force, Taking Care of Home or Family. 3 Yr Avg.

FIGURE 4.2. Report on stay-at-home moms. (Lehner, 2014)

descriptions, tables, and graphs or draw diagrams of important features and relationships, graph data, and search for regularity or trends."[5] To make sense of many policy discussions, purchases (such as homes with mortgages), and ensuing decisions, citizens must understand something about how change is described mathematically. How might a .25% change in interest rate impact the cost of a home over time? What if you paid off the home 5 years early? What if you made extra payments early in the loan versus in the middle of a loan? Understanding the mathematical nature of change is important for consumers.

There are multiple elements in learning about analysis of change and graphing in an Algebra course that can be the source of potential student errors. For example, to begin to ***Understand Rate of Change***, students must understand that rate is a numeric relationship between two changing quantities. A weak connection to that concept is at the heart of many student errors and misconceptions. Research indicates that students' experience of the real world has a significant influence on how they understand graphs. Some misconceptions are based on students seeing ***Graphs as Literal Pictures.*** Generally, one of the main objectives in an Algebra course is to develop students' skills in ***Interpreting Graphs***. In this chapter, we share research on the many common ways that students struggle with making sense of visual information and the connection to the data and symbolic representation. Finally, one particular aspect of learning about change is ***Understanding Speed as Rate of Change*** and working with ***Motion Graphs.***

RATE AND PROPORTIONAL REASONING

Common Core Standards: 7.RP.2, 8.EE.5, 6

"Reasoning with ratios and proportions requires 'understanding the multiplicative relationships between rational quantities.' This type of reasoning is widely regarded as a critical bridge between the numerical, concrete mathematics of arithmetic and the abstraction that follows in algebra and higher mathematics."[6] Proportional reasoning forms the basis for understanding rates of change and slope. It "plays such a critical role in a student's mathematical development that it has been described as a watershed concept, a cornerstone of higher mathematics, and the capstone of elementary concepts."[7] Many books have been written on this topic, so we address a few key issues in this section.

Comprehension of linear functions hinges on an understanding of proportional thinking. Thinking about ratios and proportions appears in many everyday applications such as speed, mixture, density, scaling, unit conversion, consumption, percentages, and pricing, but it is one of the more difficult ideas for students to grasp.[8] For instance, consider the problems in Table 4.1 from the National Assessment of Educational Progress (NAEP) for eighth-grade students. On the first, simple proportion problem, 57% of students answered incorrectly; on the second algebraic proportion problem, 80% of students answered incorrectly, indicating

TABLE 4.1. Ratio and Proportion Problems

Problem	Response
Problem 1	
Rima and Eric have earned a total of 135 tokens to buy items at the school store. The ratio of the number of tokens that Rima has to the number of tokens Eric has is 8 to 7. How many tokens does Rima have?	
A. 8 B. 15 C. 56 D. 72 E. 120	A. 7% B. 9% C. 24% D. 43%✓ E. 14% Didn't respond: 3%
Problem 2	
In which of the following equations does the value of y increase by 6 units when x increases by 2 units?	
A. $y = 3x$ B. $y = 4x$ C. $y = 6x$ D. $y = 4x + 2$ E. $y = 6x + 2$	A. 20%✓ B. 4% C. 8% D. 13% E. 53% Didn't respond: 2%

Source: Institute of Education Sciences (2013)

TABLE 4.2. Proportion Problem #3

Problem 3
Find x:
$\dfrac{6}{14} = \dfrac{x}{35}$

that there is much work to be done to develop students' ratio and proportional thinking.

Proportionality is a mathematical relationship between two quantities—a statement of equality of two ratios, i.e., $\dfrac{a}{b} = \dfrac{c}{d}$.[9] Researchers describe proportional reasoning as "reasoning in a system of two variables between which there is a functional relationship (that) leads to conclusions about a situation or phenomenon that can be characterized by a constant ratio."[10] The key for students is to learn the nature of that relationship and how it is different than other math relationships they have learned. Consider how students solve Problem 3 in Table 4.2.

Rather than seek multiplicative relationships, students often find the difference between 6 and 14 and then subtract it from 35 in order to get an answer of $x = 27$.[11] Prior knowledge can influence students' thinking about proportions. In this case, students are used to thinking in terms of additive solution methods, so that is how they tend to approach the problem. One major study found that 25–40% of errors on proportion problems were due to students using additive strategies.[12] Other incorrect strategies include students simply ignoring some of the data in a problem, as in the following problem:

Problem 4: Metal Alloy

In a particular metal alloy, there is 1 part mercury to 5 parts copper and 3 parts tin to 10 parts copper. How many parts mercury to would you need for how many parts of tin?

TABLE 4.3. Interview for Metal Alloy Proportion Problem

Student:	It's there, one part mercury and three parts tin.
Interviewer:	Doesn't it make any difference that there are five parts copper and ten parts copper?
Student:	Yes, it does, I suppose.
Interviewer:	What are you going to do about it?
Student:	Well, you can't work it out like that.

Source: Hart (1981, p. 95)

Table 4.3 provides an interview with a student indicating a struggle with taking into account multiple aspects of the problem simultaneously. Generally, in order of increasing difficulty, students are challenged by 1:2 ratios, 1:*n* ratios, *m*:*n* ratios (in which neither term is 1), and non-integer ratios.[13]

There are four primary difficulty factors for students solving proportion problems.[14]

- *First*, the presence or lack of integer ratios. Students sometimes acquire facility with proportions as long as the relevant data is integral. When non-integers are used, students become confused and apply incorrect techniques (e.g., constant differences rather than constant ratios, as in the additive strategy used in Problem 3 above).[15] To overcome this, teachers should provide plenty of examples in which the data is non-integral and help the students see that the thinking is the same regardless of the kinds of numbers used.
- *Second*, the location of the variable in the proportion with respect to the other three numbers of the proportion. For instance, whether the variable is in the numerator ($\frac{x}{4} = \frac{28}{9}$) or denominator ($\frac{9}{4} = \frac{28}{x}$).
- *Third*, the size of the numbers involved in the ratio.
- *Finally*, the context of the proportional reasoning problem can also affect the difficulty. If students are familiar with the setting for the problem, they are more likely to be successful.[16]

As teachers design or make use of proportion problems, it is important to vary these factors to develop students' proficiency more broadly.

One approach that teachers often use with their students is the rote algorithm of cross multiplication. That is, in Problem 3 students would multiply 6*35 and 14*x to get 210 = 14x and divide each side by 14 to get x = 15 (or more generally, $\frac{A}{B} = \frac{x}{D}$, AD = Bx, $x = \frac{AD}{B}$). However, research indicates that this algorithm does not promote proportional reasoning on the part of the student for the following reasons:[17]

- The algorithm is poorly understood by students,
- Using the algorithm is seldom a naturally generated solution method, and
- The algorithm is used by students to avoid proportional reasoning.

This algorithm does not help students make connections between the relationships of the quantities. One concern is that students only encounter "missing value problems where three of four values in two rate pairs [are] given and the fourth is to be found"[18] and the cross-multiplication method is the only strategy that students learn to solve them.[19] Students apply the method to similar problems rather than being able to compare different problem types and approaches to solving proportions.[20] Instead, students should be given the opportunity to use informal reasoning to explore relationships between the objects and the internal and external ratios. Teachers should engage students in "activities that help them discover that proportion is the variation of two quantities related to each other."[21]

One of two primary strategies is used when students solve proportion problems accurately: either the multiplicative or building-up approach.[22] When solving Problem 3, a student using the multiplicative approach might think, "$\frac{6}{14}$ reduces to $\frac{3}{7}$ and 7 times 5 is 35 so 3 times 5 would be 15." Using the building-up approach, a student might think, "I get 3 for every 7 so 6 for every 14, 9 for every 21, 12 for every 28, and 15 for every 35." Both strategies can help students solve proportional thinking problems, but the building-up approach only works well for students when dealing with integers.[23] Furthermore, researchers believe that the build-up strategy "is a relatively weak indicator of proportional reasoning."[24]

Using either of those strategies, there are multiple ways that researchers have described how students need to think about proportional relationships.[25] The first has been described as the *internal ratio*, the "*within* relationship," or *scalar reasoning*. When using this way of thinking with Problem 3, students consider the relationship between 6 and 14, which can be reduced to 3 to 7, and use that relationship to solve the problem. Or put another way, suppose I can make 16 origami paper cranes in two hours. I can find how many cranes I can make in 4 hours by noticing that 2 hours by 8 cranes per hour yields 16 paper cranes, so multiplying 4 hours by 8 cranes per hour yields 32 cranes. The second way of thinking has been described as the *external ratio*, the "*between* relationship" across the equal sign, or *functional reasoning*. That is, in Problem 3, the relationship between 14 and 35 can be compared to get the relationship between 6 and *x*. In the cookie problem, I see that an input of two hours gives me 16 cookies, so doubling the input means doubling the output, and I can make 2 times 16, or 32 cookies. When students can reason in both ways, they are gaining mathematical power. Students should be given the opportunity to use informal reasoning when exploring problems of proportionality. It is important for students to discuss the reasonableness of their solutions and use mathematical arguments to justify their findings.

There is quite a bit of evidence indicating that visual representations have the potential to enhance students' understanding of mathematics.[26] However, students must understand how the visual representation "depicts information about the content."[27] Teachers often use double number lines,[28] tape diagrams,[29] and ratio tables[30] as models for proportional reasoning (see Figures 4.3, 4.4, and 4.5). These representations offer systematic ways of writing down the relationship between two variables. The position of the points on a double number line is meaningful, and the columns in a ratio table can be in any order. Japanese teachers use tape diagrams extensively to "visually illustrate relationships among quantities in a problem."[31] Each model functions as a tool to break down complicated calculations into smaller, more manageable steps. Understanding relationships through

Time (days)

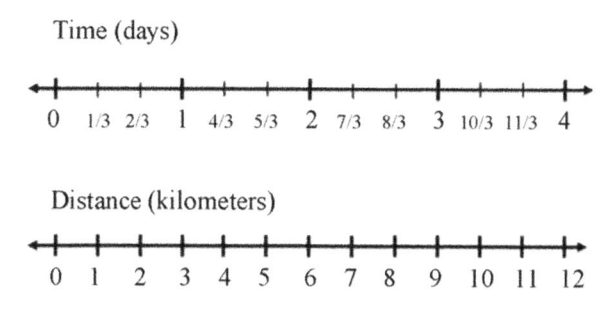

```
0  1/3  2/3  1  4/3  5/3  2  7/3  8/3  3  10/3  11/3  4
```

Distance (kilometers)

```
0  1  2  3  4  5  6  7  8  9  10  11  12
```

FIGURE 4.3. Double Number Line.

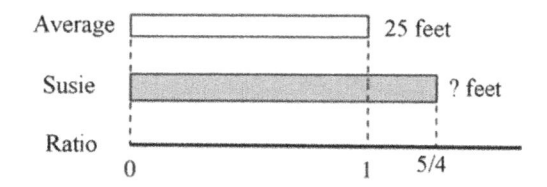

```
Average  [           25 feet        ]
Susie    [      ? feet        ]
Ratio    |----------|----------|
         0          1         5/4
```

Susie and her friends were tossing Frisbees. The average distance was 25 feet. Susie's throw was 5/4 of the average distance. How many feet did Susie toss her Frisbee?

FIGURE 4.4. Tape Diagram: Throwing Frisbees. Based on Murata (2008, p. 381)

these models can lead to successful thinking about proportions through graphs and their symbolic representation (see Figure 4.6).[32]

Proportionality is a simple example of a mathematical function that can be represented by a linear equation. It is a useful bridge between numerical and abstract relationships such as $y = mx$. Teachers should work with students to help them comprehend how algebraic ideas with proportional relationships can be

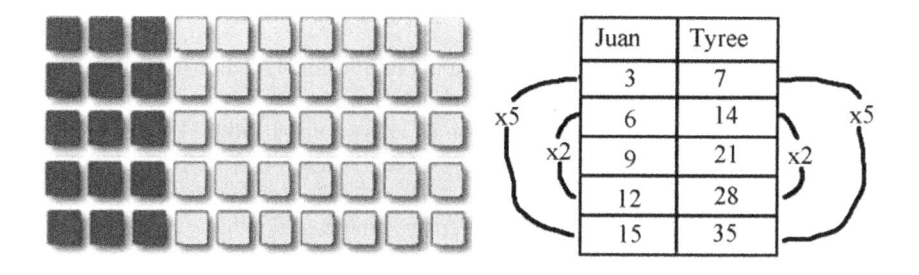

Juan	Tyree
3	7
6	14
9	21
12	28
15	35

FIGURE 4.5. Ratio Table: Blue and Yellow Boxes. Based on http://www.mathvillage.info/node/91

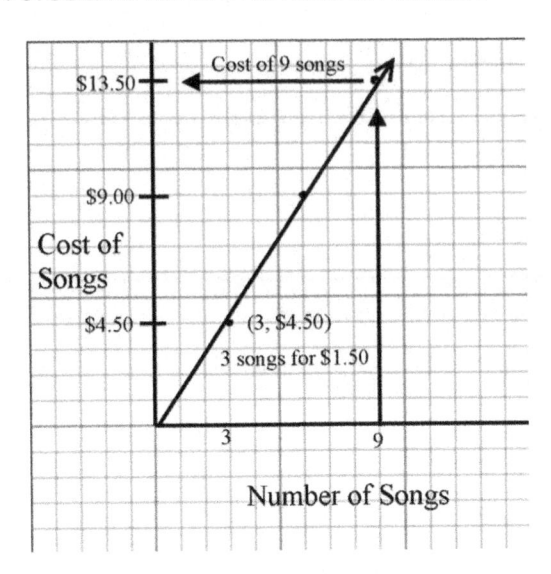

FIGURE 4.6. Graphical Representation of Proportion.
Based on Post, Behr, & Lesh (1988, p. 86)

seen in tables, graphs, equations, and diagrams. The teacher's goal for students' proportional reasoning should be more qualitative than quantitative. That is, students should be able to compare without specific values. For instance, consider the problem: "If Nicki ran fewer laps in more time than she did yesterday, would her running speed be faster, slower, the same, or you can't tell?"[33] When the context is familiar, students should be able to interpret the meaning of two ratios in this problem without needing numbers to understand the dynamic. Facility with proportional reasoning in this way will support later thinking with concepts such as slope in Algebra courses.

Teachers

- Help students understand both the within relationship and between relationship in a proportion.
- Use visuals to help students understand proportional relationships and connect symbolic representations.
- Don't rely on rote procedures such as cross multiplication. Help students see the conceptual ideas and proportional relationships behind symbolic manipulation.
- Be sure to use non-integers to increase students' flexibility with proportional reasoning.

UNDERSTANDING RATE OF CHANGE

Common Core Standards: 6.EE.9, 7.RP.2, 8EE.5, 6, 8.F.4, 5, 8.SP.3

Rate of change is a challenging concept for students to comprehend. Calculus students often struggle to understand rate.[34] To understand rate of change, students must understand four critical aspects:[35]

1. Rate as a relationship between two changing quantities.
2. Rate as a relationship, which may vary, between two changing quantities.
3. Rate as a numerical relationship, which may vary, between two changing quantities.
4. Rate as a numerical relationship, which may vary, between any two changing quantities.

Students must also understand how graphs can represent rate of change (See ***Understanding Slope***) and how speed is an important example of rate of change (See ***Understanding Speed as Rate of Change***). Table 4.4 includes information about the varied ways students experience and think about rate.

TABLE 4.4. Development of the Concept of Rate

Brief Description	Illustrative Quotes
A word rating, a quality	Interviewer: Can you give me an example? Student 1: Ratings on movies and stuff.
A word associated with a numeric value	Interviewer: Can you give me an example of rate? Student 2: Well there is that bank rate thing. They add rate if you don't pay your bills or something on time they keep adding.
The result of a formula calculation with little meaning	Interviewer: Can you give an example of rate? Student 3: If you gave me a problem [formula], I could substitute the numbers in it and then I would be able to give you a rate.
A single quantity	Interviewer: Can you tell me about the rate the clown is walking? Student 4: He's walking at about 8 seconds.
A relationship between two changing quantities	Student 5: As the area of sunlight increases, the height of the blind also increases.
A constant numeric relationship between two changing quantities	Student 6: The area increases by a rate of 3.2 for every 0.5.
A numeric relationship between two changing quantities of distance and time, i.e. speed	Interviewer: What is rate? Student 7: He [the clown] walks 22 meters in 7 seconds and then he [the frog] only makes 7 meters in 7 seconds.
A numeric relationship between any two changing quantities	Student 8: That's increasing so that it [area] starts off small, 1, and it goes up to 2.6, so that's 1.6, the difference, and then the next one is 2.2 the difference and the one after that is 2.8 the difference. Student 9: He's walking at 4 meters per second. So for every second, he walks 4 meters.

Based on Herbert & Pierce (2009, p. 220)

As is evident in the interviews in Table 4.4, students may have trouble coordinating the two quantities required for a rate. The idea of "rate" typically develops in stages.[36] In the first stage, students do not have a mathematical conception of rate, and instead describe rate based upon everyday usage. For example, in the interview, Student 1 refers to movie "ratings." At a further stage of development, students see rate as a single numerical quantity. For example, Student 4 says that the clown is walking at a rate of 8 seconds, and not 8 feet per second. Next, students may see a relationship between two quantities, but may be unable to describe that relationship numerically. For example, Student 5 does not use numbers, but describes the relationship generically—that as the area of sunlight increases, the height of the window shade also increases. Finally, students might see that rate is the numerical relationship between two changing quantities. Student 9 identifies a specific relationship by explaining that the rate the man is walking is 4 meters per second, because for every second the man walks four meters. The first four descriptions indicate gaps in students' understanding. Comprehension of rate of change must include some sense of two changing quantities and the relationship between them. A more sophisticated understanding of rate of change would ultimately involve reasoning that is independent from the context, the concept of variable rates of change, and developing potential for abstraction into symbols.

Another concern with students' thinking regarding rate of change is that students may appear to understand one representation of rate but not be able to transfer that understanding to a different representation or context.[37] In particular, students "do not transfer their understanding of rate demonstrated in tables and graphs to the corresponding symbolic representation."[38] Different representations provide different rate-related information for different students. Students have greater success when they are familiar with the setting of the reasoning task.[39] For instance, students often understand the concept of speed, but many are not able to relate that understanding to a context not involving speed.

Finally, body language can give teachers helpful insights into student thinking when talking about rates.[40] When a student's words and gestures match, it is likely the student has a clear understanding of the concept. Sometimes a student's gestures and verbal descriptions do not agree. For example, a student might say that the shadow is getting bigger while their hand motion indicates that the height of the shadow is decreasing. The mismatch between words and gestures indicates that the student does not have a good understanding of the concept. When a student cannot verbalize a concept, but correctly demonstrates a concept through body language, the teacher can help the student focus on vocabulary development and symbolic representation.

Teachers

- Try to determine the level of a student's development of the concept of rate so you know what aspects of the concept they need help with.
- Use multiple representations of rate in different contexts with integers and non-integers to help students develop a comprehensive understanding.

CONSTRUCTING GRAPHS FROM REAL-WORLD DATA
Common Core Standards: 6.EE.9, 8.F.5

Real-world data provides a rich source of activities for students to graph, offering a more meaningful context for developing algebraic thinking. As noted in the introduction to this chapter, when teaching about graphing, it is best to embed students' understanding of graphing in real-world examples. When students are familiar with the context, such as the amount (rate) of orange juice concentrate to water, and the way the rate is used, then representation of the mathematical relationship is easier for students.[41] "A representation is 'first and foremost, something that stands for something else.'...this description implies the existence of two 'worlds,' a 'represented world' and a 'representing world,' with correspondences between these worlds."[42] One researcher suggests that there are five aspects of representation that students must consider (see Table 4.5).

Teachers need to ensure that students understand each of these aspects of the representation in order to comprehend the concepts that need to be portrayed. For example, teaching students the mechanical skills of graphing—such as plotting points, drawing the axis, and labeling—is reasonably straightforward, although there are common misconceptions that may develop (for instance, see *Scaling* below). What is more challenging is to help them understand the representation of the relationship between the variables they are plotting. Comprehending graphing involves both constructing and interpreting. In this section we focus on constructing, and in a later section we address *Interpreting Graphs.*

Graphs may take on many different forms that help communicate the ideas of change. However, research typically focuses on the difficulties students have with traditional representation of mathematical ideas. When we "provide [students] with a specific representation and ask them to use it, we are greatly narrowing what we can learn about" their knowledge of concepts.[43] Instead, some research suggests that teachers be flexible with how they let students communicate their understanding of motion by allowing them to invent representations.

TABLE 4.5. Aspects of Representing

Aspect	Example
1. Represented world	Mix juice concentrate and water
2. Representing world	3:1
3. Modeled aspects of represented world	Ratio of water to concentrate
4. Aspects of representing world doing the modeling	Ratio of 3:1
5. Correspondence between the two worlds	• 3 corresponds to cups of water, 1 corresponds to the cup of concentrate • ratio expresses relationship between

Source: Palmer (1977)

178 • HOW STUDENTS THINK WHEN DOING ALGEBRA

Along that line of reasoning, in order to discuss the way students develop in their thinking about graphically representing real-world data, we first share the following problem:

Problem 5: Motorist

A motorist is speeding along a highway and he's very thirsty. When he sees a store, he stops to get a drink. Then he gets back in his car and drives slowly away.[44]

In one study, middle and high school students were asked to create their own representations to depict the motion described in Problem 5. As a result, students created a wide variety of representational forms. The work of two students, Jason and Clare, is shown in Figures 4.7 and 4.8, and the explanations of their responses to the problem are in Table 4.6.

Students are not likely to choose a coordinate graph to represent a situation if they have not had experience with that type of model. The first set of interviews (see Table 4.6) show how students can find other ways to represent a contextual situation. In each of these students' representations, you can get a sense of their perception of rate of change and the base of knowledge they have to work with as they move toward more sophisticated understanding. They typically use existing representations from their experience (or "constructive resources" as one researcher put it) as the fodder for their new representations.[45] In Clare's case, she "uses arrows of various lengths to show the magnitude of the speed, and dots to

FIGURE 4.7. Jason's Representation of the Motion Problem. Based on Sherin (2000, p. 412)

FIGURE 4.8. Clare's Representation of the Motion Problem. Based on Sherin (2000, p. 415)

TABLE 4.6. Interviews for the Motorist Problem174

Interviewer:	Tell me about your picture.
Jason:	It said the car was speeding, then he stopped at a store and got a drink and then went slowly after that. I drew a car to show each part of the trip and how fast he was going. I had to make up the speeds. So, you can see on my picture where he stopped to drink because his speed is 0 and I drew a person by the store.
Clare:	The man starts out on the left and the arrow shows that he is going fast. The dot shows he stopped. Then the short arrows mean he went slowly and then got faster.

Note: Author created interview to articulate a potential student description.

indicate a stop."[46] She visually demonstrates understanding of some of the qualitative dynamics of the problem. This can be a foundation for moving toward more traditional graphic representations in the future.

When it comes to constructing more traditional graphs, students must learn many mathematical conventions and processes. "Construction involves going from raw data (or abstract function) through the process of selection and labeling of axes, selection of scales, identification of units, and plotting...construction requires generating new parts that are not given."[47] One study suggests the set of categories in Table 4.8 for creating a graph. In that study, students were given a set of data and asked to graph the data (see Table 4.7).

Besides the different levels of understanding listed in Table 4.8, a big part of constructing graphs is determining the appropriate representation. Should it be a set of points? A line? A curve? A bar chart? In order for students to make that choice, they need to understand the use of each representation.[48] Students generally have considerable difficulty in deciding which visual format matches the

TABLE 4.7. Data for Graphing

Temperature (degrees Celsius)	Length (cm.)
351	1416
346	1377
337	1340
325	1305
312	1270
301	1237
290	1206
283	1176
254	1092
225	1020
212	998
200	976
194	956
186	937

Source: Wavering (1985)

180 • HOW STUDENTS THINK WHEN DOING ALGEBRA

TABLE 4.8. Categories of Students' Understanding of Graphing

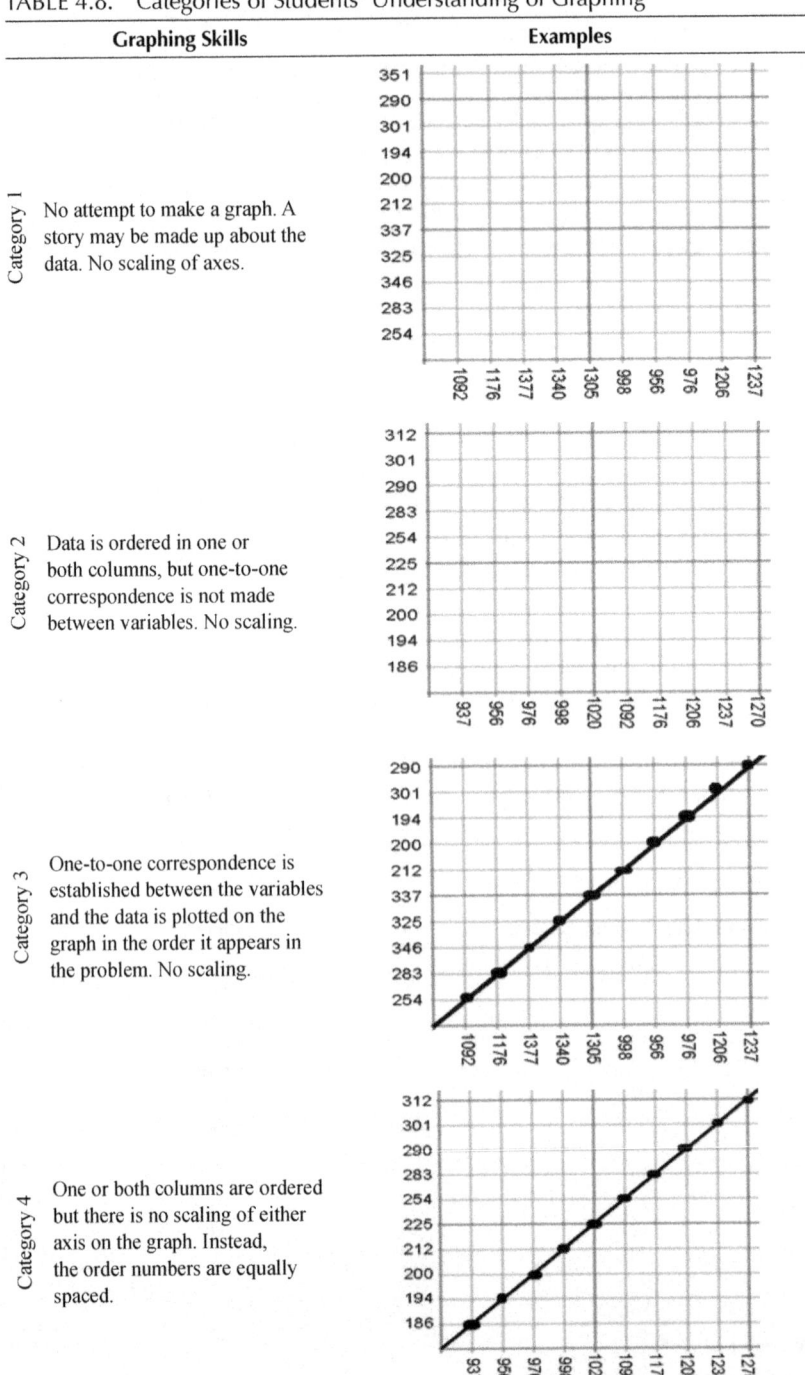

	Graphing Skills	Examples
Category 1	No attempt to make a graph. A story may be made up about the data. No scaling of axes.	Data values: 351, 290, 301, 194, 200, 212, 337, 325, 346, 283, 254. Empty graph with x-axis labels: 1092, 1176, 1377, 1340, 1305, 998, 966, 976, 1206, 1237.
Category 2	Data is ordered in one or both columns, but one-to-one correspondence is not made between variables. No scaling.	Data values: 312, 301, 290, 283, 254, 225, 212, 200, 194, 186. Empty graph with x-axis labels: 937, 966, 976, 998, 1020, 1092, 1176, 1206, 1237, 1270.
Category 3	One-to-one correspondence is established between the variables and the data is plotted on the graph in the order it appears in the problem. No scaling.	Data values: 290, 301, 194, 200, 212, 337, 325, 346, 283, 254. Graph with points plotted in a line with x-axis labels: 1092, 1176, 1377, 1340, 1305, 998, 966, 976, 1206, 1237.
Category 4	One or both columns are ordered but there is no scaling of either axis on the graph. Instead, the order numbers are equally spaced.	Data values: 312, 301, 290, 283, 254, 225, 212, 200, 194, 186. Graph with points plotted in a line with x-axis labels: 937, 966, 976, 998, 1020, 1092, 1176, 1206, 1237, 1270.

TABLE 4.8. Continued

	Graphing Skills	Examples
Category 5	One axis is ordered and the other axis is scaled.	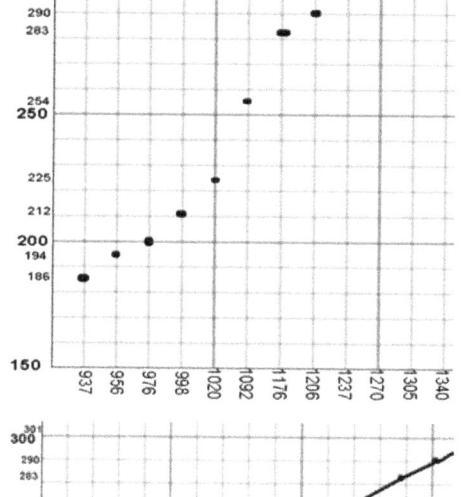
Category 6	Data in both columns is rearranged from low to high and a new one-to-one correspondence is established. Both axes scaled.	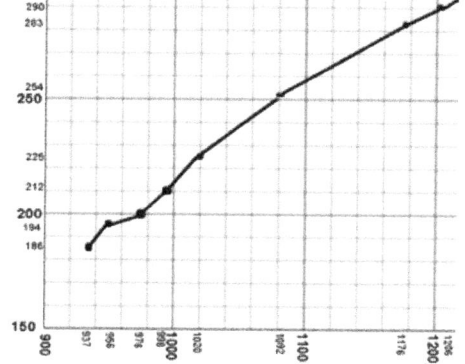
Category 7	Axes on graphs are correctly scaled but student does not see a relationship in the data as graphed.	
	Student states the correct relationship between the variables.	
Category 8	Student understands the characteristics of an exponential curve, recognizing that as one variable increases, so does the other and it is not a straight line relationship.	

Adapted from Wavering (1985)

needs of the situation. One study gave 92 eighth-grade students a problem similar to the following about the relationship between three students' time studying and their scores on tests:

Problem 6: Relationship between studying and test scores
a. Taylor said the more she studies, the better her grades are.
b. Bree said no matter how long she studies, she always gets the same grade.
c. Fina said that when she studies up to two hours, her grades are better; but, beyond two hours, she becomes tired and her grades are lower
d. Bridget said that the more she studies her grades go down.

Each student had a different relationship: increasing, constant, curvilinear, and decreasing. While 60% of students were able to construct at least one correct graph, only 28.9% were able to construct all four graphs.[49] Students often focus on only one aspect of the relationship in their visual representation—in this problem, the amount of time (see Figure 4.9).[50] Thirty-six percent of students created a series of similar graphs focusing on one dimension. This way of thinking aligns with Category 2. Thirty-seven percent of students argued that a single point represented at least one of the relationships (for example, see Student 2, Figure 4.10). In Figures 4.9 and 4.10, students' responses reflect similar, narrow focus upon some aspects of the dynamic and not others.

When students are engaged with data, they often assume there is a linear relationship, in the form of $y = mx + b$, with m > 0 as in Problems 6b and 6c.[51] In this study, while one problem is constant and the other a parabola, students used the information to graph a steep line, reflecting the thinking in categories 3–5 (see Figure 4.11). They use the data the best way that they know how to communicate the information. An important focus for teachers here is teaching mathematical conventions of graphing as well as the meaning of each point in the graph.

If the teacher over-emphasizes one type of relationship (e.g., an increasing relationship) or one type of graph (e.g., line graph), it may lead to a student perception

Student 1

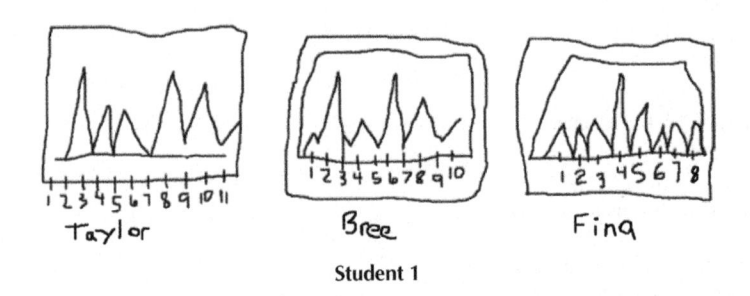

FIGURE 4.9. Student-Constructed Graphs for Problem 6.
Source: Mevarech & Kramarsky, 1997

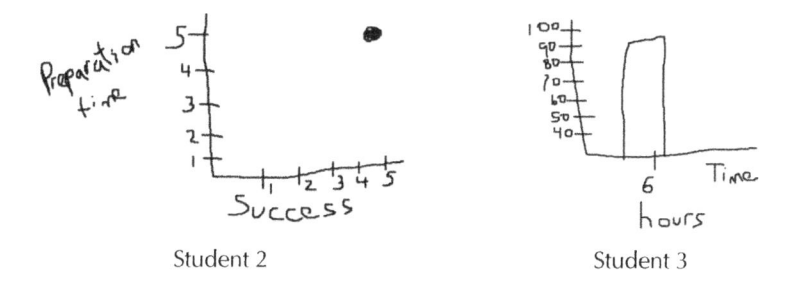

FIGURE 4.10. More Student-Constructed Graphs for Problem 6.
Source: Mevarech & Kramarsky (1997)

that all graphs have that form. Teachers should make sure that students are exposed to different types of graphs (e.g., linear, increasing, decreasing, non-linear, those that go through the origin, and those that don't). Only using graphs that are increasing in nature may lead to the common mistake of making all graphs increasing, even when this does not fit the situation.

Traditional approaches to teaching graphing that begin the process of producing a graph by first breaking it into step-by-step items in a procedure may, in fact, perpetuate the perceived problem that graphing is a difficult topic. Instead, present students with a purposeful task in a familiar context, and students will be able to act intuitively to use line graphs. Using computers/graphing calculators to help construct the graphs can help students make the connection between the specific procedures of creating the graph, parameters of the function, and the meaning of the graph as it pertains to the data.[52] More constructivist approaches toward using

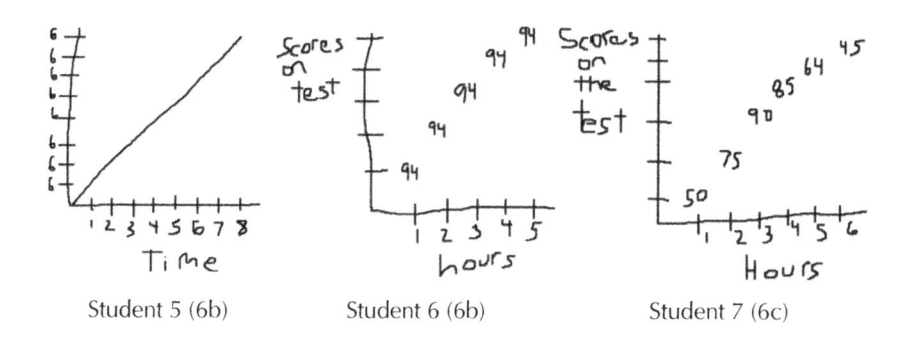

FIGURE 4.11. Tendency to Graph a Sloped Linear Relationship for Problems 6b & 6c. Source: Mevarech & Kramarsky (1997)

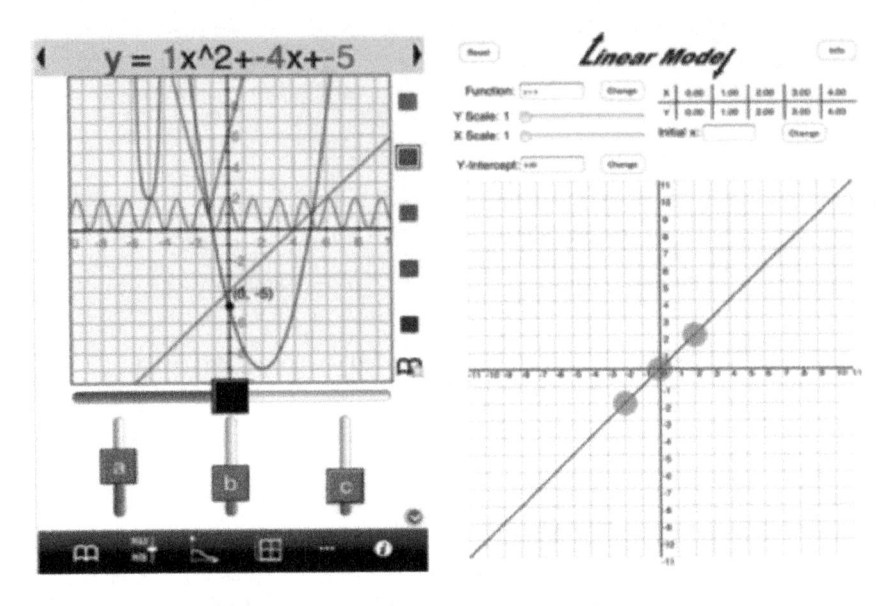

FIGURE 4.12. Math Flyer and Linear Model iOS Apps.

technology in mathematics instruction have been shown to be effective in helping students learn[53]. Virtual manipulatives from the Center for Algebraic Thinking, such as *Math Flyer* and *Linear Model* (see Figure 4.12), which connect dynamic visual images to the abstract symbols of algebra, can be especially helpful for developing students' understanding.

"The resources students employ, even if they lead to unconventional or suboptimal scales in some contexts, are ideas that can be built upon."[54] Students bring intellectual resources with them to help them problem solve. It is important for teachers to see students' attempts to construct visual representations about change as logical constructions with rationales. As teachers explore those rationales with students in comparison with conventional approaches and reasoning, students are more likely to fill the gaps in their thinking and address misconceptions. Using problem situations intentionally designed to evoke common errors and create discussion is more effective than teaching styles that avoid or ignore student misconceptions. For instance, if students agree that the graph of the speed of a bicycle going over a hill should look like a hill, consider starting with this representation and ask students to think about how their speed would change (draw on their experience with bicycles). It may help to summarize their thinking in words: "So what I am hearing is that you would start out with some speed going into the hill, but then you would slow down as you climb the hill. Then at the top, you would coast a little and then speed up as you go down the other side. Is that correct? So let's take a look at this graph to see if it matches what you told me..."

Teachers

- Encourage informal graphing prior to formal graphing as a way for students to get on paper their ideas about change.
- Help students see the relationship between the two variables in the dynamic of the graph. That is, how are the changes in the graph reflected in changes in the variables?
- Use students' attempts at constructing a graph to model a real-world situation as a basis for conversation about the need to account for multiple factors in a graph.

SCALING

Common Core Standards: 8.EE.5 8.F.5

Here we extend the discussion we initiated in the last section (***Constructing Graphs from Real-World Data***) on the difficulty students had in understanding scaling when creating axes for graphs (see also ***Patterns & Functions: Graphing Functions***). "A complete understanding of graphical representation means realizing what visual features of the graph will not change under the change of scales (e.g., the *x*- and *y*-intercepts) and what features change when the scales are altered (e.g., the geometrical angles that the line creates with each of the axes)."[55] Consider Figure 4.13: If a student constructed one of those graphs, what would

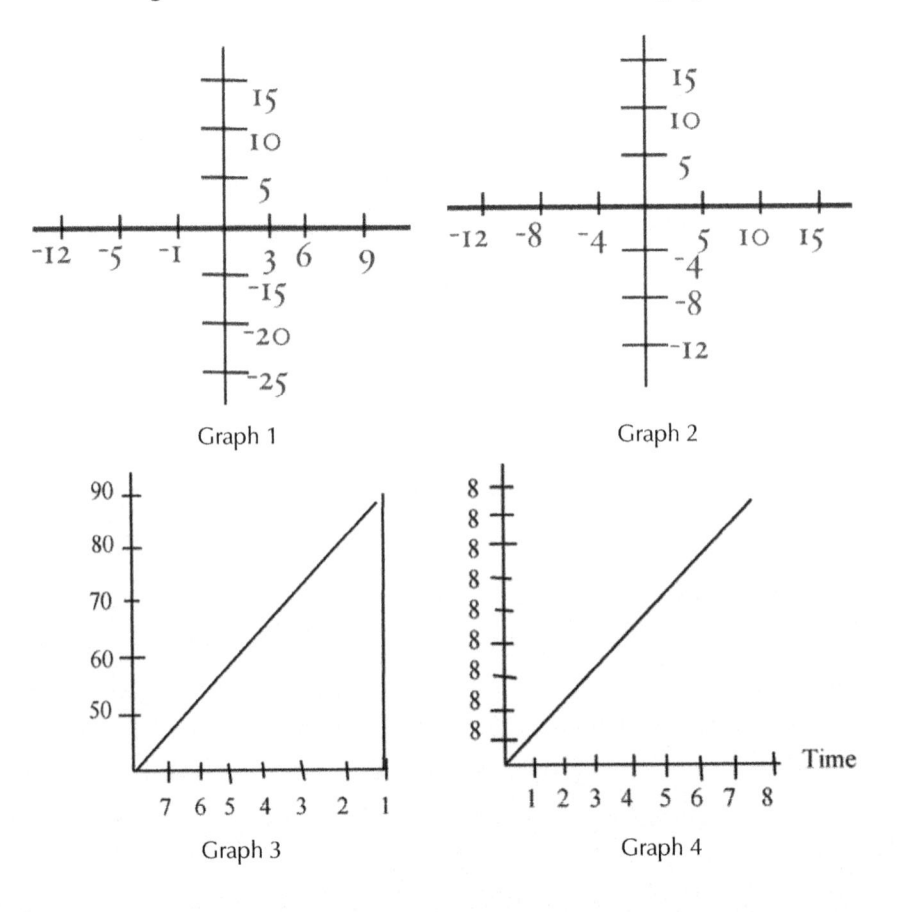

FIGURE 4.13. Struggles with Scaling.
Note: Graphs 1 & 2 from Center for Algebraic Thinking, graphs 3 & 4 from Mevarech & Kramarsky (1997)

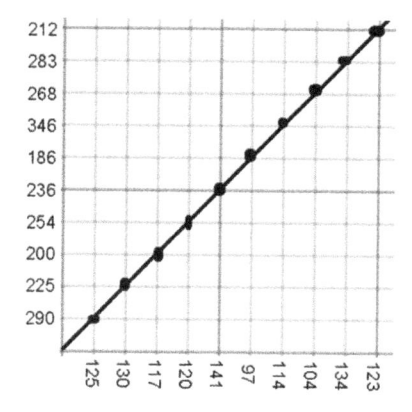

FIGURE 4.14. Discrete Points on Axes, Rather Than Continuous. Based on the data from Wavering (1985)

be the difficulty with their understanding? In each case, students struggle with the conventions of how to label the axes. In graph 2, for instance, students thought that it was possible to have different scales on the positive and negative parts of the axes.[56]

Looking at each of these graphs, it becomes clear that in order to understand the axes of a graph, students need to comprehend:

- Consistent intervals along an axis (e.g., counting by 2's, unlike Graph 1),
- Numerical order/continuous (unlike Graph 1 or 4),
- Scaling must be the same along each axis but can be different between axes (unlike Graph 2), and
- Numbers grow from left to right or from bottom to top (unlike graph 3).

Some students believe that the points on the axes are discrete rather than continuous, thinking that there aren't any points between the marked points on the axis, perhaps reflected in an image such as Figure 4.14.[57] Students may simply graph the data available to them. Other students "may place data points on successive, evenly spaced intervals regardless of their magnitude."[58]

When students don't understand scaling, interpretation of graphs can be heavily influenced by the intervals chosen for the graph.[59] Consider the graphs in Figure 4.15. They each imply a very different story of the state of the stock market. However, the data is exactly the same. The only thing that has changed is the scale of the y axis. Similarly, in Table 4.9, students are asked to view three graphs and determine which two showed the same information. About one-third of students didn't realize that the b graph was the same data but differently scaled. In order to be good consumers and conveyors of information, students must comprehend the role scaling plays in the construction of graphs.

Dow Jones Industrial Average

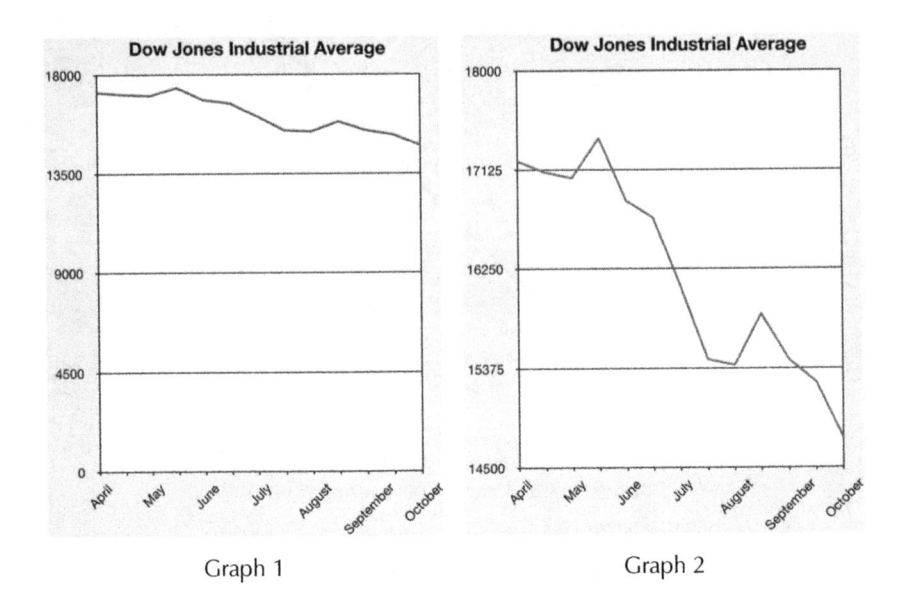

Graph 1

Dow Jones Industrial Average

Graph 2

FIGURE 4.15. Two Graphs: Same Data, Different Scales.

There are unique challenges with scaling when using graphing calculators—as opposed to apps or computer software—that can be viewed as limitations of the technology or great opportunities for conversation. Students often do not pay attention to scale and the format of the window of a graphing calculator. Researchers working with students on Problem 8 in Table 4.10 found that less than 16% recognized that the visual representation of the slope of a linear graph depends on the scaling of the axes.[60] Simply changing the scale on a graph can have a great impact on the difficulty of the task for students. In many cases, students continue

TABLE 4.9. Problem 7: Scaling Problem--Which Two Show the Same Information?

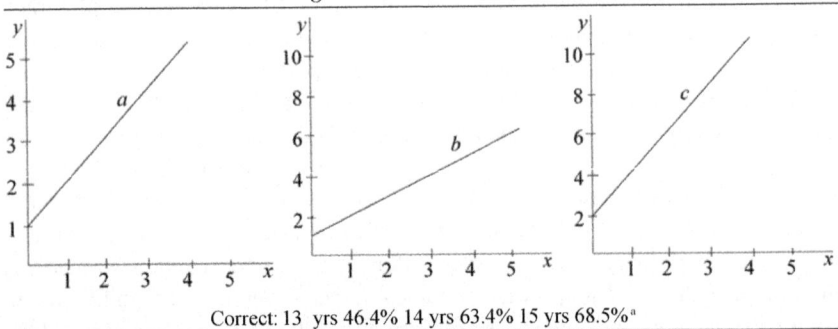

Correct: 13 yrs 46.4% 14 yrs 63.4% 15 yrs 68.5%[a]

[a]Percent of 1789 13- to 15-year-old students in Kerslake (2004)

TABLE 4.10. Problem 8: Graphing Calculator

a. Using a graphing calculator, plot $y = 2x+3$ and $y = -0.5x - 2.5$

b. Explain why the graphs of the lines do not appear at right angles on the screen. What could you do to make the lines look more perpendicular?

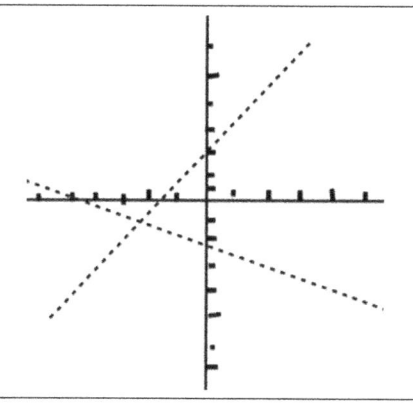

Source: Mitchelmore & Cavanagh (2000)

to read the scale as if it was in single units even when it has been changed to count by 2s, 5s, or other increments.

Encouraging students to use graphing software to enter data and make several similar graphs can promote discussion and understanding regarding which type of graph and scale best portrays the trends and patterns associated with the task being investigated. Researchers have found some student difficulties when using graphing calculators, including challenges in interpreting scale changes when zooming;[61] the appearance that perpendicular lines are not perpendicular, depending on the scale (see Table 4.11 for a student interview);[62] and the tendency to accept whatever picture appears as the initial graph on the screen without zooming in or out.[63] These technology tools can improve student understanding of scale as well as distort perceptions;[64] therefore, care needs to be taken in how teachers address student misconceptions based on the technology. Students may benefit from confronting the limitations of the technology and attempting to explain them. Tablet apps such as *Math Flyer* allow students to pinch zoom in and out to quickly change the scales and see how those manipulations alter the visual representation of the graph.

TABLE 4.11. Interview About Problem 8

Interviewer:	Why do the lines not appear at right angles on the screen?
Student:	Because the lines are not perpendicular to each other.
Interviewer:	What is the slope of the two lines?
Student:	2 and -½. Oh...those should be perpendicular. I wonder why they aren't
Interviewer:	Can you change the scale on the calculator to make them perpendicular?

Source: Mitchelmore & Cavanagh (2000)

Some researchers suggest that it might be useful to differentiate between *relative* and *absolute* understanding of scale. "The former correctly regards scale as a ratio of distance to value, while the latter interprets scale solely as either the distance between adjacent marks or the value represented by this distance."[65] Students tend to develop an *absolute* understanding of scale in which they can work with a fixed context, but find it difficult to explain how graphs change as a result of changes in scale.[66]

Finally, similar to our discussion regarding **Constructing Graphs from Real-World Data,** research indicates that it may be beneficial to give students the opportunity to invent scales and graphs and then discuss and critique them regarding their benefits and constraints in light of mathematical conventions.[67]

Teachers

- Do not always choose functions that fit neatly into a grid that spans from -10 to 10 on the x and y-axes. Students need to learn that the shape of the graph depends on the window you are viewing it through.
- Work with students on graphs in which the axes are not scaled equally.
- Encourage students to manipulate the scales and intervals of graphs and interpret the visual change in comparison with the data and symbolic representation.
- Teachers may want to avoid linear functions when helping students understand the shape and scale of graphs because linear functions are easy to graph, but do not contain enough features to develop students' visual intuition. It may be more effective to analyze the graphs of linear functions after more complicated algebraic functions.

CONNECTING GRAPHS TO ALGEBRAIC RELATIONSHIPS

Common Core Standards: 6.EE.9 8.F.1, 2, 3, & 4

Given a linear or non-linear equation, students should be able to make a connection between that algebraic equation (e.g., $y = 3x + 4$) and a graphical representation. Similarly, given a graphical representation, students should be able connect to and write a descriptive algebraic equation, assuming the graph is amenable to an algebraic description. We examine this second issue by addressing two topics in this section: *The Missing x-Intercept in "$y = mx + b$"* and *The Value of b in "$y = mx + b$."* We will examine how students are **Understanding Slope** in the next section.

The missing x-intercept in "$y = mx + b$"

When working with linear functions, students frequently believe that the value of the x-intercept should be represented in the equation "$y = mx + b$."[68] This is not the case, though: "m" is the slope of the function and "b" is the y-intercept of the function. Many teachers might disregard this as a simple error, but this error might represent a complex misconception or "transitional conception," as some researchers refer to it.[69] The line in Problem 9 (see Table 4.12) has a y-intercept of 1 and a slope of 2, so the correct solution is "$y = 2x + 1$." However, common incorrect solutions by students placed the x-intercept value into the equation for

TABLE 4.12. Problem 9: Graph of $y = 2x + 1$

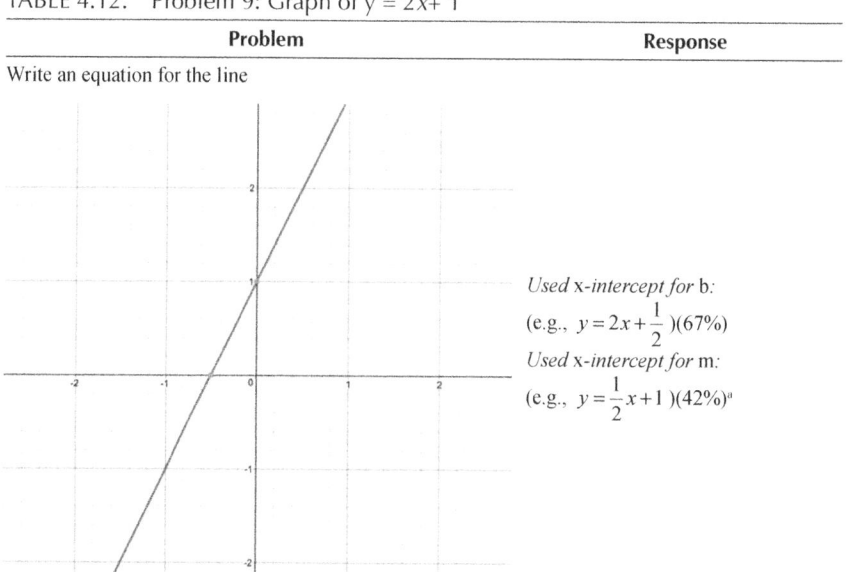

Problem	Response
Write an equation for the line	Used *x-intercept for* b: (e.g., $y = 2x + \frac{1}{2}$)(67%) Used *x-intercept for* m: (e.g., $y = \frac{1}{2}x + 1$)(42%)[a]

[a]Percent of 18 9th- and 10th-grade students in Moschkovich's (1999) study.

either the *y*-intercept value or the slope. Data from research interviews with students indicate that "the use of the *x*-intercept was more than a superficial mistake from which students could recuperate."[70] Students had well-thought-out justifications for their actions, often having a "three-slot-schema" for equations in which they look for three pieces: the slope, *y*-intercept, and *x*-intercept.[71] Students may expect the *x*-intercept to be just as salient as the *y*-intercept in the equation as it is in the graph.

The idea that the value of the *x*-intercept should appear in the equation "$y = mx + b$" is persistent even with instructional intervention.[72] One way to help students change this misconception is to guide them towards understanding the actual relationship between the *x*-intercept and the equation $y = mx + b$. Students should explore the case where $m = 1$, and see the connection between how the changing value of *b* affects the *x*-intercept. Next, students should explore cases where the slope is not 1. As the value of the slope changes, students can see how the shift in *m* relates to the steepness of the slope of the graph, and a shift of the *x*-intercept toward the origin. This may guide students away from seeing the *x*-intercept as a reflection of a change in *b* and seeing it as also depending on *m*.[73]

The Value of *b* in "y = mEl + b"

For many students, the y-intercept may seem like a less complex concept when compared with slope. However, research indicates that it can be challenging for students when they need to translate between different representations. "For example, when students are presented with a line in slope-intercept form, identification of the *y*-intercept is relatively straightforward. However, within a table and a graph, students must manage the value of the *y*-intercept with the *x*-coordinate as well."[74] Students tend to have good intuition about slope because of their life experiences with steepness, but they are not as familiar with *y*-intercept because there are not clear connections with their experience.[75] Students can easily identify the *y*-intercept in the $y = mx + b$ equation,[76] but have a more difficult time understanding its meaning. Researchers suggest that an understanding of *y*-intercept includes four dimensions (see Table 4.13).[77] Each dimension conveys a different aspect of how to think about the *y*-intercept.

TABLE 4.13. Dimensions of Understanding y-Intercept

Symbolic	Graphical	Numerical	Context
Constant term of the equation. Can either be a letter or number.	*y*-intercept. A coordinate of a point on the line.	Value of *y* when $x = 0$.	Meaning of *b* in a real-world context (e.g., 32°F in temperature conversion).

Source: Bardini and Stacey (2006, p. 115)

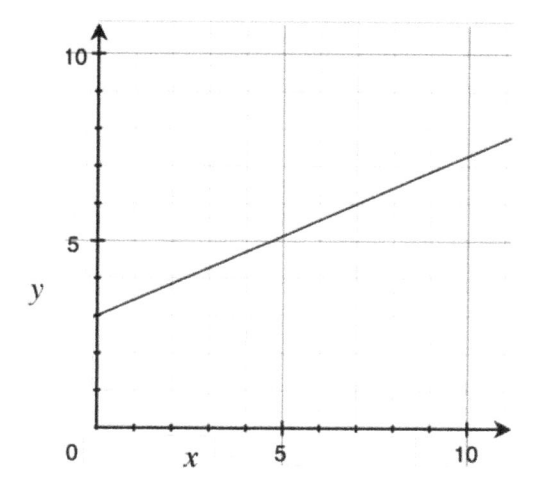

FIGURE 4.16. y-Intercept as Starting Point.

In order to help students understand the connection between the equation of a line and data, the *y*-intercept is often referred to in classrooms and textbooks as the "starting point" or "initial value" of a graph of a line.[78] In Figure 4.16, the line "starts" at (0,3). The Core-Plus Mathematics program even changes the order of the $y = mx + b$ equation to the $y = b + ax$ form to emphasize the *y* intercept as the beginning point of data in a real-world context.[79] This approach may help students understand the context when the data has a limited domain ($x \geq 0$), but also may get them into habits of rote replies to questions, rather than thoughtful consideration of the problem situation. For instance, in Table 4.14, the student states that the *y*-intercept is the "starting price" when it is actually the loss of $450 if they don't sell any tickets. Further, it appears that students do not always understand how the *y*-intercept affects the function, particularly when students think of *b* as the starting point of the graph.[80]

Similarly, confusion is possible when the data is linear to a point. Consider the data in Table 4.15 and the resulting graph in Figure 4.17. Students came up with the line of best fit as $y = -0.1x + 11.7$. What does the *y*-intercept represent in this case? How does a line of best fit make sense when the data points are discrete? The interview in Table 4.16 indicates that students struggle to think about the

TABLE 4.14. Interview for y-Intercept as Starting Point

Teacher:	Profit at the Palace Theater is a function of number of tickets sold, according to the rule $P = -450 + 2.5T$. What is –450 in this case?
Student:	The starting price.

Source: Davis (2007)

TABLE 4.15. TV Audience Data

Rank	Show	Average Weekly Audience (millions of households)
10	*The Voice*, NBC	10.9
20	*60 Minutes*, CBS	9.1
30	*The Simpsons*, FOX	8.4
40	*Chicago Fire*, NBC	7.6
49	*Scandal*, ABC	6.7
59	*Law & Order*, NBC	5.8

Source: Coxford et al. (1998, p. 162)

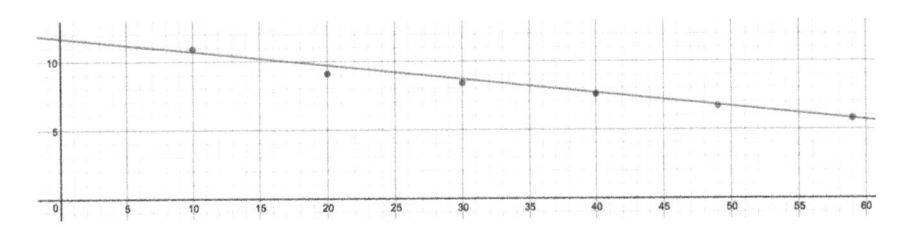

FIGURE 4.17.   Graph of TV Audience Data.

meaning of the *y*-intercept in relation to the real-world data. While *m* will always represent the rate of change, *b* may not have a practical interpretation.[81] It is important that students realize the nature of a line as infinitely progressing in two directions and the *y*-intercept as the point (0, *b*) at which the line intersects with the *y*-axis. Data can often be interpreted as going in either direction (e.g., the more tickets you sell, the more money you make or the less tickets you sell, the less money you make) so the idea of "starting point" may create misconceptions with students regarding direction of data. The discussion of a linear relationship and

TABLE 4.16.   Interview About TV Audience Data

Teacher:	What is the equation for the line of best fit?
Student 1:	Negative 0.1x + 11.7.
Teacher:	What is that 11.7?
Student 2:	Starting point.
Teacher:	Our starting point would be the show that was ranked what?
Student 2:	Zero.
Teacher:	Zero, that would be the best you could ever do. The show that is ranked best is ranked what?
Student 2:	One.

Adapted from Davis (2007)

y-intercept can be complex, depending on whether the data is discrete or continuous, as in the TV audience example. The relationship of the data to the line and *y*-intercept is important for students to reflect on and can be complicated, depending on the problem situation.

Other research indicates that students were more successful graphing slope than *y*-intercept when they were asked to create a graphical representation of a function that was described verbally.[82] Students also tend to begin lines at the origin, possibly "out of a default tendency because they do not understand how to represent a change in *y*-intercept on the graph or because they do not understand the difference between variables being linearly related and proportionally related."[83]

When examining graphs of parallel lines (see Figure 4.18) and thinking about the role of the *y*-intercept, students frequently describe the lines as being horizontal shifts of one another, rather than vertical shifts.[84] A horizontal shift of a graph is a shift to the left or the right (e.g., line *b* moves over 9 in Figure 4.18), whereas a vertical shift of a graph is a shift straight up or straight down (e.g., line *b* moves up 3 in Figure 4.18). This conception is not incorrect, because lines translate infinitely—they are both vertical and horizontal translations of one another. In fact, some students see the movement as diagonal, or a simultaneous movement of both horizontal and vertical.[85] Dynamic computer tools (e.g., *Desmos*) and apps (e.g., *Math Flyer*) allow students to manipulate values for *b* in the $y = mx + b$ equation to see how they affect the graph. By visual observation, the result could easily be interpreted as a horizontal or diagonal, rather than vertical, change. However, the standard representation of lines is in the form $y = mx + b$, so it is more helpful for students to see parallel lines as vertical translations of one another. In particular,

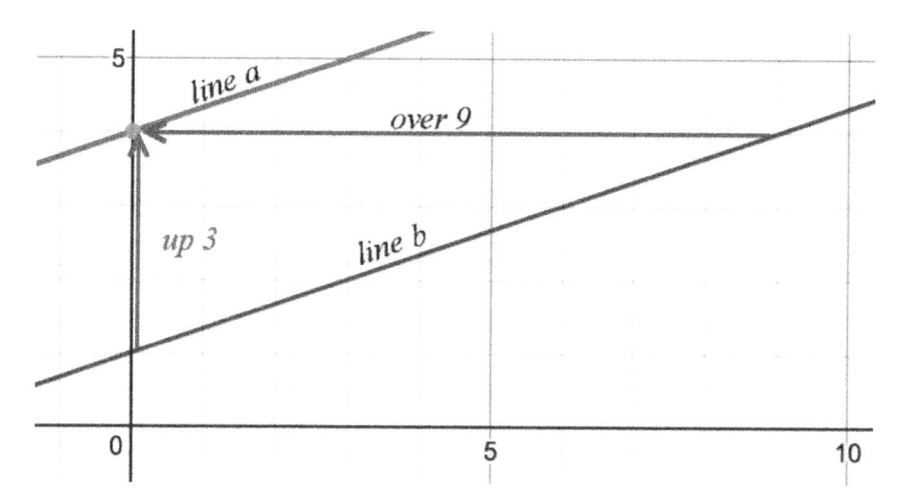

FIGURE 4.18. Graph of Parallel Lines.

the b value represents exactly the amount that the line has been shifted vertically. In the form $y = mx + b$, a change in b shifts the graph horizontally $-\frac{b}{m}$ units, but the horizontal shift is impacted by both the b value and the slope of the line.

Technology can play an important role in helping students see the connection across representations, particularly between graphs and equations. Graphing calculators have historically been a useful tool for allowing students to problem solve, develop conceptual understanding, quickly graph multiple functions on the same set of axes, and consider patterns.[86] However, students and teachers often struggle to understand how to use the devices effectively.[87] The new-generation mathematical apps are much more dynamic and allow for seeing hundreds of variations in seconds. Apps mentioned earlier, such as *Desmos* and *Math Flyer*, use sliders to allow students to vary the values of m and b, and explore how the graphical representation of the function is influenced by the coefficients in the algebraic equation. The Center for Algebraic Thinking's *Linear Model* allows students to manipulate a table, equation, or graph to see changes in the other two simultaneously. The app *Lion Grapher* is similar to the former program *Green Globs*, in which players must adjust the $y = mx + b$ equation to make a line go into a lion's mouth. These apps encourage students to think about how manipulations of values for equations of a line will influence the path of lines.

Teachers

- Place a greater emphasis on qualitative graphs when helping students make initial connections with algebraic symbolism, particularly with the meaning of the y-intercept.
- Use a range of discrete and continuous data as a resource for identifying algebraic relationships and discuss the nature of the data in relation to the algebraic symbolism.
- Be careful with using informal terminology such as "starting point" when helping students learn the idea of y-intercept, as misconceptions may develop about the nature of the data and relation to the algebraic symbols.
- Help students understand how the slope of a line is related to the x-intercept.
- It is important to honor how students see and understand changes in graphs, connect that perspective to the algebraic relationships, and discuss efficiencies of perspective such as seeing line movement as vertical change because b is easily isolated in $y = mx + b$.

UNDERSTANDING SLOPE

Common Core Standards: 6.EE.9, 8.F.5

"Understanding and graphing slope requires that a student be able to track two variables and the nature of their co-varying relationship."[88] Slope describes both the direction (increasing or decreasing) and steepness of a line, and is the visual representation of rate of change. On one hand, slope may be easy for students to comprehend because of their experience with the steepness of a hill or ratios such as two oranges for a dollar. On the other hand, having to consider at least two different points and the relationship between them can be challenging for students. Many students think of slope as a geometric ratio—"I go up this much and I go over this much." Or they may have a limited algebraic understanding of slope—simply using the linear equation $y = mx + b$ to identify m as the slope. Generally, there is quite a bit of research discussing students' difficulties understanding slope.[89]

Researchers suggest that there are four dimensions of understanding the concept of slope (see Table 4.17). A *symbolic* understanding simply regards where m resides in the linear equation, as the coefficient of x. A *graphical* understanding is one of visual interpretation: that m communicates the steepness of the line. A *numerical* understanding involves the computation of the change in the two y values of two points, over the change in the two x values. Lastly, understanding of the slope in *context* is realizing the meaning of m as the rate of change, which is a ratio of two values such as 27 miles per one gallon. This framework was extended by other researchers to include all the concepts in Table 4.18.

Students are generally successful in identifying which term is slope in an equation, demonstrating symbolic understanding (see Table 4.19, Problem 10), although they have less success when it is not written in $y = mx + b$ form (see Table 4.19, Problems 11 & 12) or when m is written as a decimal.[90] It is interesting to note that so many students opted for the ratio answers, likely because they have extensive experience thinking about slope as rise over run and struggle to think of a number such as 3 as $\frac{3}{1}$.

Students are less successful in determining whether two lines might be parallel (Determining property). Consider Table 4.20, Problem 13, in which students are asked to decide whether PQ is parallel to the line $y = 3x - 5$. Students had

TABLE 4.17. Dimensions of Understanding Slope

Symbolic	Graphical	Numerical	Context
Coefficient x in $y = mx + b$. It can be a letter or a number.	Gradient of the graph $y = mx + b$. The slope of a line.	The ratio $\frac{\Delta y}{\Delta x}$	The meaning of m in a real-world context (e.g. a speed or rate of pay).

Source: Bardini & Stacey (2006, p. 115)

TABLE 4.18. Conceptions of Slope

	Conceptions of Slope
Geometric ratio	Slope as rise over run, vertical change over horizontal change as a geometric property.
Algebraic ratio	Slope as the change in y over the change in x in an algebraic formula, such as $m = \dfrac{y_2 - y_1}{x_2 - x_1}$ or $m = \dfrac{\Delta y}{\Delta x}$.
Physical property	Slope through words such as "slant," "steepness," "incline," "pitch," and "angle."
Functional property	Slope as the rate of change between two variables, representations including tables and verbal descriptions (e.g., when x increases by 2, y increases by 3).
Parametric coefficient	Slope referenced by m in the equation $y = mx + b$, the coefficient of x.
Real-world situation	Slope in a static or physical situation, such as a wheelchair ramp or dynamic, functional situation (e.g., distance vs. time).
Trigonometric conception	Slope as the tangent of the angle of inclination.
Calculus conception	Slope as it relates to the derivative, tangent line to a curve at a point.
Determining property	Slope used to determine parallel and perpendicular lines, or a line if given a point and slope.
Behavior indicator	The number for slope indicates increasing versus decreasing, as well as magnitude of steepness.
Linear constant	Slope as a constant property, regardless of representation, unique to "straight" figures, a constant property independent of representation.

Source: Moore-Russo, Conner, & Rugg (2011, p. 9), Stanton & Moore-Russo (2012, p. 271)

considerable difficulty, with less than 10% giving an accurate answer, saying the lines weren't parallel because "one is longer than the other" or "because the line is diagonal" and saying the lines were parallel "because they both go together" or "because the lines have 3 marks on the y axis all the way along."[91]

One common error found in the research is slope-height confusion.[92] Students frequently don't know whether information from the slope or the height of the graph is useful for solving a problem. For example, in Problem 14a (see Table 4.21), students were asked to recognize that the slopes of the lines represent the speed of the objects and Line A rises more steeply than Line B, so Object A is traveling faster than Object B. In one study, 79.6% of 7th- and 8th-grade students answered this type of question incorrectly.[93] Of those who answered incorrectly, researchers found that many students chose the higher line "due to failure to realize that information about the velocity cannot be extracted from the height."[94] Students also tended to state that the two objects had the same speed at $t = 7$, not realizing that the two lines have different slopes, so they would never have the same speed.[95] Some students simply made graph reading errors or had reasonable

Analysis of Change (Graphing) • **199**

TABLE 4.19. Linear Equation Problems

Problem	Response[a]
10. The slope of the line of $y = 3x + 2$ is:	a. 3 (65%)
	b. 2 (11%) ✓
	c. $\frac{3}{2}$ (22%)
	d. -2 (1%)
11. The slope of the line $y = 2 - 3x$ is:	a. 3 (8%)
	b. 2 (14%)
	c. $-\frac{3}{2}$ (16%)
	d. -3 (56%) ✓
12. The slope of the line $2y = 6x + 5$ is:	a. 6 (14%)
	b. $\frac{6}{5}$ (11%)
	c. 2 (4%)
	d. 3 (52%) ✓

[a]Percent of 252 12th-grade students in Barr (1981).

TABLE 4.20. Problem 13: Parallel Lines Graph

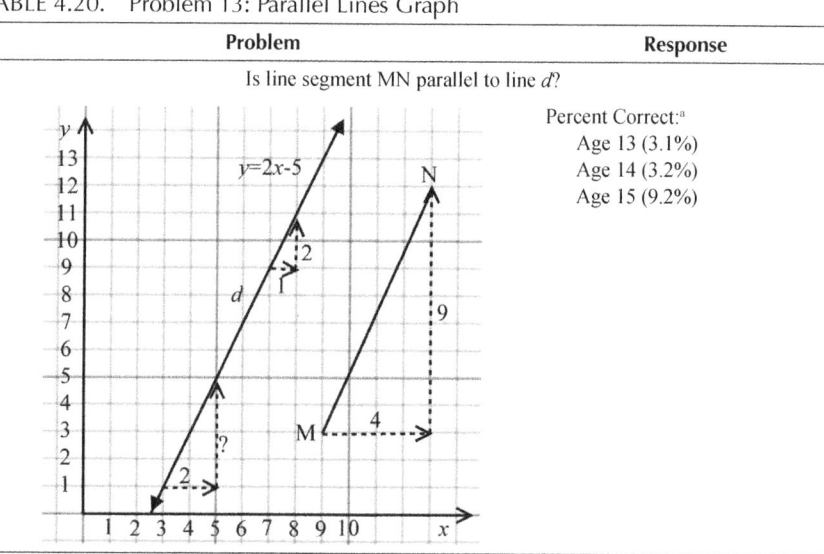

Problem	Response
Is line segment MN parallel to line *d*?	

Percent Correct:[a]
Age 13 (3.1%)
Age 14 (3.2%)
Age 15 (9.2%)

[a]Percent of 1798 13- to 15-year-old students in Hart et al. (1981, 2004)

TABLE 4.21.   Problem 14: Position Versus Time

The position versus time graph shows the motions of two objects, A and B, moving along the same path.

a) At the instant $t = 2s$, is the speed of Object A greater than, less than, or equal to the speed of Object B? Explain.

b) Do Objects A and B ever have the same speed? If so, when? Explain.

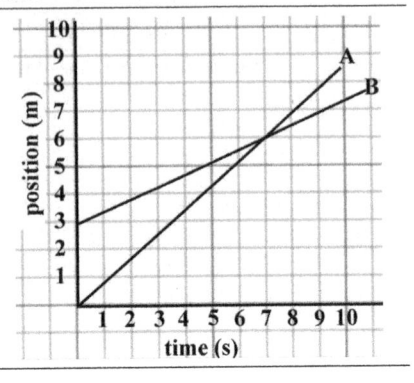

Based on McDermott, Rosenquist, van Zee (1987)

but faulty logic for their choice, rather than having held any misconceptions about the "picture," so it is important for teachers to probe students' thinking.[96] Other research indicates that this confusion can be effectively addressed by instruction, particularly with technology.[97]

Slope can represent real-life situations such as the steepness of a hill, or slope can measure the rate of change of two varying quantities. When dealing with physical situations of steepness, students rarely use proportional reasoning (see **Rate and Proportional Reasoning**), comparing the height of an object to the base of the object. Only 22.7% of students in a high school physics class used a ratio to explain their thinking for Problem 15 (Table 4.22).[98] Instead, students focused on angles ("Measure the angle at the top of the ramp") or a single attribute, such as the length of the ramp ("I would find the length of the hypotenuse") or used whatever they knew ("One-half base times height").[99]

Students are often taught slope as a fraction, with the change in y over the change in x. With only this instrumental understanding of slope, students may have a difficult time with the way a line is positioned in a plane or the idea of rate of change.[100]

TABLE 4.22.   Problem 15: Ski Ramp

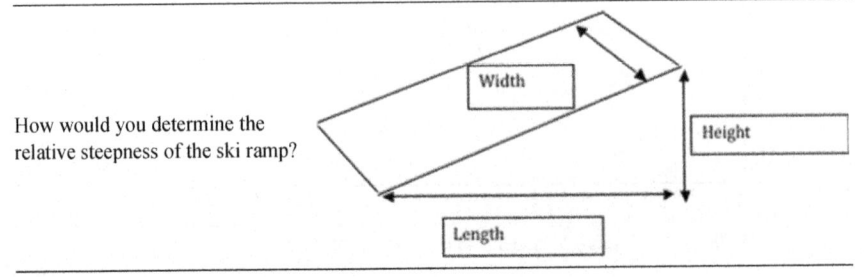

How would you determine the relative steepness of the ski ramp?

x	y
0	0
1	3
2	6
3	9
4	12

(+3 between each y value)

×3

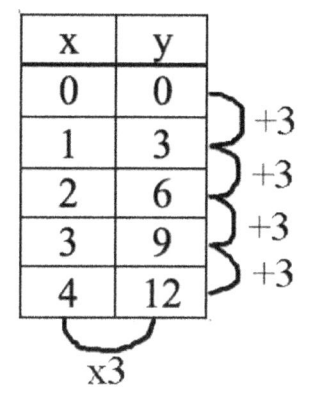

FIGURE 4.19. Additive and Multiplicative Relationships.

They may struggle to understand the idea that the slope is a ratio or rate of change (functional relationship) and to think of ratio as a measurement of steepness (physical relationship).[101] For instance, when the slope is written as a decimal ($y = 3.4x + 2$) students have difficulty thinking of slope as a rate of change.[102] Researchers find that students tend to have a better understanding of slope as a rate of change than as a measure of steepness.[103] These variations on the idea of slope are based in students' understanding of fractions, decimals, and percentages; they are more likely to use the idea of rate or ratio when dealing with symbolic equations than in real-world situations. Students are most successful in calculating rate as a function of time, which in many ways is the most intuitive form of rate, rather than measures of steepness.[104] If a student knows that they have traveled 200 miles in four hours, they are able to identify the rate as 50 miles in one hour.

An important reason for students to understand slope as rate is to enhance their understanding of the relationship between the points on a line. Students can look at the relationship between the points in Figure 4.19 in two ways; first, in an additive relationship in which each x coordinate is 3 more than the previous one; and second, that the x coordinate is 3 times the y coordinate. Research indicates that as students gain understanding of slope, they may have difficulties recognizing multiplicative relationships.[105] This is evident from Problem 16 in Table 4.23, as students struggle to understand the relationship between the points in order to determine additional points. There is also quite a bit of research indicating that students don't understand that there are infinitely many possible ordered pairs making up a line.[106] In Problem 17, less than 20% of students understand that lines contain an infinite number of points.[107]

One way to encourage students to develop an understanding of slope, and particularly the multiplicative relationship between the infinite number of points on a line, is the Center for Algebraic Thinking's app, *Point Plotter* (see Figure 4.20).

202 • HOW STUDENTS THINK WHEN DOING ALGEBRA

TABLE 4.23. Problems for Points on a Line

Problem	Response[a]
16. Fill in the table so that the points with coordinates (x, y) all lie on the same line. <table><tr><th>x</th><th>y</th></tr><tr><td>1</td><td>3</td></tr><tr><td>2</td><td></td></tr><tr><td>3</td><td></td></tr><tr><td>4</td><td>15</td></tr></table>	Correct: 33% Incorrect: 66%
17. Draw a line through the points (2,5), (3,7), and (5,11). How many points do you think lie on the line altogether?	Some finite number (e.g., 4, 5, 8) (63.8%, 69.2%, 51.6%) "Hundreds," "lots," etc. (12.6%, 5.1%, 9.7%) "Infinite," "as many as you like," etc. (6.7%, 6.2%, 19.6%)

[a]Percent of 1789 13-, 14-, and 15-year-old students in Hart, et al. (2004), respectively.

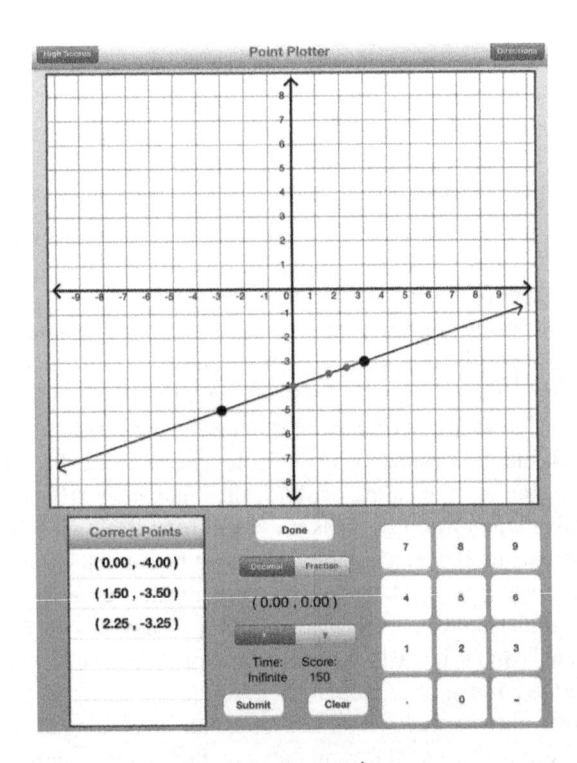

FIGURE 4.20. Point Plotter App.

In this game, students must determine as many points as possible between the two given points in a timed or untimed version. The "easy" points at the grid crossings are quickly done and leave students to figure out the relationship of the points between points. In order to be successful in this game, students need to develop an understanding of the slope and the multiplicative relationships between the points to determine new points on the line. As students use strategies such as taking half the distance between two points (developing/using the midpoint formula) or coming up with the slope and equation of a line to calculate new points, they come to realize the infinite possibilities.

Teachers

- There is often a disconnect between being able to calculate slope and actually knowing what slope represents, so teachers need to ask students to explain the meaning of the slope in a given situation. One way to prompt student thinking is to ask, "What does the slope represent in the context of this situation?"
- Researchers suggest that, when preparing lessons to teach the concept of slope to students, a teacher should reflect on the following questions:[108]
    a. What representations of slope are in my concept definitions?
    b. Do I have a flexible understanding of representations for slope?
    c. What representations of slope do I use?
    d. What real-world situations do I use to illustrate the concept of slope?
    e. What is my knowledge of student difficulties with slope?
- Both the *Submariner* and *Lion Grapher* apps from the Center for Algebraic Thinking create a context for students to think about how to manipulate the slope and y-intercept of a linear equation to achieve a goal. As students play either game, they develop an intuitive understanding of the steepness and direction of a line by changing the slope and/or y-intercept.
- One activity to promote an understanding of steepness as a ratio is to have students physically measure real-world handrails, sidewalks, and ramps. Models of ramps that have slopes that are gentle, moderate, and steep can also be used. Compare the ratio of the rise of the ramp with the run. Students should realize that the ratio with the greatest value relates to the ramp with the steepest slope.[109]

GRAPHS AS LITERAL PICTURES

Common Core Standard: 8.F.5

To understand graphing, students must understand that a coordinate graph is a visual representation of a numeric relationship between two changing quantities. Students' understanding and experience of reality has a strong influence on their thinking as they examine graphs. There is a strong tendency for students to view graphs as pictures[110] and this tendency persists into college.[111] Often, they incorrectly interpret the interaction between the axes because their experience leads them to interpret the image as a representation of the reality visually, rather than information from the action. Researchers refer to this phenomenon as *iconic translation*.[112] Instruction, particularly with computers, can effectively help students understand the limitations of that perspective.[113]

To illustrate this idea, Graph 1 in Figure 4.21 displays the data for a ball thrown vertically in the air that comes straight back down right where it left. Students, however, often see the graph as a ball thrown from one person to the other.[114] In Graph 2, at $t = 4$, many students explain that Car A is passing Car B.[115] Multiple researchers have studied a "bicyclist on a hill" problem and asked students to graph speed versus position or time. Up to half of students drew the hill in response.[116] Even when students are given the detailed graphic in Figure 4.22 and asked to graph the speed versus distance, half of the students drew the hill.[117] Many students tend to draw the movement paths of objects, rather than the distance versus time graph.[118] In another study, students were given Problem 18 in Table 4.24. Based on the rate of correct responses, it is clear that students had difficulty separating the dynamic between the axes and the "picture" of the situation. Students' explanations included that the person was (a) "going east, then

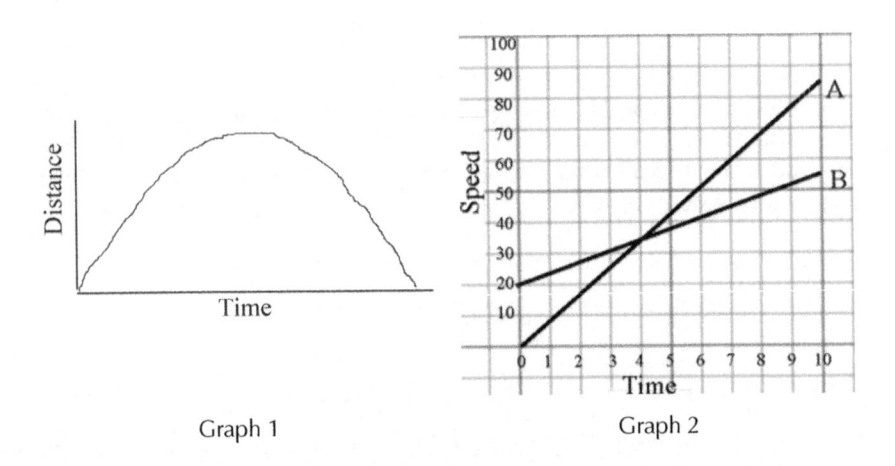

Graph 1

Graph 2

FIGURE 4.21. Iconic Translation Graphs

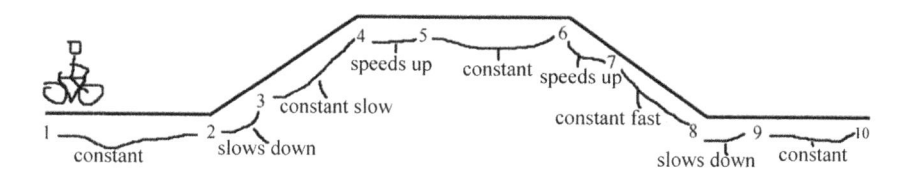

FIGURE 4.22. Bicyclist on a Hill.
Source: Barclay & TERC (1986)

due north, then east," "went along a corridor, then up in a lift, then along another corridor;" (b) "going along, then turning left," "going back the way he came;" and (c) "climbing a mountain," and "going up, going down, then up again."[119]

There are two particular technological tools that may help students make the connection between reality and the data on a graph. First, computer-based motion detectors provide an opportunity for students to have a physical, experiential connection to the data regarding position versus time that appears on a graph (see Figure 4.23). Motion detectors can be used with moving objects as well as with moving students; activities can include matching a pre-made graph or exploring the interaction between movement toward and away from the detector and the resulting graph. Research suggests that motion detectors encourage students to evaluate their perspectives regarding the connection between their bodily movement and what appears on a screen.[120] Conversation about their experience and connecting their bodily motion with mathematical symbolism can enhance learning.[121] Also, due to interaction with a motion detector, students' abilities were significantly more developed in interpreting, modeling, and transforming graphical data.[122]

The second helpful tool is the Center for Algebraic Thinking's *Action Grapher* app (Figure 4.24) that focuses directly on the difficulty students have in under-

TABLE 4.24. Problem 18: Graphs of Potential Journeys

Problem	Response
Which of the graphs below represent journeys? Describe what happens in each case. Why do you think that?	Correct Responses[a] a) 9.5%, 8.4%, 15% b) 11.1%, 9.3%, 15.7% c) 14.7%, 17.2%, 25.2%

[a]Percent of 1798 13-, 14-, and 15-year-old students in Kerslake (2004), respectively.
Source: Kerslake (2004)

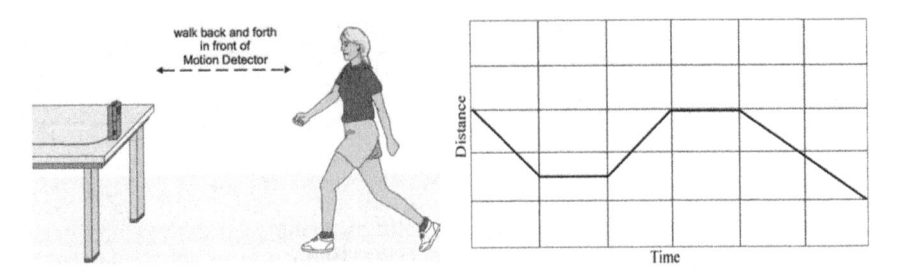

FIGURE 4.23. Motion Detector Activity With Pre-Made Matching Graph. From http://www.vernier.com/

standing the connection between a graph and the motion of a bicyclist on a hill. The student takes a bike on a journey up and down a hill and sees three different graphical representations of what is going on, based on elevation, speed, and distance versus time. First, the student sketches what she thinks each of the graphs might look like. Next, she plays the animation to see how her hypothesis compares to each of the actual graphs. The student can change the terrain when ready to test new hypotheses.

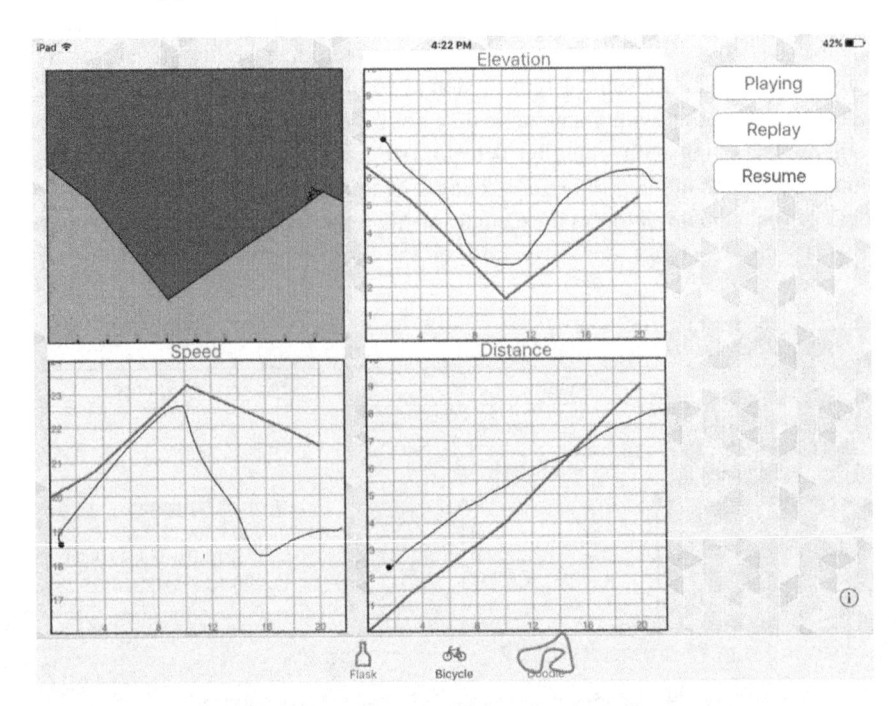

FIGURE 4.24. Action Grapher App: Biker.

Teachers

- Use computers whenever possible to develop students' understanding of interpreting graphs. Graphs can be formed instantaneously as students create, input, or manipulate data, as graphs are less likely to be seen by students as "static pictures and more likely to be seen as involving dynamic relationships."[123]
- Instruction may include tasks in which students collect their own data, create a graph, and are then asked questions about the graph they created (e.g., Where on the graph does it show...?).
- Consider having students start learning to read graphs within contexts that are familiar to them, and then move toward traditional curricular topics once internal graph reading competency has been established. For example, students can collect data on the drop height and bounce height of a ball, or height of a plant over time.

INTERPRETING GRAPHS

Common Core Standards: 8.EE. 5 & 6, 8.F. 1, 2, 3, 4, & 5

"A graph is useful only to those who know where to focus their attention."[124] Information often comes to us in the form of graphs, whether it is about which ketchup pours more slowly or how much our mortgage will cost us over time. "Our society is becoming more reliant on the representation of information in graphical forms as traditional communication and literacy demands change and adapt to what could be considered a burgeoning information age."[125] We need to interpret information in order to be entertained or informed. This ability to interpret is sometimes termed *graph sense* or *graphicacy* (based on "graphical literacy"), meaning the skill to read graphs and learn from that data.[126] The skills required for graph sense are:[127]

1. Recognizing components of graphs, the interrelationships among these components, and the effect of these components on the presentation of information in graphs.
2. Speaking the language of specific graphs when reasoning about information displayed in graphical form.
3. Understanding the relationships among a table, a graph, and the data being analyzed.
4. Responding to different levels of questions associated with graph comprehension, or more generally, interpreting information displayed in graphs. The three levels of questioning are:
   a. extracting data from a graph,
   b. interpolating and finding relationships in the data as shown on a graph, and
   c. extrapolating from the data and interpreting the relationships identified from a graph.
5. Recognizing when one graph is more useful than another on the basis of the judgment tasks involved and the kind(s) of data being represented, and considering both the nature of the data and the purposes for analysis.
6. Being aware of one's relationship to the context of the graph, with the goal of interpretation to make sense of what is presented by the data in the graph.

Many students have difficulty interpreting graphs. Consider the problems in Table 4.25: students can generally find a specific point on a graph and plot points, so many students were able to respond to Problem 19 with a date for questions (a) and (b).[128] However, when it came to questions (c) and (d), students had significant difficulty explaining about growth and the relationship of this graph to a linear equation. Similarly, on Problem 20, many students struggled to interpret the dynamic between the two axes and the resulting graph.

TABLE 4.25. NAEP Problems on Graphing

Problem	Response[a]

Problem 19

HOUSEHOLDS WITH TELEVISION SETS IN THE UNITED STATES

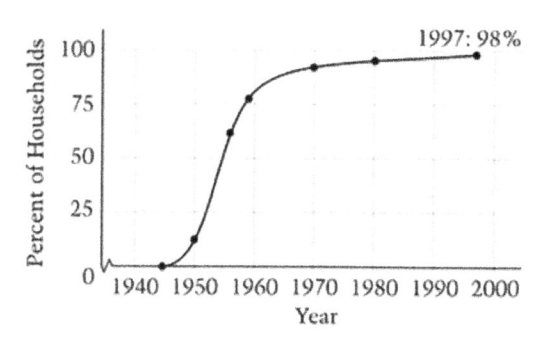

[Graph showing Percent of Households vs Year from 1940 to 2000, with an S-shaped curve reaching 1997: 98%]

(a) In which year were television sets first available for purchase in stores?
(b) In which year did 50 percent of the households own at least one television set?
(c) Write a sentence or two comparing growth in the percentage of households with television sets across the three decades of the 50s, 60s, and 70s.
(d) The points (1950, 10) and (1970, 90) lie on the graph above. They are also solutions of $y = 4x - 7790$. However, if the graph of $y = 4x - 7790$ were drawn for 1940 to 1997, it would not look like the graph shown. Explain why not.

Satisfactory: 10%
Incorrect: 21%
Minimal: 35%
Partial: 29%
Omitted: 6%

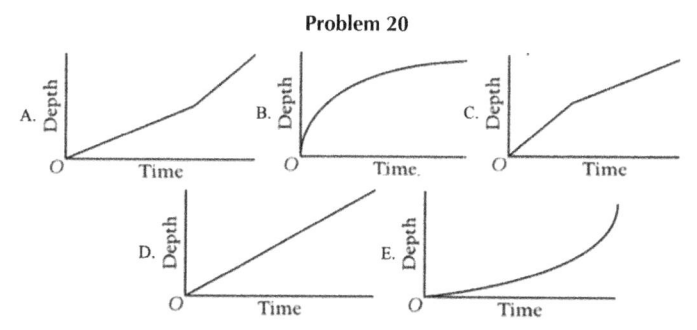

Problem 20

[Five graphs labeled A–E, each plotting Depth vs Time with different shapes]

Martine is filling a rectangular fish tank using two hoses that fill the tank at the same flow rate. When the tank is about half full, she turns off one hose but does not change the flow rate of the other hose. Which of the following graphs best represents how the depth of the water in the tank changes over time?

A) 10%
B) 18%
C) 52% ✓
D) 15%
E) 4%

[a] Percent of students responding on NAEP (2013)

TABLE 4.26. Levels of Graphical Information Processing

Level 1: Explicit
Superficial reading that focuses on the details of the graph such as title, type and names of variables, type of phenomenon represented. This type of processing does not involve interpreting meaning for the data represented.
Level 2: Implicit
This involves interpretation of the graph beyond the reading of discrete elements of the graph. It emphasizes the translating of information that is present into symbols. It involves knowing certain "codes" that the reader must have mastered. It involves going beyond the reading of coordinate points in isolation and includes noticing patterns between the values.
Level 3: Conceptual Information
This level emphasizes the relationships between the variables and also draws on prior knowledge to interpret the phenomenon being represented. Students base the conceptual information on analyzing the structure of the graph and use their knowledge of the content to produce interpretations, explanations, or predictions.

Source: Based on Postigo & Pozo (2004)

Research describes different stages of student learning about interpreting graphs. One study proposed three levels of analysis that explain the way students read or process graphical information (see Table 4.26).[129] Although these three levels of processing are increasingly more complex, they should not always be considered sequential. The way students respond to a graph will also depend on the questions or tasks they are given in association with the graph. The interviews in Table 4.27 provide examples of student thinking within these three levels. Student 1 (Explicit) focuses on the visual details of the graph, making no interpretation. Student 2 (Implicit) makes some observations about the motion of the graph and translates that into information, but does not extend that direct translation. Student 3 (Conceptual) makes connections with outside knowledge (e.g., linear relationship) to help interpret the graph.

TABLE 4.27. Interviews Indicating Three Levels of Graphical Information Processing

Interviewer:	What can you tell me about this graph (Figure 4.25)?
Student 1:	Time is on the x axis and Distance is on the y axis. It has straight lines and they go up. (Explicit)
Student 2:	It appears that this object traveled at a constant rate over time, stopped briefly, continued at the same rate as before, stopped again for a short time and then continued at the same rate. (Implicit)
Student 3:	This is basically a linear relationship, meaning the time and distance seem to remain constant except for the two places there is no distance being covered for a short amount of time. Whatever the "object" is, it appears to be able to go from 0 to its travel speed instantaneously since I don't see any place on the graph indicating a speeding up or slowing down. (Conceptual)

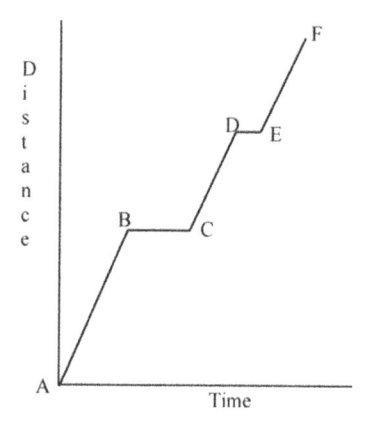

FIGURE 4.25. Problem for Interview on Levels.

Based on these three levels, teaching graph comprehension might take place at three successive stages.[130] At the *initial stage,* students should be able to extract information from the data, or "read the data." Teacher questions such as "How many cars were sold in 1980?" is an appropriate cue for this level. To be successful at the initial stage, students must understand the conventions of the graph design, including the scale and labels.

At the *intermediate stage,* students find relationships within the data, or "read between the data." Teacher questions to encourage this type of thinking might include condensing data categories (such as "How many boxes have more than 30 raisins?"), identification of trends in parts of the data ("What is the relationship between car sales and engine size between 1970 and 1985?"), and observing relationships in visual displays without reference to the contextual meaning ("How do the changes in these two curves compare?"). At the intermediate stage, students must be able to make comparisons and perform the appropriate computations.

The *final stage* "reads beyond the data," and requires interpolation/extrapolation such as prediction or generation. Teacher prompts to develop this type of thinking might include reduction of the data to a single statement ("From June 15 to June 30, what was the trend in the value of stock X?"), extending the data to form predictions ("If students opened another box of raisins, how many might they expect to find?"), and using the data as evidence to support arguments ("If this graph was offered as a piece of evidence to prove true the statement, 'The economy is improving,' how would you describe the connection between the graph and the attempt to prove the statement true?"). To be successful in the final stage, students must also be able to relate the data to the context of the situation.

An important aspect of graph interpretation is the ability to understand the connection between an equation and a visual representation. Researchers call this the *Cartesian Connection*: that each point on a line represents an ordered

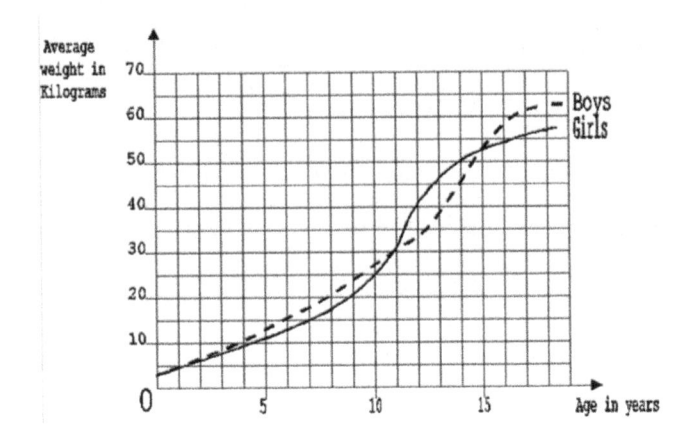

FIGURE 4.26. Average Weight of Boys Versus Girls.
Hadjidemetriou & Williams (2001)

pair that satisfies the equation representing the line.[131] If the equation of the line is $y = 3x + 2$, for instance, the point on the line (1, 5) satisfies the equation because the ordered pair makes the equation true: $5 = 3(1) + 2$. "Understanding both the equation-to-graph and graph-to-equation connections is considered fundamental in developing the flexibility to move fluently among representations."[132] In a study looking at both directions of process, the majority of students had particular difficulty moving from the graph to the equation—not understanding that any point on the graph would be a solution to the equation.[133] Students' difficulties in this area seem to be caused more by instructional rather than cognitive factors.[134] Teachers can help by moving back and forth between algebraic and graphical representations so that students can clearly understand the connection.[135]

Another area in which students struggle when interpreting graphs is understanding when a point or an interval is the most useful information.[136] For instance, students were asked to determine whether boys or girls are growing faster in Figure 4.26, and an example of their responses is in Table 4.28. In this conversation, it is clear that the student believes that the highest point (see discussion of slope-height confusion in **Understanding Slope**) is what is important in answering the question. She fails to understand that the rate of change at that point is what will answer the question, which requires her to look at the interval around that point. "Instead of identifying the largest increase, students identified the largest value."[137]

When studying students' eye movement as they worked on graph problems, researchers found that the format of the graph (line, bar, pie, etc.) also had an influence on students' understanding and the way they interpreted and inferred information,[138] applying different types of cognition to different kinds of graphs.[139]

TABLE 4.28. Interview for Average Weight of Boys Versus Girls Graph

Student:	Em, I wrote Girls.
Interviewer:	Why did you write girls?
Student 1:	Because they weigh more than boys and this means that they could be taller than ... they could be eating more.
Interviewer:	Ok. So, she is saying that "girls" because they weigh more...
Student 1:	Yeah, they weigh more at that point. [Indicating where the boys and girls are 14.]
Interviewer:	At that point.
Student 2:	But the gradient is steeper.
Student 1:	Yeah, I know, but it is not an area, it is just for one point.

Source: Hadjidemetriou & Williams (2001)

The visual characteristics, as well as the context for the information, impacted students' ability to make inferences. Students are also often confused about when data is discrete or continuous.[140] They have difficulty in determining the meaning of a line graph, often only "seeing" the marked points (see Figure 4.27).[141] In one large study, for Problem 21, many students said that there were no points between two marked points, while others explained that there was just one, probably meaning the midpoint (see Table 4.29).[142] In another study, students believed that points had some mass or size rather than being an abstract entity, which may lead them to believe they can only "fit" so many in the space.[143] They often connect discrete points when it is not appropriate, indicating that they struggle to understand what contexts dictate what type of data.[144]

An integrated approach that emphasizes both the mechanical interpretation of graphs *and* the power of graphs as a model for studying relationships between variables is necessary. Student difficulties appear to arise because variable relationships presented symbolically are too abstract for most students in the 11–13 age range.[145] Students should be involved in the entire process—gathering the data, creating the graph, and analyzing variables. It is possible to teach graphing skills in isolation if all we expect is for students to perform algorithmically. However, if we want them to interpret visual data in a way that indicates they comprehend meaning and subtleties, then that requires understanding the whole process.

TABLE 4.29. Number of Points Between Two Points

Problem 21	Response[a]
How many points lie between two points?	None (15.4%, 17.7%, 13.1%)
	One (36.8%, 41.8%, 34.4%)
	Some finite number (e.g. 4, 5, 8)(27.5%, 22.1%, 21.3%)
	"Infinite," "as many as you like," etc. (4.1%, 4.0%, 11.1%)

[a] Percent of 1789 13- to 15-year-old students in Kerslake, 2004.

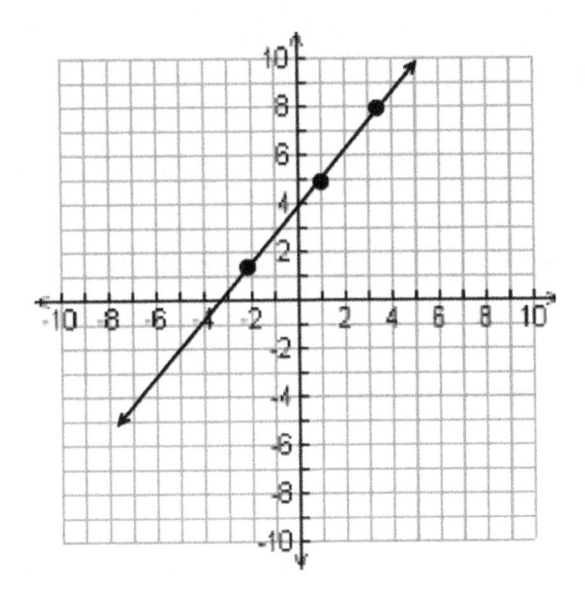

FIGURE 4.27. Line Graph.

Lastly, one of the questions addressed by researchers is whether teachers should teach graph interpretation or construction first.[146] Some argue that the initial emphasis should be on analyzing graphs because "If you teach design before data analysis it is harder for students to understand why design matters."[147] Others argue that, since they "are interconnected, there can be no one simple linear sequence for learning and teaching."[148] Ultimately, the answer is likely teaching both at the same time, as it is important to highlight aspects of graph construction as you work with students to interpret the visual representation of data.

Teachers

- Graphs presented with only qualitative features (i.e., no numerical values such as Problem 20 in Table 4.25) might help students draw on their common sense and reality-checking strategies.[149]
- Research indicates that spreadsheets have power in helping students understand the connection between data and their graphs.[150]
- Students should be involved in both graph construction and interpretation right from the beginning in order for them to make sense of graphs.

UNDERSTANDING SPEED AS RATE OF CHANGE

Common Core Standards: 6.EE.9, 8.F.5

Speed is commonly used to introduce students to the idea of rates of change. It is the quantification of motion—"the proportional correspondence between distance and time."[151] Students have experience with speed that they can bring to the context of a problem to help them interpret situations,[152] but that experience does not always help them understand the concepts.[153] It is sometimes difficult for students to understand that speed is a measure of the rate of change of two varying quantities: time and distance. Students' experience and thinking about speed is a resource that can be improved upon.

At the beginning of their understanding of speed, students often conceive of speed as distance.[154] Prospective teachers have been found to have the same confusion.[155] For example, a speed of 60 feet per second is understood as a chunk of 60 feet that can be traveled in 1 second. Although this is not incorrect, it only allows students to reason in one direction: given a speed and an amount of time, the student can add up "speed-lengths" or "rate-lengths" to find the distance. However, in order to find the amount of time it would take to cover a given distance, the only method available for students with that mindset would be to break up the distance into speed-lengths and count the number of speed-lengths in that distance.

For example, in one study, a student said "the turtle will always be 10 feet behind the rabbit" when the turtle was moving at 20 ft/sec and the rabbit at 30 ft/sec.[156] The student was thinking in terms of speed-lengths, in which distance was the primary quantity to consider and time only existed implicitly (see left side of Figure 4.28). That is, the student counted segments of 30 feet as she computed how long it would take to travel 100 feet. The speed-lengths accumulate and time follows. This way of thinking was problematic when she encountered a problem in which she needed to determine the speed with which the rabbit would go a given distance in a given amount of time (see right side of Figure 4.28). A better

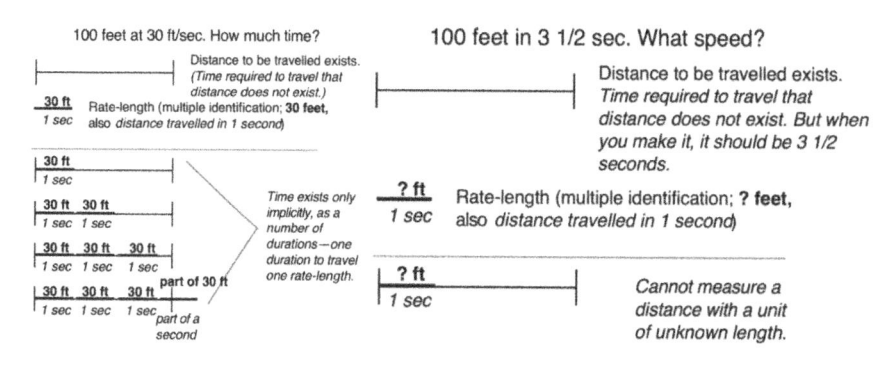

FIGURE 4.28. Rate-Lengths in Relation to Speed.
Source: Thompson (1994, p. 49)

216 • HOW STUDENTS THINK WHEN DOING ALGEBRA

understanding of speed includes comprehending that there is a direct proportional relationship between distance and speed. Partitioning the total time to travel a given distance implies a proportional partition of the distance traveled, and vice versa.

Researchers suggest that the development of an understanding of speed may follow along this trajectory for some students:[157]

- Speed as a distance and time as a ratio;
- Speed-lengths accumulate, time follows;
- Distance and time accumulate simultaneously; and
- Speed as a rate.

This comprehension might then be extended to include concepts of instantaneous speed, average speed, and acceleration for a more sophisticated understanding of rates of change.

In one study, 60% of prospective middle and elementary teachers were confused about the difference between instantaneous and average speeds.[158] Instantaneous speed is the rate of change of the object at a point, while average speed is the rate of change over an interval. In thinking about average speed, it is easiest to think in terms of one person making a trip at varying speeds and a second person making the same trip at a constant speed, such that they would tie.

In regard to acceleration, a student in one study, Carl, was asked to consider the graph in Figure 4.29. Based on some questioning, it was evident that Carl was operating on two main principles:

- *The acceleration principle:* If the green car is behind the blue car and is accelerating very rapidly while the blue car is not accelerating at all, then the green car will get closer to the blue car.

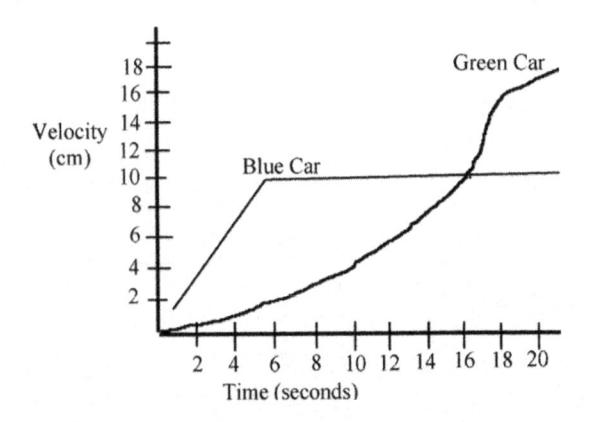

FIGURE 4.29.   Velocity Versus Time Graphs for Two Cars.
Source: Hale (2000) from Monk (1994)

- *The speed principle:* If the blue car is ahead of the green car at time t_0 and the blue car is going faster than the green car, then the blue car will stay ahead of the green car and will actually get farther ahead of the green car.

These are logical principles based on his experience. However, they don't generalize to more complex situations, such as a car entering a freeway on-ramp.

The red car could be on the freeway traveling at 70 MPH while the blue car is entering the freeway and accelerating rapidly from 30 MPH to 65 MPH. In that situation, the blue car would not be getting closer to the red car; in fact, the red car is moving farther away from the blue car. The speed principle does not generalize because it considers only the cars' velocities at a particular point in time and does not take acceleration into account.[159]

A complex understanding of speed as a rate of change includes comprehending the difference between speed and acceleration. Each of the ideas of instantaneous speed, average speed, and acceleration are integrally tied to the context, so it is important that students begin to get a sense of the relative comparisons based on contexts, such as getting on the freeway versus a constant speed on the freeway. When students have a strong understanding of these ideas, they are more likely to be successful when the context moves to mixture types of problems.

Teachers

- Computer-based activities can be very effective in helping students differentiate their understanding between velocity and acceleration.[160] Graphs of realistic representations of motion can "help students confront paradoxes that arise when their new view of motion conflicts with their previous undifferentiated view."[161]
- Researchers found that having students interact with motion detectors and moving objects, then seeing the resulting graphs in real time was helpful in repairing students' misconceptions about speed and acceleration.[162] From one student's perspective: "Doing that one lab where we actually had to come up with the scenarios and then kind of play them out to see whether they worked—that helped out the most, I think."[163] When computers are used to link physical motion to graphical shape, the tools become resources for understanding. "Graphs created by a motion detector in response to a student's own body motion embody a link between a mathematical representation and body motion."[164]
- Computer simulations like the Center for Algebraic Thinking app *Tortoise and the Hare Algebra* (Figure 4.30) can be used to provide examples that help students explore the effects of speed and different rates of change on the classic race between the tortoise and the hare. In this app, students manipulate how many feet (or meters) per second each racer travels—including a little nap time for the hare—and then watch the animated race. They can also change perspectives between seeing the overall map of the race

FIGURE 4.30. Tortoise and the Hare Algebra App.

and watching individual racers up close to see how the rates of change for each racer look in each context. This provides additional experience with the concept of speed as a rate of change as students try to solve challenges posed in the app, such as how fast would the tortoise need to travel to finish the 100-yard race in 8 seconds? Or, if rabbit travels at 50 ft/sec and tortoise travels at 40 ft/sec, how far behind will turtle be when the race ends?

MOTION GRAPHS

Common Core Standards: 6.EE.9, 8.F.5

Interpreting a Cartesian graph of objects in motion is a process of understanding the representation of the relationship between selected elements. The research describes how students attempt to make sense of graphs related to problems of relative motion. In one study, researchers asked 10th-grade students to interpret a graph of students in motion (see Table 4.30).[165] Many students had difficulty interpreting what was happening at points A, B, C, and D in the graph for Problem 22 because both students were moving. In the first part of the interview (see Table 4.31), Student 1 saw the horizontal segment of the graph as representing that one of the people stopped, when in fact it meant that the distance between the two people stayed the same while they were walking.

In the interview, each of the students is at a different place in his or her thinking, struggling to make sense of the graph. The diverse perspectives keep them in conversation trying to resolve their differences. Their primary challenge is to

TABLE 4.30. Problem 22: Two Students in Motion

| Marna and Peter are two students who are standing one meter apart. They start walking in a straight line one behind the other. Marna is walking behind Peter carrying a calculator and CBR to measure the distance between them. They walk for 7 seconds. Interpret the graph. | *(Graph: Distance vs. Time, showing segment from A rising to B, flat from B to C, descending to D)* |

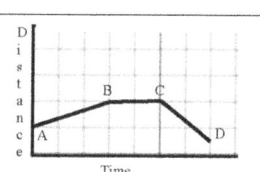

Source: Radford, Miranda, & Guzman (2008)

TABLE 4.31. Interview with Students for Problem 22

S1:	Peter moves away from Marna for 3 seconds, then he stops for 2 seconds, and then he moves closer to Marna for 2 seconds.
S2:	Well, even though he moves away, he moves back again, I don't know.
S3:	Well, if she walks with him, the graph really doesn't make sense.

(Researcher's comments in italics)

S1's interpretation is based on Peter moving and Marna remaining in one place. Peter moves away, stops and moves back. S2 makes an unsuccessful attempt to interpret the graph. S3 reminds the group that Marna is also moving.

S1:	So from A to B, Peter is walking faster than Marna. Right?
S2:	That could be and then Peter stops from B to C.
S3:	No, they are the same distance apart from B to C so they are walking at the same speed.

The interpretation of the graph has changed to Peter walking faster than Marna, to Peter stopping, to Peter and Marna walking at the same speed. The interpretation that Peter is walking faster than Marna could be coming from the positive slope of AB.

Source: Radford, Miranda, & Guzman (2008)

TABLE 4.32. Interview with Students for Problem 22: Instructor Intervention

I:	If Peter is walking faster than Marna, will the distance between them remain the same?
S1:	No. What happens between B and C?
S2:	Peter stops.
I:	Is the CBR still moving?
S3:	Yes...No, I don't know. Did Marna stop? She has the CBR.
The conversation is changing from relative speed to relative distance.	
I:	What does the point A represent?
S2:	1 meter.
I:	Is that a distance or a speed?
S1:	A distance.
I:	What is the distance measuring?
S2:	No, from B to C they are walking at the same speed because the distance between them is staying the same.
A reasonable explanation for BC has been made by S2.	
S3:	Then from C to D the distance between Peter and Marna is decreasing so Marna would be walking faster than Peter. Does that make sense?
S1 and S2 agree. *Students have come to a reasonable interpretation of the graph that makes sense.*	

think about the difference between absolute and relative motion. The dynamic changes as the instructor enters into the conversation and helps them focus their attention on particular aspects of the story (see Table 4.32).

Cartesian graphs are used to describe real world phenomena and convey complex mathematical meanings. One of the challenges for students working with motion graphs is that "space and time have to be understood in relational terms. Both space and time have to be measured numerically against an arbitrary starting point."[166] Researchers suggest that the logic of interpreting a Cartesian representation of relative motion can become apparent through discussion as opposed to direct instruction. The discussion is a "social process related to the manner in which students become progressively aware, through personal deeds and interpretations, of the cultural logic of mathematical entities."[167] One researcher described the process as "progressive disentanglement."[168] The role of the teacher in helping students move toward particular mathematical ideas was significant. The interviewer for this problem used a number of questions to probe students' thinking. Sometimes it is useful for the teacher to help move discussion forward by breaking down the dynamic of motion in the graph with questions such as the following:

1. What does point A represent? (The distance between Marna and Peter at time t=0.)

2. What is happening over segment AB? (The distance between Marna and Peter is increasing.)
3. What is happening over segment BC? (The distance between Marna and Peter is staying the same.)
4. What is happening over the segment CD? (The distance between Marna and Peter is decreasing.)

Another important part of the process was students' gestures to communicate their thinking. Researchers find that gestures can help students process their ideas as well as provide teachers with additional formative assessment of students' thinking, particularly with motion graphs.[169] "Gestures help construct better mental representations and mental operations to solve problems."[170] Gestures can be physical, bodily movements as well as "inked gestures" of drawings on paper.[171]

Teachers

- Students can bring real-world digital photos and video of motion into mathematical software such as *The Geometer's Sketchpad* and Vernier's *Video Physics*. Using pictures from students' experience can personalize mathematics, engage students, and integrate math with other subjects to help students makes sense of the mathematical dynamics.[172]
- Research indicates that encouraging students' use of gestures with hands or body may help them communicate and think about motion.[173]
- Another approach to helping students understand motion graphs is the Center for Algebraic Thinking's *Action Grapher* app (see Figure 4.31). It was built in response to research on students' thinking with motion graphs. The biking activity in the app was described earlier. Another activity is fill-

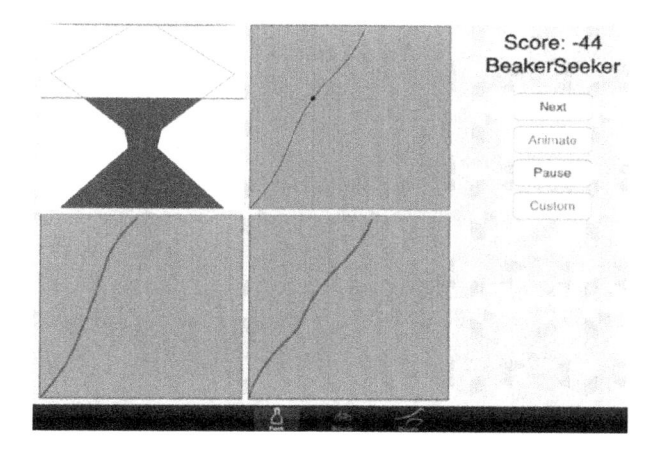

FIGURE 4.31. Action Grapher App: Flask.

ing flasks with water. The student watches a flask fill with water and must determine which of three representations accurately describes that motion. The activity includes a "custom" option in which students can test their hypotheses by designing their own flasks and filling them with water. The app automatically constructs three different representations for the student to choose from as the correct representation.

ENDNOTES

[1] NCTM, 2000
[2] Demana, Schoen, & Waits, 1993; Leinhardt, Zaslavsky, & Stein, 1990
[3] Dugdale, 1993; Leinhardt, Zaslavsky, & Stein, 1990; Mitchelmore & Cavanagh, 2000; Moschkovich, Schoenfeld, & Arcavi, 1993
[4] Cisneros, 2014
[5] National Governors Association Center for Best Practices, Council of Chief State School Officers (2010).
[6] Jitendra, et al., 2009, p. 250, based on Boyer, Levine, & Huttenlocher, 2008, p. 1478, and Fuson & Abrahamson, 2005; Lamon, 2007; Post, Behr, & Lesh, 1988
[7] Lamon, 1994, p. 90 based on Lesh, Post, & Behr, 1988 and reflected as well in Kilpatrick, Swafford & Findell, 2001
[8] Post, Behr, & Lesh, 1988
[9] Tourniaire & Pulos, 1985, p. 181
[10] Karplus, Pulos, & Stage, 1983b, p. 219
[11] Fuson & Abrahamson, 2005, p. 213; Hart, 1981
[12] Hart, 1984, p. 5
[13] Tourniaire & Pulos, 1985
[14] Tourniare & Pulos, 1985
[15] Hart, 1981; Noelting, 1980a, 1980b
[16] Tourniare & Pulos, 1985
[17] Lesh, Post, & Behr, 1988, p. 93-94
[18] Post, Behr, & Lesh, 1988, p. 78
[19] Karplus, Pulos, & Stage, 1983; Noelting, 1980a
[20] Jitendra, et al., 2009
[21] Silvestre & da Ponte, 2012, p. 74, based on Stanley, McGowan, & Hull, 2003
[22] Hart, 1981; Tourniaire & Pulos, 1985
[23] Hart, 1981
[24] Lesh, Post, & Behr, 1988, pp. 104-105
[25] Lamon, 1994; Tourniaire & Pulos, 1985
[26] Hembree, 1992; Koedinger & Terao, 2002; Mayer, 1989, 2005
[27] Rau, 2016, p. 1. Also Booth & Koedinger, 2012
[28] Orrill & Brown, 2012; Watanabe, 2015
[29] Murata, 2008
[30] Middleton & van den Heuvel-Panhuizen, 1995
[31] Murata, 2008, p. 379
[32] Post, Behr, & Lesh, 1988, p. 86
[33] Post, Behr, & Lesh, 1988, p. 79
[34] Orton, 1983; Ubuz, 2007
[35] Herbert, 2010; Herbert & Pence, 2009
[36] Thompson, 1996
[37] Karplus, Pulos, & Stage, 1983; Vergnaud, 1980

[38] Herbert, 2010, p. 246
[39] Saunders & Jesunathadas, 1988
[40] Herbert & Pierce, 2007
[41] Tourniaire, 1984
[42] Sherin, 2000, p. 403 quoting Palmer, 1977
[43] Sherin, 2000, p. 400
[44] Based on Sherin, 2000, p. 408
[45] Sherin, 2000
[46] Sherin, 2000, p. 415
[47] Leinhardt, Zaslavsky, & Stein, 1990, p. 12
[48] diSessa, Hammer, Sherin, & Kolpakowski, 1991
[49] Mevarech & Kramarsky, 1997
[50] Bell, Brekke, & Swann, 1987a, 1987b; Mevarech & Kramarsky, 1997
[51] Dreyfus and Eisenberg, 1982; Markovits, Eylon, & Brukheimer, 1983; and Zaslavsky, 1997
[52] Barclay, 1987; Burrill, et al., 2002; Ellington, 2003; Hall, & Chamblee, 2013; McClaran, 2013; Roschelle, Pea, Hoadley, Gordin, & Means, 2000
[53] Li & Ma, 2011
[54] Delgado & Lucero, 2015, p. 652
[55] Leinhardt, Zaslavsky, & Stein, 1990, p. 19
[56] Hart, 2004
[57] Leinhardt, Zaslavsky, & Stein, 1990
[58] Delgado & Lucero, 2015, p. 652 based on Adams & Shrum, 1990; Brasell, 1990; Mevarech & Kramarsky, 1997; Vergnaud & Errecalde, 1980; Wavering, 1989
[59] American Association for the Advancement of Science, 1993
[60] Mitchelmore & Cavanagh, 2000
[61] Goldenberg, 1988
[62] Vonder Embse & Engebretsen, 1996
[63] Steele, 1994
[64] Burrill, et al., 2002; Ellington, 2003; Mitchelmore & Cavanagh, 2000
[65] Mitchelmore & Cavanagh, 2000, p. 262
[66] Goldenberg and Kliman 1988; Mitchelmore & Cavanagh, 2000
[67] Delgado & Lucero, 2015; diSessa, Hammer, Sherin, & Kolpakowski, 1991
[68] Schoenfeld, Smith, & Arcavi, 1993; Moschkovich, 1989, 1990, 1992
[69] Moschkovich, 1999
[70] Moschkovich, 1999, p. 170
[71] Schoenfeld, Smith, & Arcavi, 1993
[72] Moschkovich, 1999
[73] Moschkovich, 1999
[74] Davis, 2007, p. 388 based on Smith, Arcavi, & Schoenfeld, 1989
[75] Hattikudur, Prather, Asquith, Alibali, Knuth, & Nathan, 2012
[76] Barr, 1980
[77] Bardini & Stacey, 2006, p. 115
[78] Davis, 2007
[79] Davis, 2007
[80] Pierce, Stacey, & Bardini, 2010
[81] Pierce, Stacey, & Bardini, 2010
[82] Hattikudur, Prather, Asquith, Alibali, Knuth, & Nathan, 2012
[83] Hattikudur, Prather, Asquith, Alibali, Knuth, & Nathan, 2012, p. 236
[84] Goldenberg, 1988; Moschkovich, 1990, 1999
[85] Chiu, Kessel, Moschkovich, & Munoz-Nunez, 2001
[86] Boers-van Oosterum, 1990; Demana, Schoen, & Waits, 1993; Dunham & Dick, 1994; Ellington, 2003, 2006; Hennessy, Fung, & Scanlon, 2001; Kenney, 2014; Slavit, 1994,

[87] Dewey, Singletary, & Kinzel, 2009; Kenney, 2014
[88] Hattikudur, Prather, Asquith, Alibali, Knuth, & Nathan, 2012
[89] Bell & Janvier, 1981; Janvier, 1981a; McDermott, Rosenquist, & van Zee, 1987; Orton, 1984; Simon & Blume, 1994; Stump, 2001
[90] Barr, 1980, 1981
[91] Hart, et al., 2004
[92] Beichner, 1993; Brassel and Rowe, 1993; Clement, Mokros, & Schultz, 1986; Hadjidemetriou & Williams, 2001; Janvier, 1978; Leinhardt, Zaslavsky, & Stein, 1990; McDermott, Rosenquist, van Zee, 1983; McDermott, Rosenquist, Popp, & van Zee, 1986
[93] Clement, Mokros, & Shultz, 1986
[94] Mcdermott, Rosenquist, & van Zee, 1987, p. 504
[95] Mcdermott, Rosenquist, & van Zee, 1987
[96] Hale, 2000; Monk, 1994
[97] Mokros & Tinker, 1987
[98] Stump, 2001
[99] Stump, 2001, p. 82-83
[100] Skemp, 1976; Walter & Gerson, 2007
[101] Stump, 2001; Swafford & Langrall, 2000
[102] Barr 1980, 1981
[103] Simon & Blume, 1994; Stump, 2001
[104] Simon & Blume, 1994; Stump, 2001
[105] Simon and Blume, 1994
[106] Hart, 2004; Mevarech & Kramarsky, 1997
[107] Hart, 2004
[108] Based on Stump, 1997
[109] Based on Stump, 2001
[110] Berg & Phillips, 1994; Clement, 1985, 1986; Janvier, 1998; Leinhardt, Zaslavsky, & Stein, 1990; McDermott, Rosenquist, & van Zee, 1987; Monk, 1992; Nemirovsky & Rubin, 1992; Noble & Nemirovsky, 1995
[111] McDermott, Rosenquist, Popp, & van Zee, 1983
[112] Monk, 1992
[113] Mokros & Tinker, 1987
[114] Kerslake, 2004
[115] Clement, 1989
[116] Monk, 1992; Schultz, Clement, & Mokros, 1986
[117] Barclay & TERC, 1986; Schultz, Clement, & Mokros, 1986
[118] Based on Brassel and Rowe, 1993
[119] Kerslake, 2004, p. 129
[120] Nemirovsky, Tierney, & Wright, 1998
[121] Ferrara, 2014; Herman, Laumakis, Dover, & Doyle, 2008; Robutti, 2006
[122] Kwon, 2002; Phillips, 1997
[123] Glazer, 2011, p. 200
[124] Knuth, 2000a, p. 504 based on Larkin & Simon, 1987
[125] From Lowrie, Diezmann, & Logan, 2011, p. 1, based on Leu, Kinzer, Coiro, & Cammack, 2004
[126] Friel & Bright, 1996; Friel, Curcio, & Bright, 2001; Postigo & Pozo, 2004
[127] Friel, Curcio, & Bright, 2001, p. 146
[128] Padilla, McKenzie, & Shaw, 1986
[129] Postigo & Pozo, 2004, pp. 627-629
[130] Based on Glazer, 2011; Friel, Curcio, & Bright, 2001; and Wainer, 1992
[131] Knuth, 2000a; Moschkovich, Schoenfeld, & Arcavi, 1993; Schoenfeld et al., 1993; Smith et al., 1989
[132] Knuth, 2000a, p. 501

[133] Knuth, 2000a
[134] Knuth, 2000a
[135] Knuth, 2000a
[136] Hadjidemetriou & Williams, 2001; Leinhardt, Zaslavsky, & Stein, 1990; Mevarech & Kramarsky, 1997
[137] Billings & Klanderman, 2000, p. 441 based on Bell & Janvier, 1981; Janvier, 1981b
[138] Carpenter & Shah, 1998
[139] Jackson, Edwards, & Berger, 1993
[140] Leinhardt, Zaslavsky, & Stein, 1990
[141] Janvier, 1983; Kerslake, 2004; Mansfield, 1985; Stein & Leinhardt, 1989
[142] Kerslake, 2004
[143] Mansfield, 1985
[144] Kerslake, 2004; Markovits, Eylon, & Bruckheimer, 1986
[145] Knuth, 2000b
[146] Cobb & Moore, 1997; DiSessa, Hammer, Sherin, & Kolpakowski, 1991; Roth & McGinn, 1997.
[147] Cobb & Moore, 1997, p. 816
[148] Glazer, 2011, p. 204
[149] From Hattikudur, et al., 2012, p. 232 and based on Goldenberg, 1987 and Krabbendam, 1982
[150] Ainley, 2000; Ainley, Nardi, & Pratt, 2000
[151] Thompson, 1994, p. 37
[152] Lobato & Thanheiser, 1999
[153] Hale, 2000; Lobato & Thanheiser, 1999; Monk, 1994
[154] Thompson, 1992; Thompson, 1994
[155] Billings & Klanderman, 2000
[156] Thompson, 1994, p. 28
[157] Thompson & Thompson, 1996
[158] Billings & Klanderman, 2000
[159] Hale, 2000, p. 415
[160] Dykstra, Boyle, & Monarch, 1992
[161] Hale, 2000, p. 416
[162] Arzarello, Pezzi, & Robutti, 2007; Hale, 2000
[163] Hale, 2000, p. 416
[164] Noble & TERC, 2003, p. 3 based on Meira, 1995; Moschkovitch, 1996; Nemirovsky & Noble, 1997; Nemirovsky et al. 1998; Noble, Nemirovsky, Tierney & Wright, 1999; Noble, Nemirovsky, Wright, & Tierney, 2001
[165] Radford, Miranda, & Guzman, 2008
[166] Radford, 2009
[167] Radford, Miranda, & Guzman, 2008, p. 166
[168] Radford, 2009, p. 12
[169] Cook, Yip, & Goldin-Meadow, 2012; Noble, 2004; Radford, Miranda, & Guzman, 2008
[170] Segal, 2011
[171] Radford, 2009, p. 14
[172] Pierce, Stacey, & Ball, 2005
[173] Herbert & Pierce, 2007; Noble, Nemirovsky, DiMattia, & Wright, 2004
[174] Author created interview to articulate a potential student description.

REFERENCES

Adams, D. D., & Shrum, J. W. (1990). The effects of microcomputer-based laboratory exercises on the acquisition of line graph construction and interpretation skills by high school biology students. *Journal of Research in Science Teaching, 27*(8), 777–787.

Ainley, J. (2000). *Exploring the transparency of graphs and graphing.* Paper presented at the Twenty Fourth Annual Conference of the International Group for the Psychology of Mathematics Education, Hiroshima, Japan.

Ainley, J., Nardi, E., & Pratt, D. (2000). The construction of meanings for trend in active graphing. *International Journal of Computers for Mathematical Learning, 5*, 85–114.

American Association for the Advancement of Science (1993). *Benchmarks for science literacy.* New York, NY: Oxford University Press.

Arzarello, F., Pezzi, G., & Robutti, O. (2007). Modelling body motion: An approach to functions using measuring instruments. *Modeling and Applications in Mathematics Education, 10*, 129–136.

Barclay, T. (1987). A graph is worth how many words? *Classroom Computer Learning, 7*, 46–50.

Barclay, W. L., & Technical Education Research Center, C. M. A. (1986). *Graphing misconceptions and possible remedies using microcomputer-based labs.* Retrieved from http://search.ebscohost.com/login.aspx?direct=true&db=eric&AN=ED264129&site=ehost-live.

Bardini, C., & Stacey, K. (2006). *Students' conceptions of m and c: How to tune a linear function.* Paper presented at the 30th Conference of the International Group for the Psychology of Mathematics Education, Prague.

Barr, G. (1980). Graphs, gradients and intercepts. *Mathematics in School, 9*(1), 5–6.

Barr, G. (1981). Some student ideas on the concept of gradient. *Mathematics in School, 10*(1), 14,16–17.

Beichner, R. J. (1993). *Testing student interpretation of kinematics graphs.* Paper presented at the 5th European Association for Research on Learning and Instruction, France.

Bell, A., Brekke, G., & Swan, M. (1987a). Diagnostic teaching: 4 graphical interpretations. *Mathematics Teaching, 119*, 56–60.

Bell, A., Brekke, G., & Swann, M. (1987b). Diagnostic teaching: 5 graphical interpretation teaching styles and their effects. *Mathematics Teaching, 120*, 50–57.

Bell, A., & Janvier, C. (1981). The interpretation of graphs representing situations. *For the Learning of Mathematics, 2*(1), 34–42.

Berg, C. A., & Phillips, D. G. (1994). An investigation of the relationship between logical thinking structures and the ability to construct and interpret line graphs. *Journal of Research in Science Teaching, 31*(4), 323–344.

Billings, E., & Klanderman, D. (2000). Graphical representations of speed: obstacles preservice K–8 teachers experience. *School Science and Mathematics, 100*(8), 440–450.

Boers-van Oosterum, M. A. M. (1990). Understanding of variables and their uses acquired by students in traditional and computer-intensive algebra. Unpublished doctoral dissertation, University of Maryland. *Dissertation Abstracts International, 51*(5), 1538.

Booth, J. L., & Koedinger, K. R. (2012). Are diagrams always helpful tools? Developmental and individual differences in the effect of presentation format on student problem solving. *British Journal of Educational Psychology, 82*(3), 492–511.

Boyer, T. W., Levine, S. C., & Huttenlocher, J. (2008). Development of proportional reasoning: Where young children go wrong. *Developmental Psychology, 44*, 1478–1490.

Brasell, H. M. (1990). Graphs, graphing, and graphers. In M. B. Rowe (Ed.), *The process of knowing*. Washington, DC: National Science Teachers Association.

Brassel, H. M., & Rowe, M. B. (1993). Graphing skills among high school physics students. *School Science and Mathematics, 93*, 63–71.

Burrill, G., Allison, J., Breaux, G., Kastberg, S., Leatham, K., & Sanchez, W. (2002). *Handheld graphing technology in secondary mathematics.* East Lansing, MI: Michigan State University.

Carpenter, P. A., & Shah, P. (1988). A model of the perceptual and conceptual processes in graph comprehension. *Journal of Experimental Psychology: Applied, 4*, 75–100.

Chiu, M. M., Kessel, C., Moschkovich, J., & Munoz-Nunez, A. (2001). Learning to graph linear functions: A case study of conceptual change. *Cognition and Instruction, 19*(2), 215–252.

Cisneros, S. (2014, June 20). *New report finds that 1 in 5 mothers stay home to care for family.* Retrieved from http://www.opb.org/news/article/new-report-finds-1-in-5-mothers-stay-home-to-care-for-family/.

Clement, J. (1985). *Misconceptions in graphing.* Paper presented at the Ninth International Conference for the Psychology of Mathematics Education, Utrecht, The Netherlands.

Clement, J. (1989). The concept of variation and misconceptions in Cartesian graphing. *Focus on Learning Problems in Mathematics, 11*, 77–87.

Clement, J., Mokros, J. R., & Schultz, K. (1986). *Adolescents' graphing skills: A descriptive analysis.* Paper presented at the Annual Meeting of the American Educational Research Association.

Cobb, G., & Moore, D. (1997). Mathematics, statistics, and teaching. *The American Mathematical Monthly, 104*(9), 801–823.

Cook, S. W., Yip, T. K., & Goldin-Meadow, S. (2012). Gestures, but not meaningless movements, lighten working memory load when explaining math. *Language and Cognitive Processes, 27*(4), 594–610.

Coxford, A. F., Fey, J. T., Hirsch, C. R., Schoen, H. L., Burrill, G., Hart, E. W., & Watkins, A. E. (1998). *Contemporary mathematics in context: A unified approach.* New York: Glencoe/McGraw-Hill.

Davis, J. D. (2007). Real-world contexts, multiple representations, student-invented terminology, and y-intercept. *Mathematical Thinking and Learning, 9*(4), 387–418.

Delgado, C., & Lucero, M. M. (2015). Scale construction for graphing: An investigation of students' resources. *Journal of Research in Science Teaching, 52*(5), 633–658.

Demana, F., Schoen, H., & Waits, B. (1993). Graphing in the K–12 curriculum: The impact of the graphing calculator. In T. Romberg, E. Fennema, & T. Carpenter (Eds.), *Integrating research on the graphical representation of functions* (pp. 11–39). Hillsdale, NJ: Lawrence Erlbaum.

Dewey, B. L., Singletary, T. J., & Kinzel, M. T. (2009). Graphing calculator use in algebra teaching. *School Science and Mathematics, 109*(7), 383–393.

diSessa, A. A., Hammer, D., Sherin, B., & Kolpakowski, T. (1991). Inventing graphing: Meta-representational expertise in children. *Journal of Mathematical Behavior, 10*(2), 117–160.

Dreyfus, T., & Eisenberg, T. (1982). The function concept in college students: Linearity, smoothness and periodicity. *Focus on Learning Problems in Mathematics, 5,* 119–132.

Dugdale, S. (1993). Functions and graphs: Perspectives on student thinking. In T. A. Romberg, E. Fennema, & T. P. Carpenter (Eds.), *Integrating research on the graphical representation of functions* (pp. 101–130). Hillsdale, NJ: Lawrence Erlbaum.

Dunham, P. H., & Dick, T. P. (1994). Research on graphing calculators. *The Mathematics Teacher, 87,* 440–445.

Dykstra, D. I., Boyle, F., & Monarch, I.A. (1992). Studying conceptual change in learning physics. *Science Education, 76,* 615–52.

Ellington, A. J. (2003). A meta-analysis of the effects of calculators on students' achievement and attitude levels in precollege mathematics classes. *Journal for Research in Mathematics Education, 34*(5), 433–463.

Ellington, A. J. (2006). The effects of non-CAS graphing calculators on student achievement and attitude levels in mathematics: A meta-analysis. *School Science and Mathematics, 106,* 16–26.

Ferrara, F. (2014). How multimodality works in mathematical activity: Young children graphing motion. *International Journal of Science and Mathematics Education, 12*(4), 917–939.

Friel, S. N., & Bright, G. W. (1996). *Building a theory of graphicacy: How do students read graphs?* Paper presented at the Annual Meeting of the American Educational Research Association., New York, NY.

Friel, S. N., Curcio, F. R., & Bright, G. W. (2001). Making sense of graphs: Critical factors influencing comprehension and instructional implications. *Journal for Research in Mathematics Education, 32*(2), 124–158.

Fuson, K. C., & Abrahamson, D. (2005). Understanding ratio and proportion as an example of the apprehending zone and conceptual-phase problem-solving models. In J. Campbell (Ed.), *Handbook of mathematical cognition* (pp. 213–234). New York, NY: Psychology Press.

Glazer, N. (2011) Challenges with graph interpretation: a review of the literature. *Studies in Science Education, 47*(2), 183–210.

Goldenberg, E. P. (1987). Believing is seeing: How preconceptions influence the perception of graphs. In J. C. Bergeron, N. Herscovics, and C. Kieran (Eds.), *Proceedings of the 11th International Conference for the Psychology of Mathematics Education* (pp. 197–203), Montreal, Canada.

Gravemeijer, K., van Galen, F., & Keijzer, R. (2005). *Designing instruction on proportional reasoning with average speed.* Paper presented at the 39th International Conference of the International Group for the Psychology of Mathematics Education.

Hadjidemetriou, C., & Williams, J. (2001). Children's graphical conceptions. In M. van den Heuvel-Panhuizen (ed). *Proceedings of the 25th Conference of the International Group for the Psychology of Mathematics Education* (Vol. 3, p. 89). Utrecht: Netherlands.

Hale, P. (2000). Kinematics and graphs: Students' difficulties and CBLs. *Mathematics Teacher, 93*(5), 414–417.

Hall, J., & Chamblee, G. (2013). Teaching algebra and geometry with GeoGebra: Preparing pre-service teachers for middle Grades/Secondary mathematics classrooms. *Computers in the Schools, 30*(1–2), 12.

Hart, K. (2004). *Children's understanding of mathematics: 11–16.* Eastbourne, United Kingdom: Anthony Rowe Ltd.

Hart, K. (1984). *Ratio: Children's strategies and errors.* Windsor, UK: NFER-NELSON.

Hattikudur, S., Prather, R. W., Asquith, P., Alibali, M. W., Knuth, E. J., & Nathan, M. (2012). Constructing graphical representations: Middle schoolers' intuitions and developing knowledge about slope and *y*-intercept. *School Science and Mathematics, 112*(4), 230–240.

Hembree, R. (1992). Experiments and relational studies in problem solving: A meta-analysis. *Journal for Research in Mathematics Education, 23,* 242–273.

Hennessy, S., Fung, P., & Scanlon, E. (2001). The role of the graphic calculator in mediating graphing activity. *International Journal of Mathematical Education in Science and Technology, 32*(2), 267–290.

Herbert, S. (2010). *Impact of context and representation on Year 10 students' expression of conceptions of rate.* Paper presented at the 33rd Annual Conference of the Mathematics Education Research Group of Australasia, Freemantle, Australia.

Herbert, S., & Pierce, R. (2007). *Video evidence: What gestures tell us about students' understanding of rate of change.* Paper presented at the 30th annual conference of the Mathematics Education Research Group of Australasia. Hobart, Tasmania, Australia.

Herbert, S., & Pierce, R. (2009). Revealing conceptions of rate of change. In R. Hunter, B. Bicknell, & T. Burgess (Eds.), *Crossing divides: Proceedings of the 32nd annual conference of the Mathematics Education Research Group of Australasia* (pp. 217–224). Palmerston North, NZ: MERGA.

Herman, M., Laumakis, P., Dover, R., & Doyle, D. (2008). Using the CBR to enhance graphical understanding. *The Mathematics Teacher, 102*(5), 383–389.

Jackson, D. F., Edwards, B. J., & Berger, C. F. (1993). Teaching the design and interpretation of graphs through computer-aided graphical data analysis. *Journal of Research in Science Teaching, 30*(5), 483–501.

Janvier, C. (1978). *The interpretation of complex Cartesian graphs representing situations—Studies and teaching experiments.* University of Nottingham, England, Unpublished Doctoral dissertation.

Janvier, C. (1981a). Difficulties related to the concept of variable presented graphically. In C. Comiti & G. Vergnaud (Eds.), *Proceedings of the Fifth International Conference for the Psychology of Mathematics Education* (pp. 189–192). Grenoble, France: Laboratoire IMAG.

Janvier, C. (1981b). Use of situations in mathematics education. *Educational Studies in Mathematics, 12,* 113–122.

Janvier, C. (1983). Teaching the concept of function. *Mathematical Education for Teaching, 4*(2), 48–60.

Jitendra, A. K., Star, J. R., Starosta, K., Leh, J. M., Sood, S., Caskie, G., Hughes, C., & Mack, T. R. (2009). Improving seventh grade students' learning of ratio and proportion: The role of schema-based instruction. *Contemporary Educational Psychology, 34*(3), 250–264.

Karplus, R., Pulos, S., & Stage, E. K. (1983a). Proportional reasoning in early adolescents. In R. Lesh & M. Landau (Eds.), *Acquisition of mathematics concepts and processes.* New York, NY: Academic Press.

Karplus, R., Pulos, S., & Stage, E. K. (1983b). Early adolescents' proportional reasoning on 'rate' problems. *Educational Studies in Mathematics, 14*, 219–233.

Kenney, R. H. (2014). Investigating a link between pre-calculus students' uses of graphing calculators and their understanding of mathematical symbols. *International Journal for Technology in Mathematics Education, 21*(4), 157–166.

Kerslake, D. (2004). Graphs. In K. Hart (Ed.), *Children's understanding of mathematics: 11–16* (pp. 120–136). Eastbourne, UK: Antony Rowe Publishing.

Kilpatrick, J., Swafford, J., & Findell, B. (2001). *Adding it up: Helping children learn mathematics.* Washington, DC: National Academy Press.

Knuth, E. J. (2000a). Student understanding of the Cartesian connection: An exploratory study. *Journal for Research in Mathematics Education, 31*, 500–507.

Knuth, E. J. (2000b). Understanding connections between equations and graphs. *Mathematics Teacher, 93*(1), 48–53.

Koedinger, K. R., & Terao, A. (2002). A cognitive task analysis of using pictures to support pre-algebraic reasoning. In C. D. Schunn & W. Gray (Eds.), *Proceedings of the Twenty-Fourth Annual Conference of the Cognitive Science Society.* Mahwah, NJ: Lawrence Erlbaum Associated, Inc.

Krabbendam, H. (1982). The non-qualitative way of describing relations and the role of graphs: Some experiments. In G. Van Barnveld & H. Krabbendam (Eds.), *Conference on functions* (pp. 125–146). Enshede, The Netherlands: Foundation for Curriculum Development.

Kwon, O. N. (2002). The effect of calculator-based ranger activities on students' graphing ability. *School Science and Mathematics, 102*(2), 57–67.

Lamon, S. (1994). Ratio and proportion: Cognitive foundations in unitizing and norming. In G. Harel & J. Confrey (Eds.), *The development of multiplicative reasoning in the learning of mathematics* (pp. 89–120). Albany, NY: State University of New York Press.

Lamon, S. J. (2007). Rational numbers and proportional reasoning: Toward a theoretical framework for research. In F. K. Lester, Jr. (Ed.), *Second handbook of research on mathematics teaching and learning* (pp. 629–668). Charlotte, NC: Information Age Publishing.

Larkin, J. H., & Simon, H. A. (1987). Why a diagram is (sometimes) worth ten thousand words. *Cognitive Science, 11*, 65–99.

Lehner, J. (2014). *Oregon's stay at home parents.* Department of Administrative Services, Office of Economic Analysis. Retrieved from https://oregoneconomicanalysis.files.wordpress.com/2014/06/oregons-stay-at-home-parents-062314.pdf.

Leinhardt, G., Zaslavsky, O., & Stein, M. K. (1990). Functions, graphs, and graphing: Tasks, learning, and teaching. *Review of Educational Research, 60*(1), 1–64.

Lesh, R., Post, T. R., & Behr, M. J. (1988). Proportional reasoning. In J. Hiebert & M. Behr (Eds.), *Number concepts and operations in the middle grades* (pp. 93–118). Reston, VA: NCTM.

Leu, D. J., Jr., Kinzer, C. K., Coiro, J. L., & Cammack, D. W. (2004). Toward a theory of new literacies emerging from the Internet and other information and communication technologies. In R. B. Ruddell & N. J. Unrau (Eds.), *Theoretical models and processes of reading* (5th ed., pp. 1570–1613).

Li, Q., & Ma, X. (2011). A meta-analysis of the effects of computer technology on school students' mathematics learning. *Educational Psychology Review, 22*, 215–243.

Lobato, J., & Thanheiser, E. (1999). *Rethinking slope from quantitative and phenomenological perspectives*. Presented at the North American Chapter of the Psychology of Mathematics Education Conference. Morelos, Mexico.

Lowrie, T., Diezmann, C., & Logan, T. (2011) Understanding graphicacy: Students' making sense of graphics in mathematics assessment tasks. *International Journal for Mathematics Teaching and Learning*, 1–21. Retrieved from http://eprints.qut.edu.au/48802.

Mansfield, H. (1985). Points, lines, and their representations. *For the Learning of Mathematics*, 5(3), 2–6.

Markovits, Z., Eylon, B., & Bruckheimer, M. (1986). Functions today and yesterday. *For the Learning of Mathematics*, 6(2), 18–28.

Markovitz, Z., Eylon, B.,,,,, & Brukheimer, M. (1983). Functions: Linearity unconstrained. In R. Hershkowitz (Ed.), *Proceedings of the Seventh International Conference of the International Group for the PME* (pp. 271–277). Rehovot, Israel.

Mayer, R. (2005). Cognitive theory of multimedia learning. In R. Mayer (Ed.) *The Cambridge handbook of multimedia learning* (pp. 31–48). Cambridge, UK: Cambridge University Press.

Mayer, R. E. (1989). Systematic thinking fostered by illustrations in scientific text. *Journal of Educational Psychology*, 81, 240–246.

McClaran, R. R. (2013). *Investigating the impact of interactive applets on students' understanding of parameter changes to parent functions: An explanatory mixed methods study*. Theses and Dissertations—Science, Technology, Engineering, and Mathematics (STEM) Education. Paper 2. Retrieved from: http://uknowledge.uky.edu/stem_etds/2.

McDermott, L. C., Rosenquist, M. L., Popp, B., & van Zee, E. H. (1983). *Student difficulties in connecting graphs, concepts, and physical phenomena*. Paper presented at the Annual Meeting of the American Educational Research Association, Montreal, Canada.

McDermott, L., Rosenquist, M., & van Zee, E. (1987). Student difficulties in connecting graphs and physics: Example from kinematics. *American Journal of Physics*, 55, 503–513.

Meira, L. (1995). The microevolution of mathematical representations in children's activity. *Cognition and Instruction*, 3(2), 269–313.

Mevarech, Z. R., & Kramarsky, B. (1997). From verbal descriptions to graphic representations: Stability and change in students' alternative conceptions. *Educational Studies in Mathematics*, 32(3), 229–263.

Middleton, J. A., & Van den Heuvel-Panhuizen, M. (1995). The ratio table. *Mathematics teaching in the middle school 1*(4), 282–288.

Mitchelmore, M., & Cavanagh, M. (2000). Students' difficulties in operating a graphics calculator. *Mathematics Education Research Journal*, 12(3), 254–268.

Mokros, J. & Tinker, R. (1987). The Impact of microcomputer-based labs on children's ability to interpret graphs. *Journal of Research in Science Teaching*, 24, 369–383.

Monk, S. (1992). Students' understanding of a function given by a physical model. In G. H. E. Dubinsky (Ed.), *The concept of function: Aspects of epistemology and pedagogy*. Washington, DC: Mathematical Association of America.

Monk, S. (1994). *How students and scientists change their minds*. MAA invited address at the Joint Mathematics Meetings, Cincinnati, Ohio, January.

Moore-Russo, D., Conner, A., & Rugg, K. (2011). Can slope be negative in 3-space? Studying concept image of slope through collective definition construction. *Educational Studies in Mathematics, 76*(1), 3–21.

Moschkovich, J. (1989). *Constructing a problem space through appropriation: A case study of guided computer exploration of linear functions.* Paper presented at the 1989 Annual Meeting of the American Educational Research Association, San Francisco.

Moschkovich, J. (1990). *Students' interpretations of linear equations and their graphs.* Paper presented at the Fourteenth Annual Meeting of the International Group for the Psychology of Mathematics Education, Mexico.

Moschkovich, J. (1992). *Making sense of linear equations and their graphs. An analysis of student conceptions and language use.* Unpublished Doctoral dissertation.

Moschkovitch, J. (1996). Moving up and getting steeper: Negotiating shared descriptions of linear graphs. *Journal of the Learning Sciences, 5*(3), 239–27.

Moschkovich, J. (1999). Students use of the x-intercept as an instance of a transitional conception. *Educational Studies in Mathematics, 37*(2), 169.

Moschkovich, J., Schoenfeld, A. H., & Arcavi, A. (1993). Aspects of understanding: On multiple perspectives and representations of linear relations and connections among them. In T. A. Romberg, E. Fennema & T. P. Carpenter (Eds.), *Integrating research 011 the graphical representation of functions* (pp. 69–100). Hillsdale, NJ: Lawrence Erlbaum.

Murata, A. (2008). Mathematics teaching and learning as a mediating process: The case of tape diagrams. *Mathematical Thinking and Learning: An International Journal, 10*(4), 374–406.

National Council of Teachers of Mathematics. (2000). *Principles and standards for school mathematics.* Reston, VA: Author.

National Governors Association Center for Best Practices, Council of Chief State School Officers (2010). *Common Core State Standards Mathematics.* Washington, DC: National Governors Association Center for Best Practices, Council of Chief State School Officers.

Nemirovsky, R., & Noble, T. (1997). Mathematical visualization and the place where we live. *Educational Studies of Mathematics, 3*(2), 9–131.

Nemirovsky, R. & Rubin, A. (1992). *Students' tendency to assume resemblances between a function and its derivative.* Cambridge, MA: TERC Communications, January 1992. Working Paper 2–92.

Nemirovsky, R., Tierney, C., & Wright, T. (1998). Body motion and graphing. *Cognition and Instruction, 16*(2), 19–172.

Noble, T., & Nemirovsky, R. (1995). *Graphs that go backwards.* Paper presented at the XIX Annual Meeting of the International Group for the Psychology of Mathematics Education. Recife, Brazil.

Noble, T., Nemirovsky, R., DiMattia, C., & Wright, T. (2002). *On learning to see: How do middle school students learn to make sense of visual representations in mathematics?* Cambridge, MA: TERC.

Noble, T., Nemirovsky, R., DiMattia, C., & Wright, T. (2004). Learning to See. *International Journal of Computers for Mathematical Learning, 9*(2), 109–167.

Noble, T., Nemirovsky, R.,Wright, T., & Tierney, C. (2001). Experiencing change: The mathematics of change in multiple environments. *Journal for Research in Mathematics Education, 32*(1), 85–108.

Noble, T., & TERC, C. M. A. (2003). *Gesture and the mathematics of motion.* Retrieved from http://search.ebscohost.com/login.aspx?direct=true&db=eric&AN=ED478190&site=ehost-live.

Noelting, G. (1980a). The development of proportional reasoning and the ratio concept Part I—Differentiation of stages. *Educational Studies in Mathematics, 11*(2), 217–253.

Noelting, G. (1980b). The development of proportional reasoning and the ratio concept Part II—problem-structure at successive stages; problem-solving strategies and the mechanism of adaptive restructuring. *Educational Studies in Mathematics, 11*(3), 331–363.

Orrill, C. H., & Brown, R. E. (2012). Making sense of double number lines in professional development: Exploring teachers' understandings of proportional relationships. *Journal of Mathematics Teacher Education, 15*(5), 381–403.

Orton, A. (1983). Students' understanding of differentiation. *Educational Studies in Mathematics, 14*, 235–250.

Orton, A. (1984). Understanding rate of change. *Mathematics in School, 13*(5), 23–26.

Padilla, M. J., McKenzie, D. L., & Show, E. L. Jr. (1986). An examination of the line graphing ability of students in grades seven through twelve. *School Science and Mathematics, 86*, 20–26.

Palmer, S. (1977). Fundamental aspects of cognitive representation. In E. Rosch, & B. B. Lloyd (Eds.), *Cognition and categorization,* (pp. 259–303). Hillsdale, NJ: Erlbaum.

Phillips, R. J. (1997). Can juniors read graphs? *Journal of Information Technology in Teacher Education, 6*(1), 49–58.

Pierce, R., Stacey, K., & Bardini, C. (2010). Linear functions: Teaching strategies and students' conceptions associated with $y = mx + c$. *Pedagogies, 5*(3), 202.

Pierce, R., Stacey, K., & Ball, L. (2005). Mathematics from still and moving images. *Australian Association of Mathematics Teachers, 61*(3), 26–31.

Post, T. R., Behr, M. J., & Lesh, R. (1988). Proportionality and the development of prealgebra understandings. In A. Coxford & A. Shulte (Eds.), *The ideas of algebra, K–12* (pp. 78–90). Reston, VA: National Council of Teachers of Mathematics.

Postigo, Y., & Pozo, J. (2004). On the road to graphicacy: The learning of graphical representation systems. *Educational Psychology, 24*(5), 623–644.

Radford, L. (2009). "No! He starts walking backwards!": Interpreting motion graphs and the question of space, place and distance. *ZDM, 41*(4), 467–480.

Radford, L., Miranda, I., & Guzman, J. (2008). *Relative motion, graphs, and the heteroglossic transformation of meanings: A semiotic analysis.* Paper presented at the Joint 32nd Conference of the International Group for the Psychology of Mathematics Education and the 30th North American Chapter. Morelia, Mexico.

Rau, M. A. (2016). Conditions for the effectiveness of multiple visual representations in enhancing STEM learning. *Educational Psychology Review, 29*(4), 717–761.

Robutti, O. (2006). Motion, technology, gestures in interpreting graphs. *International Journal for Technology in Mathematics Education, 13*(3), 117–125.

Roschelle, J., Pea, R., Hoadley, C., Gordin, D., & Means, B. (2000). Changing how and what children learn in school with computer-based technologies. *The Future of Children, 10*(2), 76–101.

Roth, W. M., & McGinn, M. (1997). Graphing: Cognitive ability or practice? *Science Education, 81,* 91–106.

Saunders, W. L., & Jesunathadas, J. (1988). The effect of task content upon proportional reasoning. *Journal of Research in Science Teaching, 25*(1), 59–67.

Schoenfeld, A. H., Smith. J. P., & Arcavi, A. (1993). Learning: The microgenetic analysis of one student's evolving understanding of a complex subject matter domain. In R. Glaser (Ed.), *Advances in instructional psychology* (Vol. 4, pp. 55–175). Eillsdale, NJ: Lawrence Erlbaum.

Schultz, K., Clement, J., & Mokros, J. R. (1986). *Adolescents' graphing skills: A descriptive analysis.* Paper presented at the Annual Meeting of the American Educational Research Association. San Francisco, CA.

Segal, A. (2011). *Do gestural interfaces promote thinking? Embodied interaction: Congruent gestures and direct touch promote performance in math.* ProQuest LLC. Retrieved from http://gateway.proquest.com/openurl?url_ver=Z39.88-2004&rft_val_fmt=info:ofi/fmt:kev:mtx:dissertation&res_dat=xri:pqdiss&rft_dat=xri:pqdiss:3453956.

Sherin, B. L. (2000). How students invent representations of motion: A genetic account. *The Journal of Mathematical Behavior, 19*(4), 399–441.

Silvestre, A., & da Ponte, J. (2012). Missing value and comparison problems: What pupils know before the teaching of proportion. *PNA, 6*(3), 73–83.

Simon, M. A., & Blume, G. W. (1994). Mathematical modeling as a component of understanding ratio-as-measure: A study of prospective elementary teachers. *Journal of Mathematical Behavior, 13,* 183–197.

Skemp, R. R. (1976). Relational understanding and instrumental understanding. *Mathematics Teaching, 77,* 20–26.

Slavit, D. (1994). *The effect of graphing calculations on students' conceptions of function.* Paper presented at the Annual Meeting of the American Educational Research Association, New Orleans, LA.

Smith, J., Arcavi, A., & Schoenfeld, A. H. (1989). Learning y-intercept: Assembling the pieces of an "atomic" concept. In Vergnaud, G., Rogalski, J., & Artigue, M. (Eds.), *Proceedings of the Annual Conference of the International Group for the Psychology of Mathematics Education* (pp. 176–183). Paris: International Group for the Psychology of Mathematics Education.

Stanley, D., McGowan, D., & Hull, S. H. (2003). Pitfalls of over-reliance on cross multiplication as a method to find missing values. *Texas Mathematics Teacher, 11*(1), 9–11.

Stanton, M., & Moore-Russo, D. (2012). Conceptualizations of slope: A review of state standards. *School Science and Mathematics, 112*(5), 270–277.

Steele, D. (1994). *The Wesley College Technology Enriched Graphing Project.* Unpublished Master's thesis, University of Melbourne, 1994.

Stein, M. K., & Leinhardt, G. (1989). *Interpreting graphs: An analysis of early performance and reasoning.* Unpublished manuscript, University of Pittsburgh Learning Research and Development Center, PA.

Stump, S. L. (1997). *Secondary mathematics teachers' knowledge of the concept of slope.* Paper presented at the Annual Meeting of the American Educational Research Association, Chicago, IL.

Stump, S. L. (2001). High school Precalculus students' understanding of slope as measure. *School Science and Mathematics, 101*(2), 81–89.

Swafford, J. O., & Langrall, C. W. (2000). Grade 6 students' preinstructional use of equations to describe and represent problem situations. *Journal for Research in Mathematics Education, 31*(1), 89–112.

Thompson, A. G., & Thompson, P. (1996). Talking about rates conceptually, Part II. *Journal for Research in Mathematics Education, 27*(1), 2–24.

Thompson, P. (1994). The development of the concept of speed and its relationship to concepts of rate. In G. Harel & J. Confrey (Eds.), *The development of multiplicative reasoning in the learning of mathematics* (pp. 181–234). Albany, NY: SUNY Press.

Thompson, P. W., & Thompson, A. G. (1992). *Images of rate*. Paper presented at the Annual Meeting of the American Educational Research Association, San Francisco, CA.

Tourniaire, F. (1984). *Proportional reasoning in grades three, four, and five*. Unpublished Doctoral dissertation, University of California, Berkeley.

Tourniaire, F., & Pulos, S. (1985). Proportional reasoning: A review of the literature. *Educational Studies in Mathematics, 16*(2), 181–204.

Tufte, E. (2001). *The visual display of quantitative information*. Cheshire, CT: Graphics Press

U.S. Department of Education, Institute of Education Sciences, National Center for Education Statistics. (2013). *National Assessment of Educational Progress*. Retrieved from http://nces.ed.gov/nationsreportcard/nqt/Home/LegacyGen0Bookmark?subject=mathematics&year=2013&grade=8.

Ubuz, B. (2007). Interpreting a graph and constructing its derivative graph: Stability and change in students' conceptions. *International Journal of Mathematics Education in Science and Technology, 38*(5), 609–637.

Vergnaud, G. (1980). Didactics and acquisition of multiplicative structures in secondary schools. In W. F. Archerhold, R. H. Driver, A. Orton, & C. Wood-Robinson (Eds.), *Cognitive development research in science and mathematics*. Leeds, England: University of Leeds.

Vergnaud, G., & Errecalde, P. (1980). Some steps in the understanding and the use of scales and axis by 10–13 year old students. In R. Karglus (Ed.), *Proceedings of the Fourth International Conference for the Psychology of Mathematics Education* (pp. 285–291). Berkeley, CA: University of California.

Wainer, H. (1992). Understanding graphs and tables. *Educational Researcher, 21*(1), 14–23.

Walter, J., & Gerson, H. (2007). Teachers' personal agency: Making sense of slope through additive structures. *Educational Studies in Mathematics, 65*(2), 203–233.

Watanabe, T. (2015). Visual reasoning tools in action: Double number lines, area models, and other diagrams power up students' ability to solve and make sense of various problems. *Mathematics Teaching in the Middle School, 21*(3), 152–160.

Wavering, M. (1985). *The logical reasoning necessary to make line graphs*. Paper presented at the Annual Meeting of the National Association for Research in Science Teaching, French Lick Springs, Indiana.

Wavering, M. J. (1989). Logical reasoning necessary to make line graphs. *Journal of Research in Science Teaching, 26*(5), 373–379.

Zaslavsky, O. (1997). *Conceptual obstacles in the learning of quadratic functions*. Unpublished Doctoral dissertation, Technion, Haifa, Israel.

CHAPTER 5

PATTERNS & FUNCTIONS

> Algebra students for whom the concept of function is central
> are able to identify the important properties of a situation,
> as displayed in any representation, and are better prepared
> than other algebra students to solve traditional algebraic word problems.
> —*Yerushalmy & Shternberg, 2001*[1]

INTRODUCTION

Functions are one of the central ideas in algebra and one of the most common sources of student difficulty. "Teachers' comprehensive and well-organized conceptions contribute to instruction characterized by emphases on conceptual connections, powerful representations, and meaningful discussions."[2] In contrast, teachers' limited views of function have been shown to lead to narrow instruction, with "missed opportunities to highlight important connections between concepts and representations."[3] The study of functions begins informally in very early grades when we ask students to continue or describe a pattern, and continues in higher grades by ***Developing the Concept of Function***[1] with more formal defini-

---

[1] Note: ***bold and italicized*** print indicates a reference to an entry in the book

tions and exploring more varied representations of functions. However, students often struggle to have a comprehensive understanding of functions.[4]

Formal definitions of function for the classroom have evolved over time and vary according to the level of the student. Early 17th-century definitions focused on curves described by motions. In the 18th century, with the emphasis on symbolic representation, functions were defined as analytic expressions made up of variables and constants, representing the relation between two variables. At that time, graphs of functions were assumed to have "no sharp corners" (one sees the connection to calculus here). However, modern mathematics defines a function as a univalent correspondence between two sets; that is, a relation in which every element of the domain is paired with exactly one element of the range (also known as the *Bourbaki* definition).

When first introduced to functions, students learn the Bourbaki definition with numerical domains and ranges: a function matches one group of numbers with another group of numbers. Some teachers do not have an understanding of this modern function concept, and many students also do not: even though they can provide a modern definition, it is not reflected in their understanding. The conceptual understandings of both students and teachers remain closer to 18th-century definitions, in which functions are generally considered to be formulas.

Function Machines are often used as a metaphor for introducing the modern definition. With this model, students are told to imagine inputs being inserted into a machine (the function) and the outputs are the result produced by the machine. As students progress mathematically, they begin working with multiple representations of functions, such as tables, equations, **Graphing Functions**, and verbal descriptions. Students usually begin with **Linear Functions** and become more proficient at **Transforming Functions** from one representation to another. Over time, they learn about the various properties that apply to all functions and those that distinguish one family of functions from another, moving from the idea of a function as a process to the idea of a function as an object (see **Process vs. Object**). Many begin using devices such as the **Vertical Line Test** to determine whether a relation is a function, and some teachers focus on the idea of **Covariation**.

However, there is concern among mathematics educators that students tend to apply tricks and overuse specific function types, leading to misconceptions rather than a deep conceptual understanding. Students with limited exposure to the variety of possible functions end up with misconceptions about what the salient features of a function are. Because students tend to focus on examples of ideas as the ideas themselves, it is important to help students develop **Reasoning and Interpreting** skills and have them work with a wide variety of examples so that they can imagine there are still more. For example, students looking at a graph of a step function might not recognize it as a function at all if the only examples of functions they have are continuous.

If students are only exposed to polynomial functions, they may then equate the notions of "polynomial" and "function," rather than seeing polynomials as just a

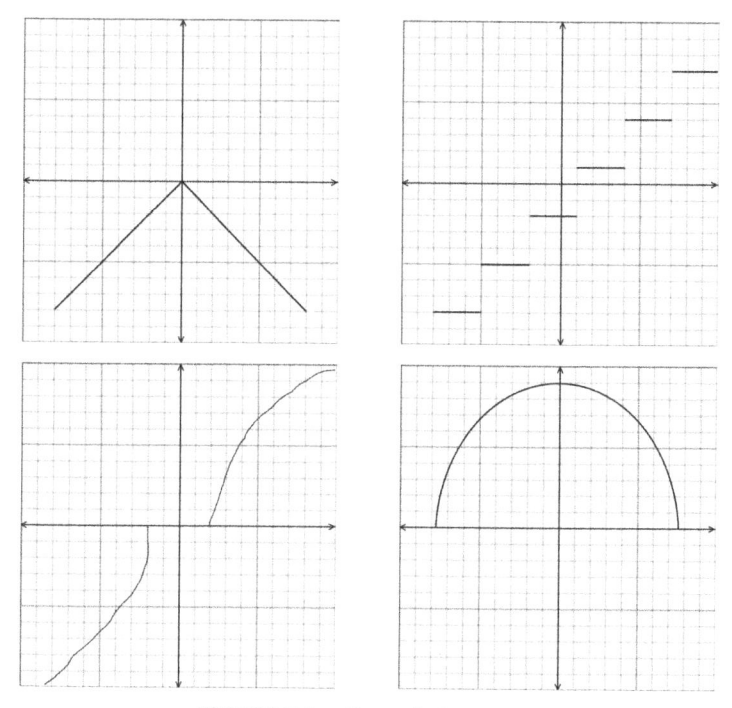

FIGURE 5.1. Example Functions

single class of examples of a broad concept. Accordingly, it is useful to provide many different kinds of examples, including discontinuous functions, functions with discrete domains (see *Sequences and Pictorial Growth Patterns*), functions with exceptional points, piecewise-defined functions, multivariable functions, and functions with non-simple domains such as $\sqrt{x^2-1}$ (see Figure 5.1).

Algebra students are primarily concerned with functions of a single variable with curves in the plane for graphs. But *Generalizing to Functions of Multi-Variables* or higher dimensions may help students see the abstract properties that make a function a function (see *Promoting & Justifying Generalization*). In fact, the whole process of extending a pattern, with which the study of functions begins, is itself a process of generalization.

In this chapter, we explore how functions are described, used, and represented; which ideas about functions students struggle with and why; and what can be done to help them grasp this essential algebraic concept.

DEVELOPING THE CONCEPT OF FUNCTION

Common Core Standards: 6.EE.5, 7, 8, 9; 7.EE.4, 8.EE7, 8.F.1, 2, 3, 4, 5, F.IF.1

Students almost universally struggle with the concept of a function, yet most have already developed a concrete and accurate idea of what a function is without even realizing it. For example, students readily attach names to their classmates' faces and can tell their parents the cost of the new games they want.[5] Each situation has a domain, a range, and a well-defined output for each input.

Nevertheless, students often begin using the term "function" without a clear understanding of what it means. Functions are often taught in schools "with just one of its representations, either symbolic or the graphical one,"[6] and secondary teachers tend to focus on graphs.[7] Many students, "even those successful in College Algebra courses, think a function must be defined by a single algebraic formula."[8] Seven in ten junior high school teachers believed that all functions could be represented by a formula.[9] Students often develop a narrow understanding of the function concept because of the limited variety of function examples they experience in school.[10] They can often understand the idea that each element in the domain corresponds to one element in the range, but can't explain why that is important or why functions are defined in that way.[11] With a limited definition, students may have difficulty understanding functions that are a set of discrete points or equations, such as $f(x) = 12$.[12]

It appears that much of the confusion also connects to some fundamental misconceptions about equality and the equation $y = f(x)$. Consider the following set of problems:

- **Problem 1:** Students are presented relations represented in various ways (i.e., set correspondence diagrams, sets of ordered pairs, equations, and graphs) and are asked to determine which are examples of functions. Which of the following are functions?
- **Problem 2**[13]**:** Give an example of a function f such that for any real numbers x, y in the domain of f the following equation holds: $f(x+y) = f(x) + f(y)$
- **Problem 3**[14]**:** Does there exist a function whose graph is (see Figure 5.2):
- **Problem 4:** Does there exist a function that assigns to every number different from 0 its square and to 0 it assigns 1?
- **Problem 5:** Does there exist a function, all of whose values are equal to each other?
- **Problem 6:** Does there exist a function whose values for integral numbers are non-integral and whose values for non-integral numbers are integral?
- **Problem 7:** What is a function, in your opinion?

Problems 2 through 6 all describe functions that have some "peculiarity" compared to most of the functions defined by formulas that students encounter. Once students realize that there are such functions, they should come to a broader un-

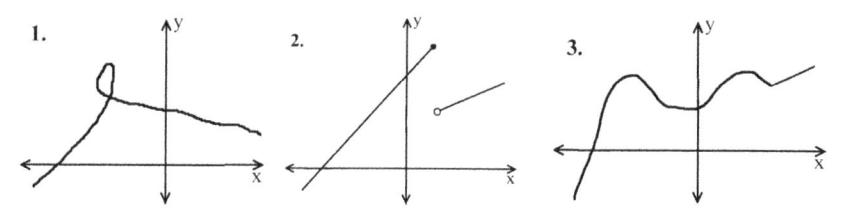

FIGURE 5.2. Example Functions.

derstanding of the function concept that is not limited to algebraic expressions. Problem 7 is an attempt to extract from students an articulation of their conclusions when considering these "unusual" examples. Ideally, there will be a teacher to help facilitate the discussion and ensure that students capture the salient features of functions.

As another potential source of confusion, functions appear in a variety of forms and guises. Sometimes they appear as processes (double the input then add three); sometimes as single, complex objects (a formula or graph); and sometimes as a list of ordered pairs—and the representations all overlap. In the student interview for $f(x + y) = f(x) + f(y)$ in Problem 8 (see Table 5.1), the student shows

TABLE 5.1. A Student's Understanding of Function Problem 8: $f(x + y) = f(x) + f(y)$

Interviewer:	How do you understand the word "function"?
Student:	It's a formula that leads us to drawing a parabola or something like that.
Interviewer:	In that case, will it be the same formula for $f(x)$ and $f(y)$?
Student:	No.
Interviewer:	And how do you understand the expression: "Function f of 3"?
Student:	Umm... It makes me think about a... a "zero of the function," or something like that... Because usually when we write, we write a formula. I mean we don't write a specific number in parentheses. So I think it is about a formula, because it would be difficult to write something like that in such a form. Now I would say something about the formula—instead of those 3 and 5... It's just that I don't know how! Just a moment... (Reads the problem once again) But this is... AN EQUATION?! (Indignantly)
Interviewer:	How do you think, is this an equation?
Student:	Not really, because when there's an f here, it can't be an equation.
Interviewer:	Why not?
Student:	Because if on the left-hand side we leave this [pointing to $f(x+y)$] and here we substitute two formulas of functions—for $f(x)$ a formula of a function and for $f(y)$ a formula, too—only then we would have it in the form of an equation.
Interviewer:	Would you substitute the same thing for $f(x)$ and $f(y)$?
Student:	No. Something different.
Interviewer:	Something completely different?
Student:	Yes, for example, $f(x + y) = (x^3 - 2x + 3) + (x^2 + 5)$.

a lack of understanding of function notation, once indicating that f was just an abbreviation for the word "function" and once saying that the letter f is the start of a function's formula. Later in the interview, we see that the student views $f(x)$ and $f(y)$ as being completely different functions. Students may consider $f(x)$ as a single symbol, in which case $f(y)$ can refer to an entirely different function, as above. To overcome this, it needs to be tackled head-on: the roles of f, x, and the parentheses in the expression $f(x)$ need to be considered separately, and teachers should be sure to use many different letters and symbols in such expressions so students don't become wedded to a particular form. In particular, if y is generally reserved for the output of a function, then when it is used in any other way, confusion results. More examples and problems using y as something other than a function output could reduce this confusion and help students see the need for declaring their variables.

Thus, while students can learn and reproduce the definition of a function, most of the examples they see are of the same few kinds and are generally given by a formula. As a result, the students end up with an "understanding" of functions as *things that have a formula*. This leaves them with an inconsistent view of functions: when asked to define "function," they are able to give the Bourbaki definition, but when faced with "unusual" functions (discontinuous, discrete, etc.), they cannot identify them as functions and have only a vague notion of what is meant by the term *function*.

For example, students were given Problems 3–7 from the table.[15] The authors grouped responses to Problem 7 into six categories and then compared what students *said* about functions to how they actually *thought* about functions. Even when a given function satisfied the definition they provided, many students said it was not a function if it had an "unusual" property such as a discontinuity or piecewise definition. The six categories of students' definitions of function are listed here, along with what the students said in response to Problem 7 (What is a function?) that put them in that category.[16]

I. *Correspondence*: A function is any correspondence between two sets that assigns to every element in the first set exactly one element in the second set (the Dirichlet-Bourbaki definition).
   - "A correspondence between two sets of elements."
   - "For every element in A there is one and only one element in B."

II. *Dependence Relation*: A function is a dependence relation between two variables (y depends on x).
   - One factor depending on the other one."
   - A dependence between two variables."
   - A connection between two magnitudes."

III. *Rule*: A function is a rule. A rule is expected to have some regularity, whereas a correspondence may be "arbitrary." The domain and the codo-

main were usually not mentioned here, contrary to Category I, where they were.
- "Something that connects the value of x with the value of y."
- "The result of a certain rule applied to a varying number."
- "A relation between x and y is a function."

IV. *Operation*: A function is an operation or a manipulation (one acts on a given number, generally by means of algebraic operations, in order to get its image).
- "An operation."
- "An operation done on certain values of x that assigns to every value of x a value of $y = f(x)$."
- "Transmitting values to other values according to certain conditions."

V. *Formula*: A function is a formula, an algebraic expression, or an equation.
- "It is an equation expressing a certain relation between two objects."
- "A mathematical expression that gives a connection between two factors."
- "An equation connecting two factors."

VI. *Representation*: The function is identified, in a possibly meaningless way, with one of its graphical or symbolic representations.
- "A graph that can be described mathematically."
- "A collection of numbers in a certain order which can be expressed in a graph."
- "$y = f(x)$."
- "$y(F) = x$."

We can see that students have very distinct ideas about what functions are. Some of the ideas show a limited understanding of functions—for example, that a function is a formula—while others, on the surface, seem to imply a deeper understanding but sometimes turn out to be rote definitions that have little or no conceptual understanding beneath them. Asking the student questions is the only way to figure out the student's understanding.

A learning trajectory for functions involves multiple aspects of the concept:[17]

1. Function definition
2. Operations
3. Composition and inverse
4. Linear, polynomial, and rational functions
5. Exponential functions
6. Sequences
7. Rate of change and characteristics of graphs

8. Quadratic functions
9. Trigonometric functions
10. Transformations
11. Modeling with functions

Notice that only 1, 7, possibly 10, and 11 get away from functions described by a formula.

Once a teacher has categorized a student's notion of function, he or she can begin to probe its depth. For example, for the student who thinks that a function must be a formula or an operation, a teacher could start with a graph of x^2 and ask if it shows a function, and then modify it just slightly (in such a way that there is no obvious formula for it) and ask whether it is still a function.

Teachers

Non-standard questions (such as those employing functional equations as in Problem 2) can test the depth of a student's understanding of functions. Because they are non-standard, however, a student is likely to need some assistance on such a problem. Students may better internalize the notion of a function if they are exposed to a variety of descriptions of functions, including non-standard problems. This forces the student to consider what a function is at a more fundamental level and may dispel misconceptions about functions. While "Skill Drills" are important for procedural mastery, they often come in large clusters with a single set of directions (e.g., "Factor each quadratic expression below"). If a similar computation arises in a different context, students may find themselves with no idea how to proceed. Plenty of practice with non-standard or differently phrased questions can help students recognize familiar algebraic steps in unfamiliar contexts.

Consider the following strategies for teaching algebra and functions:[18]

- Begin before Grade 5 and build on students' informal knowledge,
- Work on students' struggles with function notation,
- Include several different forms of algebraic thinking,
- Integrate learning algebra with learning other subject matter,
- Encourage students to reflect on what they learn and articulate what they know, and
- Encourage active learning and the construction of relationships that puts a premium on sense making and understanding.

In order for students to develop a robust understanding of the function concept, they need to see many substantially different examples of functions—including, for example, functions given by a formula (of which they see plenty), a graph, a table, a verbal description, or a diagram, as well as both continuous and discontinuous functions and functions defined on a discrete domain. For example, teachers can approach linear functions through price models, such as determining the total cost for some number of pounds of bananas priced at $3.98 per pound. Such examples bring out the ideas of domain, scale, precision, and rate of change. To bring in the concept of a *y*-intercept, teachers can change this to a commission model (for example) in which a salesperson receives a fixed salary plus a commission. This revisits all of the ideas listed previously and introduces *y*-intercepts. Many cell phone plans charge by the minute, where fractions of a minute are rounded up; this is an example of a step function. Interest charges grow exponentially, taxes use a discrete domain and have a piecewise definition, etc. Building time into the lesson for discussion of the models gives students a chance to verbalize and solidify their understanding of these ideas. Another essential component is gradually introducing more and more sophisticated ideas (e.g., linear—quadratic—exponential—trigonometric or smooth—cusped—discontinuous—piecewise-defined) over time, which requires an earlier introduction to the basic concepts.

Teachers can take advantage of everyday function situations to make strong connections between functions and the ideas students have already internalized. Rather than starting with the definition of a simple linear function, like $f(x) = 3x$, teachers can begin with a concept of function that ties immediately to the students' experiences. Abstraction to function notation and variables should come easier for students who already understand the underlying principles.

Be aware of three common misconceptions that teachers and students have about functions with respect to the notion of *arbitrariness*:[19]

1. Functions are (or can always be represented as) equations (counterexample: an arrow diagram associating each student with his or her height).
2. Graphs of functions should be "nice" (counterexample: a step function).
3. Functions are "known" (counterexample: a squiggly graph drawn on the board).

On the other hand, it is also common to assume a relation that fits these criteria is a function—a circle being a good example. Teachers can focus on helping students understand and make proper use of the **Vertical Line Test**. Too often it is used in place of a more comprehensive function definition or as a rule that students can follow to get the right answer, devoid of understanding.

Make sure students understand *univalence*: "a function is a relation such that every number in the domain is matched to one and only one number in the range" (many state it backwards). Many teachers do not know the history that led to functions being defined this way and therefore do not understand the mathematical need for univalence. Also, students inadvertently impose a constraint of "uniqueness" (i.e., functions must be one-to-one), leading them to conclude that relations such as $f(x) = 2$ are not functions.[20]

As noted above, students should be exposed to as wide a variety of function types as possible. "The ability to identify and represent the same concept in different representations, and flexibility in moving from one representation to another, are crucial in mathematics learning, as they allow students to see rich relationships, and develop deeper understanding of concepts."[21] They should see continuous and discontinuous functions, functions with algebraic descriptions and functions with no algebraic description, functions with continuous domains and functions with discrete domains, etc., and the representations of the functions should also vary. Furthermore, students should see examples of relations that are not functions.

VERTICAL LINE TEST

Common Core Standards: 8.F.1, F.IF.1

Fairly early in their study of functions, students learn about the vertical line test as a way of using the graph of a given equation or relation to determine whether it is a function. The vertical line test states that if a vertical line intersects the graph in more than one place, then the equation or relation is *not* a function. Studies with pre-service teachers indicate that overuse of the vertical line test contributes to limited conceptions of function, as well as misconceptions about functions that are not necessarily represented by an equation.[22] This is illustrated in the interview exchange in Table 5.2.[23] As a result, students often do not believe that ordered pairs,[24] tables of values, or statements such as "Mary owes $6, John owes $3, and Sue owes $2" can be functions.[25] While it is generally assumed that teachers' subject-matter knowledge and pedagogical content knowledge are interrelated, there is little research to help us better understand these relationships. The function concept is especially challenging and complex.

The following is a framework for developing a deeper understanding of function:[26]

a. Essential features—what is a function?
b. Different representations of functions
c. Alternative ways of approaching functions
d. The strength of the concept—the inverse function and the composition of functions
e. Basic repertoire—functions of the high school curriculum
f. Different kinds of knowledge and understanding of the function concept
g. Knowledge about mathematics

This framework provides a structure that will help ensure students see the variety of functions they need to in order to develop a solid understanding of them. It begins with what a function is, then continues to ways of representing them, fundamental operations on functions, a basic library of functions students should know, and a broader view of functions. For example, consider the characteristic function of the rational numbers: $g(x) = 1$ if x is rational and $g(x) = 0$ if x is irrational. One student teacher showed a rote understanding of the definition of a function but did

TABLE 5.2. Functions and Equations Interview

Student Teacher:	Yes, I think you could write all functions in terms of equations. It might be a trigonometric equation, like $\sin x$, but in every term the y value is going to be equal to some operation with an x value.
Interviewer:	So you can always describe a function using a formula or equation.
Student Teacher:	Yes, I think so.

248 • HOW STUDENTS THINK WHEN DOING ALGEBRA

not have a solid grip on the concept: "Yes, there is an assignment of a single value to each number." Later in the interview, this gave way to, "I don't know if it's a function. It fits the criteria of mapping, but it does not look pretty. It is not really graphable. It might just not be."[27]

Problems 9 and 10 approach the function concept from different directions and can be used to assess the strength of a student's understanding of functions.[28] They include functions that are discontinuous, have discrete domains, are constant, and that can't be graphed in any reasonable way. They can be used as a diagnostic tool to determine where students have gaps in their understanding of functions.

Problem 9

A student marked all of the following (see Figure 5.3) as non-functions (R is the set of all real numbers. N is the set of all the natural numbers).

a. For each case, decide whether the student was right or wrong. Give reasons for each of your decisions.
b. In cases where you think the student was wrong, try to explain what the student was thinking that could cause the mistake.

(iv) A correspondence that associates 1 with each positive number, -1 with each negative number, and 3 with zero.
(v) $g(x) = x$, if x is a rational number, or
0, if x is an irrational number
(vi) $\{(1, 4), (2, 5), (3, 9)\}$

Problem 10

When asked to draw a graph of a function that passes through the points A, B, and C, a student gave the answer in Figure 5.4:
What do you think about this answer? Is it correct? Why?

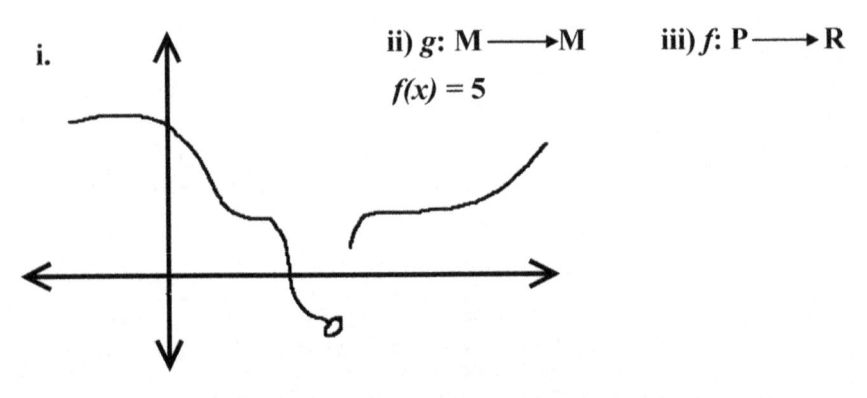

i.

ii) $g: M \longrightarrow M$
$f(x) = 5$

iii) $f: P \longrightarrow R$

FIGURE 5.3. Student Answers.
Adapted from Even (1993)

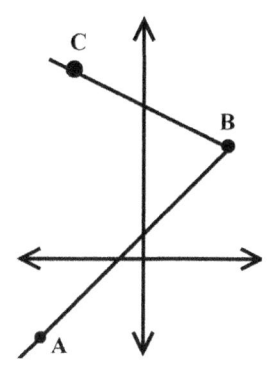

FIGURE 5.4. Student answer to Problem 10.

- If yes--is there any other correct answer? Give me an example. Can you find another one? Why do you think the student gave this answer?
- If no--What would you like to see as a correct answer? Give me an example. Another one. Why do you think the student gave this answer?

Teachers

- Avoid overreliance on the vertical line test.
- Remember that not all functions can be represented by a "formula."
- Provide examples of functions that are not readily graphed.
- Verify that students understand the connection between the vertical line test and the definition of a function.

FUNCTION MACHINES

Common Core Standards: 8.F.1, F.IF.1

Too often, students associate success in algebra with symbol manipulation. This focus on manipulating symbols leads to students having little flexibility in their algebraic thinking and often having considerable difficulty with function notation specifically.

Student self-evaluations from a course on college algebra indicated that many felt a function machine model helped them make sense of notation, organize their thinking, and generate equations from given data.[29] Students who were the most successful in the course made references to input and output as well as function machines in their written work, while the least successful students did not. One student's self-evaluation said,

> I know that function machines are good models for mathematical relationships when one wants to clearly identify and separate the input, output, and process. I also can look at function notation, $f(x)$, and understand what it is stating: that f is a function of x. I see the input (x), the output $f(x)$, and the process, when in a relationship form like $f(x) = 3x+10$ ... I know now that "solving" can refer to many different things including:
>
> 1. Solving for an unknown variable;
> 2. Solving a system of equations (where the given functions share the same input and output values); and
> 3. Solving an equation (finding the input value(s) that produce an output of 0).[30]

This was without being asked specifically to discuss function machines; the text emphasized a function machine approach for the first few chapters, then did not mention them. The prompt for the self-evaluation was "What mathematics have you learned during this time? Write a summary of what you have learned." Stronger students are able to think flexibly in this way, able to evaluate a function and to determine which input(s) correspond to a given output.

Another student said,

> I had no idea what a function is and what a relation is, but now I do. Relation is mathematical notation with at least two variables. Independent is input. Dependent is output. I know that a relation, which has only one output for each given input is a function. If I see $f(8) = x + 7$, I know I am given an input. $f(x) = 8$ gives me output. I can work with ordered pairs. In case (8, 45), I see 8 as input and 45 as output."

This student understood function notation, had a grasp of the roles of inputs and outputs, and understood multiple representations of functions.

Compare the following tasks:

Task 1: Given a table from a TI-82 graphing calculator. What are the output(s) if the input is -2? What are the input(s) if the output is -3?

Task 2: Consider the equation $y = 3x - 7$. What are the output(s) if the input is 5? What are the input(s) if the output is 0?

In Task 1, students tended to answer both questions correctly. In Task 2, students did not tend to answer both questions correctly, but they made strong gains over the course of the semester, a semester in which a function machine approach was used.[31] Although there was no control group in this experiment, students' self-evaluations indicated they attributed their gains at least partially to a function-machine approach. In Task 1, students could read inputs and outputs directly from the table, which did not rely on any algebraic skill, while in Task 2, they needed to understand the functional relationship between x and y.

A function machine representation combined with an emphasis on the process-oriented view of the notion of function helps student better organize their thinking and make sense of the notation. Of course, for maximum effect this should not be the only view presented. The Center for Algebraic Thinking's *Function Mystery Machine* app[2] is a useful tool for teaching this concept; it explicitly treats a function as a machine.

Teachers

- A function table from a graphing calculator may help students gain flexibility in thinking about functions.
- The problems above can be used as formative assessments.
- "Function machine" perspective on functions can help students understand function notation and what a function is from a process perspective.

---

[2] All Center for Algebraic Thinking apps are freely available on iTunes except Math Flyer ($0.99).

COVARIATION

Common Core Standards: 8.F.4, 5, F.IF.4, 5

The function machine approach is very much focused on functions as a process, which, while useful, is a limited perspective. A *covariation* approach to functions may help to expand students' views of functions beyond simple input-output procedures. In addition, it may be more intuitive for some students than the usual input-output approach: how do x and y change together? Furthermore, if the changes in x are unit changes, this ties in very neatly with introducing students to the concept of rate of change, including non-constant rates of change (see **Analysis of Change: Understanding Rate of Change** and **Ratio and Proportional Reasoning** as well as other sections in Chapter 4).

In their efforts to chart a visual course to the concept of function as one of covariation and a rate of change, researchers found that providing images and language for a list of function properties (see Table 5.3) helped students use the terms constant, increasing, and decreasing when describing functions. Students made the link between a situation and its visual representation but struggled to conceptually understand the link between a graph and its symbols. Consider Problem 11:

> Students are given a worksheet in a dynamic geometry package. The construction, hidden from the students, is as follows: given two points, A and B, construct the line

TABLE 5.3. Representing Visual Properties of Functions

		Quantity	
Rate of change	Increasing	Decreasing	Constant
Increasing	⌣	⌢	
Decreasing	⌢	⌣	
Constant	╱	╲	───

Source: Yerushalmy & Shternberg (2001)

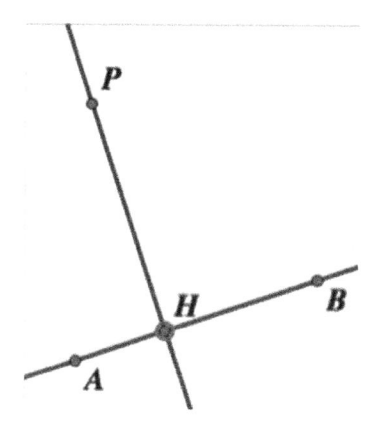

FIGURE 5.5. Sample Problem 11 Answer

through them. Take a third point, P, not on the line, and construct H, the foot of the perpendicular from P to line AB. Students can drag any of the points. The problem is to determine the relationships among the points (Figure 5.5 shows a possible answer). Ask: "What does it mean to say two functions are equal?"

Problem 11 allows any of the variables (points in the plane) to be the "independent" variable, creating a situation in which covariation is the natural way to think about functions. The interview in Table 5.4 is a conversation regarding two functions being equal, having explored Problem 11 with dynamic geometry software. Rather than just giving the students a definition of function equality, the teacher has a conversation with them about what it should mean. During the conversation, they bring up several misconceptions that are explored and then either built upon or discarded, as appropriate. In the end, the students have developed a working definition of function equality that, while it lacks rigor, nevertheless captures the principle they were working toward.

Teachers

- A covariation approach to functions can help students see beyond the process view of functions and gain a larger perspective on functions.
- A covariation perspective can help students see a *mutual* dependence between the variables.

The downside of a covariation approach is that, by its nature, there is some implication that the functions are discrete: to see how a function behaves with this approach, one tries a few values and looks for a pattern. From this perspective, a function is almost like a sequence. Teachers who adopt this method will need to be aware of such a pitfall. One way to avoid that trap is to use dynamic geometry software such as Cabri or Geometer's Sketchpad. As in Problem 11, teachers can

TABLE 5.4. Interview for Problem 11

Teacher:	We must find an agreement on a definition, which can be one of these, or an improvement of one of these, or the fusion of these... We must decide.
Andrea:	According to me, Gabriele and Marco's definition is wrong.
Teacher:	So, Andrea, according to you, Gabriele and Marco's definition is wrong. Let's read it again (she reads again). "Two functions are equal when they have the same range and (when) the same domain is fixed for both."
Andrea:	Because to get to the same range, someone could pass through... we could have several journeys; in fact, if there were a subset of the domain... we can't say that the functions are...
Teacher:	Tiziano, could you try to explain it better?
Tiziano:	Yesterday, we saw that we can, by doing the same domain, we can create the same range and this, with different functions (he means construction procedures).
Teacher:	Let's read the text. You say that if they have the same domain and the same range for each subset of the domain...
Tiziano:	But, here it's like having the same procedure.
Teacher:	Hum, and why is it like having the same procedure?
Several voices:	Because...
Gabriele:	...As we go further, the subsets of the domain and vice versa...
Teacher:	Do you agree, Andrea?
Gioia:	The domain is the plane, then you have the straight line, then a segment...
Teacher:	What are these?
Andrea:	The domain can be whatever.
Gioia:	They are subsets.
Teacher:	And then, the procedure, what does it do? That is to say, I.... Where does it start from?
Andrea:	The domain can be one point too... if we want!
Teacher:	The subset of the domain can be one point too. Oh!
Andrea:	For whatever point, we get the same point of the range.
Teacher:	And this gives us the idea to say that...
Gioia:	I'm doing the same procedure.
Andrea:	I'm doing the same procedure.
Teacher:	I'm doing the same procedure. Therefore, for whatever point of what?
Andrea:	For each point of the domain we have the same... as the result of the function, the same point of the range.
Teacher:	Do you agree? (referring to Tiziano. The students are perplexed. Silence. The teacher writes at the blackboard and reads:) "For each point of the domain, we have as the result of the function, the same point as the range."

Souce: Falcade, Laborde, & Mariotti (2007)

provide students with a predetermined construction that is hidden from the students. Students then manipulate different components of the diagram (given lines and points) and observe the resulting changes in the other parts of the diagram. This is a natural example of a continuously varying function that students can internalize at an intuitive level; the motion of the points illustrates the covariation idea since students can move any of the points and see the effect on the others.

Of course, just playing on the computer will not in itself achieve the desired result. In one study, the following structure helped students understand the concept of a function:[32]

- Students engage in lab activities, exploring a task with the software. They choose a point to move and observe how other points are affected; this establishes a relationship between the "input" (the point they chose) and the "output" (the point they observe).
- Students write about their experience with the software and any thoughts or conclusions they have.
- The class discusses the relationship together. The teacher's role in guiding this discussion appropriately is essential. She or he must be able to discern the key concepts when they arise, even if they are not well articulated.
- At some point, students will need to move from this geometric presentation of a function to numerical or algebraic functions. In order for this approach to be effective, students need to see the connection with the underlying principle of functional dependency.

With such an exercise, students experience a function that has no "formula" and whose inputs and outputs are not just numbers, expanding their concept of what can be a function. In addition, the functional relationship is visual rather than algebraic, potentially helping students whose algebra skills are weak to see what a function is.

PROCESS VERSUS OBJECT

Common Core Standards: 8.F.2, 3

The foregoing discussion of function machines and covariation leads us to one of the major difficulties students have with functions: reconciling the apparently conflicting dual natures of functions as processes and as objects (see also ***Variables & Expressions: Acceptance of Lack of Closure*** for a similar discussion of students understanding expressions as objects or products versus processes). The former corresponds to thinking of a function as a rule—generally, one that can be expressed algebraically (e.g., $y = x^2$)—and the latter corresponds to thinking of a function as a single entity (e.g., the entire graph of a function). In other words, the process perspective links the input and the output, while the object perspective holds the various representations as complete entities. It can be difficult to keep both in mind simultaneously, but with practice, students can become more fluent at changing perspective, depending on the situation. With this fluency comes a greater understanding of the nature and use of functions.

Traditional algebra curricula encourage students to develop skills for symbolic representation of their ideas prior to developing skills for graphical representation. Word problems reinforce this notion by having students apply formulas, rather than model real life. As a result, students tend not to experiment with graphical representation and, when faced with an authentic modeling situation, they attempt to represent the situation symbolically rather than graphically, even when a graphical representation would be more appropriate (see, for example, ***Modeling & Word Problems: Drawing Models as an Intermediate Step in Solving a Word Problem***).

In this section, we examine some problems that conceive of functions as objects. In particular, students are asked to solve problems that are difficult or impossible to solve algebraically, but can be solved somewhat easily when considered graphically. Compare this with the ***Vertical Line Test*** section. Consider Problem 12, addressing students' understanding of graphs.[33] More than 3/4 of the 178 students chose an algebraic rather than a geometric approach, and fewer than a third recognized a graphical approach as an option. An algebraic approach to this problem is both inefficient and difficult. However, a simple reading of the graph can give (-2, 4), (-1,1), (0,-2), and (1,-5) as solutions with minimal effort.

Problem 12: Missing x Coefficient

The graph in Figure 5.6 represents the equation

$$?x + 3y = -6$$

(we do not know the value of the coefficient of x).

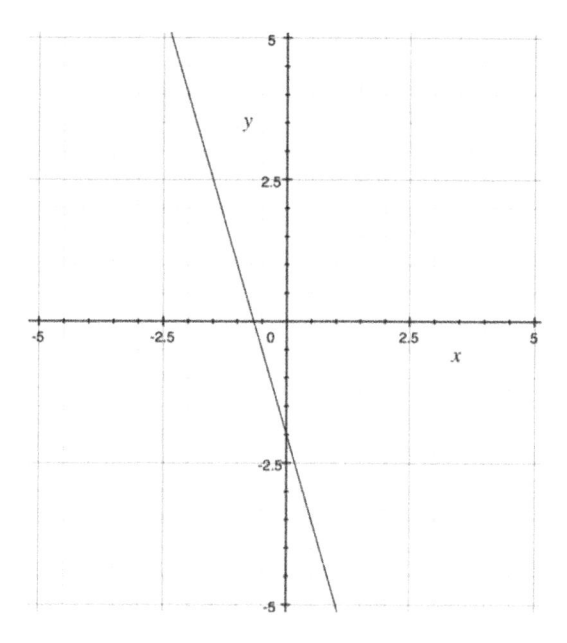

FIGURE 5.6. Graph representing $?x + 3y = -6.90$
Knuth (2000)

a. Is it possible to find a solution to the equation without the missing value? Explain your answer.
b. How could we find the missing value? Explain your answer (also provide an alternative solution).

Students were asked to make a graphical representation in the xy-plane of the following situation: "Hava loves to play with the ball. She throws it to the ground. The ball hits the ground and then it hits the wall." They did this with software that simultaneously produced graphs of x and y as functions of a parameter t (see Table 5.5).[34]

After they made their drawings, they were given graphs of how x and y changed over time and asked to interpret them according to their drawing. Students readily interpreted the y-coordinate's change over time, but had trouble interpreting the x-coordinate's change over time.[35] Some of their difficulty may stem from the fact that they are accustomed to having a horizontal x-axis, but this parametric rendering uses a vertical x-axis. Note that because the problems are presented in a qualitative way, they are naturally approached as objects; there is no standard process to apply here.

The optimization tasks on Problems 13 and 14 in Table 5.6 were intended to assess students' beginning notion of function and to lead them through a three-

TABLE 5.5. Hava Bounces a Ball

Narratives	The path of the ball plotted on xy-plane	The y-coordinate's changes in time	The x-coordinate's changes in time
We are drawing the movement of Hava's ball. Now we go up. Here is Hava's ball. It touches the floor,			
			I think that here it is Hava's Ball. She threw it to the ground.
then hits the wall,			
		So, here it hits the wall,	Here somewhere it stayed on the wall.
goes back down			
		goes back to the ground,	It hits the wall and stayed here for a while, just stayed.
and comes back to Hava.			
		and then goes back to Hava	nd then bounced back to Hava

TABLE 5.6. Optimization Tasks

Problem 13	Response
At time $t = 0$ water begins to flow through a hose into an empty tank for two minutes. At that point, the tap is closed for a little while. After a while, a pump starts up and pumps water out of the tank. Draw a graph of the volume of water in the tank against the time.	Seventh-grade students who had experienced a visually focused pre-algebra curriculum readily drew graphs indicating the volume increasing, hold constant for a short time, and then decreasing.

Problem 14	Response
A cook has a large portion of meat at room temperature that should be cooked as quickly as possible. He has at his disposal a conventional oven and a microwave. In the microwave oven meat temperature increases at a constant rate, and in the conventional oven it increases at a changing rate. Using the results of an earlier cooking trial, the cook knows that although meat temperature in the conventional oven is always higher than in the microwave, cooking time for this piece of meat would be 2 hours in either one of the ovens. Could cooking time be less than two hours using some combination of these two ovens?	A group of students who had taken algebra, pre-calculus, and calculus attempted this problem and struggled. A group of 7th graders who had experienced the Visual Math program and its visual approach to thinking about functions and authentic contexts readily solved the problem.

step process designed to develop their understanding of functions.[36] Students recognized the structure of the relationship through by-hand calculations that they could represent with a table, graph, and equation in a Technology-Rich Teaching and Learning Environment (TRTLE). The web in Figure 5.7 shows the meanings of function explored through the lessons with the TRTLE environment.

In the second step, students consider Problem 14 in Table 5.6. The students who had had algebra, pre-calculus, and calculus mostly failed to solve the problem because they attempted a symbolic approach when there was no data to work with. In contrast, seventh-grade students were highly successful by approaching the problem from a graphical—that is, object-oriented—perspective. They sketched graphs of temperature versus time for the conventional and microwave ovens, drew conclusions about their shapes, and graphically combined them to form a solution.[37] Samples of the conversation among pairs of students as they worked on the Health Clinic Task appear in Table 5.8. In the first conversation, Ben and Ken discuss the gradient (slope) as a ratio that they need to calculate. They demonstrate a fragile sense of finding the slope.

In the second conversation (Table 5.9), Kate and Meg struggled to understand what the problem was asking but demonstrated a developing understanding of gradient. Kate showed evidence of understanding what Meg was saying, as she then suggested that for a third patient the cost would be an additional $20 compared to the cost for two patients. The pair then proceeded to calculate similar

260 • HOW STUDENTS THINK WHEN DOING ALGEBRA

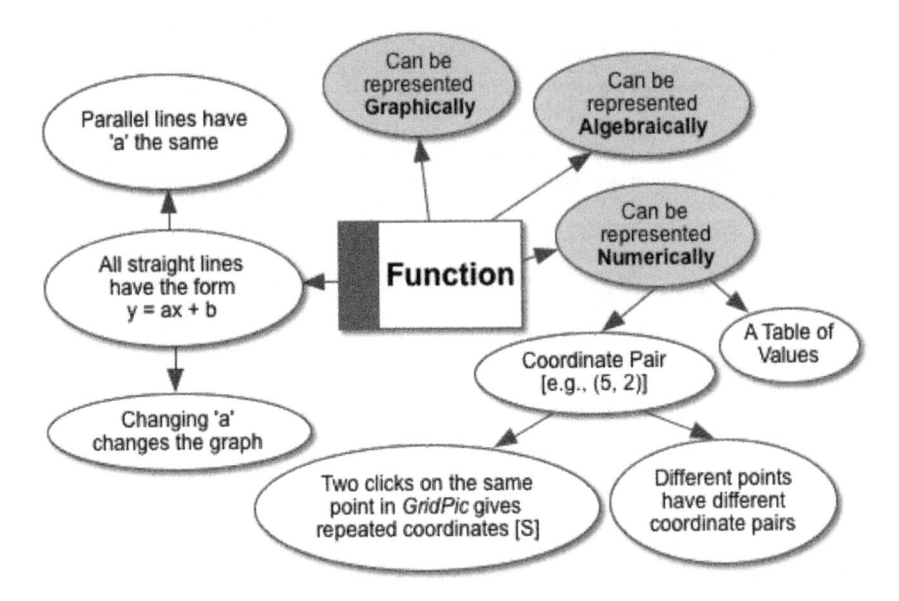

Function (central concept) connected to:
- Can be represented Graphically
- Can be represented Algebraically
- Can be represented Numerically
  - A Table of Values
  - Coordinate Pair [e.g., (5, 2)]
    - Two clicks on the same point in *GridPic* gives repeated coordinates [S]
    - Different points have different coordinate pairs
- Parallel lines have 'a' the same
- All straight lines have the form $y = ax + b$
  - Changing 'a' changes the graph

FIGURE 5.7. Webs of meaning: Conceptions related to functions raised during lesson 1. Brown (2007, p. 158)

TABLE 5.7. Problem 15: Health Clinic

A Village Health Clinic in Mali

The weekly cost of running a small village clinic at Lake Haogoundou in Mali is a function of a constant weekly value and varies with the number of patients (n) attended. The cost is $1100 when 50 patients are treated and $1740 when 90 patients are treated.

Part 1 of the task required students to:

Draw a plot showing a linear relationship for an appropriate domain. Identify the relationship decoding from the text. Identify the domain and the dependent and independent variables. Using a linear rule $C(n)$, find C given n. Write the linear relationship as an algebraic rule. Find n, given C.

Part 2 of the task required students to:

[Functions for two other clinics are given: Bamako: COST = 390 + 17.50 x number of patients, Timbuktu: COST = 115 + 19.75 x number of patients]

Find the costs, C_B and C_T, given $n = 50, 60, \ldots 200$ recording in the table of values. Focusing on one specific value of n, state which cost is cheapest, C_B or C_T. Calculate $|C_B - C_T|$ for this value of n. Determine the value of n when C_B becomes lower than C_T. Construct a table to support this result. Explain how the table of values supports these ideas. Graph the 2 functions over an appropriate domain. Identify the rate for C_B and C_T.

Source: Brown (2007, p. 158)

Patterns & Functions • 261

TABLE 5.8. Interview for Problem 15

Ken:	Well we have two points.
Ben:	Yeah, so. Village Health Clinic in Mali.
Ken:	1100 minus.
Ben:	No, 1340 - 1100. [Calculates $(1340 - 1100)/(170 - 90) = 240 \div 80$]. 240. Divided by 80. So the gradient is three.

TABLE 5.9. Second Interview for Problem 15

Meg:	What have I got? I think you just, I am going to put in just one [hesitantly].
Kate:	Mmm. And then?
Meg:	Okay, 19.75 x 1 + 115, [pause] whatever.
Kate:	How come you are doing it times 1?
Meg:	Umm, because when you find each additional patient after, like from, you go up by one.
Kate:	Oh yeah.
Meg:	It is hard to explain. Each time it goes up by. Each time it adds on to the 115 [fixed cost]. (Later, Meg suggested that she was now confident that they are being asked for more than the cost for one patient.)
Meg:	Wouldn't it be 20 [approximate difference between $C(2)$ and $C(1)$]?
Kate:	What?
Meg:	I thought it asked to do one, but to do each additional one. So what if you times it by two and then take away what you got for one. Like then it is about 20. It will go up by 20 each time you treat somebody.
Kate:	Why 20?
Meg:	Because you times. If you treated 2 patients you add, yeah you get, well, [how] much it would cost and then you take that away from one [$C(1)$]. No take 1 away from that. [$C(2) - C(1)$].
Kate:	Wait, what are you saying? You go 19[.75 x 1 + 115] [pause]. Yeah, you do that right?
Meg:	Yep, yeah.
Kate:	Then what?
Meg:	Then there is like one, and then there is like two. And there is like the difference. I think the difference there is like 20. Because if this here is 134.5 [$C(1) = 134.75$], and this one here is like 15.5, no 14.5. Wouldn't you just take them away? [$C(2) = 154.50$]
Kate:	[incoherent] It is 154.5.
Meg:	That is how many if you treat two patients, that is how much it costs.

TABLE 5.10. Third Interview for Problem 15

Amy:	[Reads] "What is the cost for treating each additional patient at Malange?"
Ben:	14.
Amy:	Yep.
Ben:	And then 12.50.
Amy:	Yeah. Gee that is intelligent having to figure that out!
Ben:	Yeah I know. [Facetiously] That was the hardest question!
Amy:	Yeah.

values for the clinic at Bamako. An error in their subtraction, which was identified as they checked their results, led to their recognition that they had found the gradient.

In contrast, for two other students who came together during the final lesson, this subtask seemed trivial as their notions of gradient were stronger (see Table 5.10). The differences among these pairs of students were not evident in the students' written work, but rather in the conversation.

Teachers

- Students who see functions only as processes or only as objects are limited by their perspective. For many students, the process perspective is the easiest to understand, so they tend to stick to it.
- Problems that require graphical reasoning can lead students to a better understanding of functions as objects.
- The ability to reason graphically puts another tool in students' toolboxes. Sometimes, a graphical argument is more natural than an algebraic one.
- Ask questions for students to explain their thinking in order to reveal the depth of their understanding of functions.

Students are generally assumed to understand the connection between algebraic and graphical representations of functions once the idea has been introduced. Accordingly, little time is devoted to this connection after the first year of algebra. However, the clear difference in understanding between the groups above illustrates how weakly some students connect the two representations and suggests that more time exploring the connections may be warranted.

Students will usually choose an algebraic technique even when a graphical one is more efficient. This appears to stem from an emphasis on teaching algebraic techniques over graphical ones. A shift to a more even emphasis could help. Furthermore, students are quite competent at taking equations and creating corresponding graphs, but they are substantially weaker at taking graphs and constructing equations from them. In particular, Knuth states that "the majority of students did not suggest that the selection of any point on the graph of a line would be a solution to the equation of the line."[38] More graph-to-equation translation problems would help to cement students' understanding of the connection.

Younger students in pre-algebra, with the proper tools and prior experience, can create graphical representations of complex situations without having any idea how to represent them symbolically. This functional approach to teaching algebra allows students to develop deep conceptual understanding before introducing traditional notation. Bridging the gap to symbolic representations may still pose difficulties, though.[39]

GRAPHING FUNCTIONS

Common Core Standards: 8.F.2, 3, 4, 5; F.IF.4, 5, 6, 7, 8 9

A graph of a function is perhaps the fastest way to convey large amounts of information about the function as a whole (or as an object). Our brains become skilled at extracting that information and interpreting it in context, but it requires familiarity and practice with graphing and reading graphs.

Problems were posed to students that required them to graph functions and/or data. The types of errors are summarized in Table 5.11. The first step in helping a student learn from a mistake is identifying the mistake. Table 5.11 can help a teacher do so. Some mistakes (such as perceptual illusion) show a fundamental lack of understanding, while others (such as transfer error) may indicate mere carelessness. Problem 16 (see Table 5.12) provides some context to examine each of these errors and the following figures demonstrate examples of each of these types of errors, as students used graphing calculators.

In Table 5.12 are student responses that illustrate some of the misconception and error types. There are a few lessons in here for students. First, over-reliance on a graphing calculator will be a problem: if any of these errors arise, the student will miss them. Second, it is important for a student to have some sense of how different standard functions behave so that she or he can decide whether the graph displayed is reasonable. Finally, a student needs to have some idea of what she or he is looking for in order to determine how best to display the graph.

TABLE 5.11. Classifications Used to Distinguish the Various Incorrect Graphs[91]

Graphing Error	Description
Pure recognition error	Student illustrates a general lack of understanding of the common behavior pattern of a specific family of functions. This category excludes recognition errors that are the direct result of a scaling error.
Transfer error	Student is unable to transfer data from the calculator to his or her paper accurately and vice versa.
Error in interchanging independent and dependent variables	Student sketches a graph that represents the inverse of the original function described in the problem.
Scaling error	Student is unable to select the appropriate dimensions of the window to illustrate the function's local and end behaviors.
Perceptual illusion	Student shows a lack of understanding of functions that exhibit asymptotic behavior or have discontinuities. Student includes the false line segment drawn by the calculator that connects points to the left and the right of the vertical asymptote to imply that the function is continuous.

Source: Pilipczuk (2008, p. 672)

TABLE 5.12. Types of Graphing Errors[92]

Problem 16
Over a two-year period, the monthly sales in thousands of units of a seasonal product are approximated by the function where t represents the time in months and $t = 1$ corresponds to January. 1. Sketch a graph of the function. 2. Use the graph to approximate the time of the year when the sales will exceed 110,000 units. $$S = 74.50 + 43.75 \sin \frac{\pi t}{6}$$

Graph Illustrating Error	Correct Graph
Scaling Issue[1]	
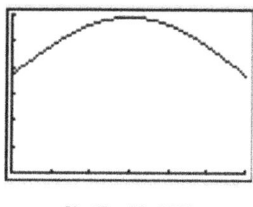 [0, 6] x [0, 120] $Xscl = 1$, $Yscl = 20$	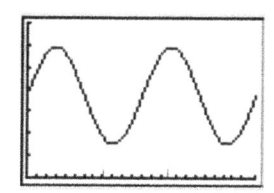 [0, 24] x [0, 140] $Xscl = 1$, $Yscl = 20$
Pure Recognition Error	
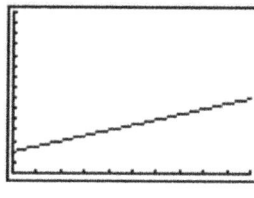 [0, 10] x [0, 150000] $Xscl = 1$, $Yscl = 10000$	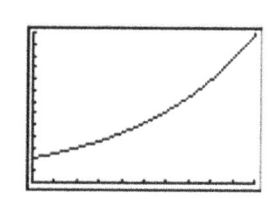 [0, 10] x [0, 130000] $Xscl = 1$, $Yscl = 10000$
Transfer Error	
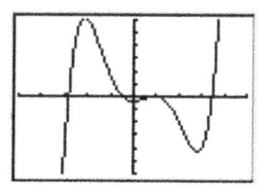 [-5, -5] x [-60, 60] $Xscl = 1$, $Yscl = 10$	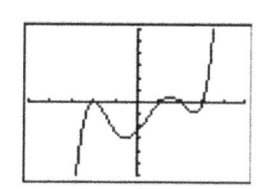 [-5, -5] x [-60, 60] $Xscl = 1$, $Yscl = 10$

Source: Pilipczuk (2008, p. 673)

Teachers

- Graphs convey large amounts of information quickly.
- Due in part to confusion about function notation, students can struggle with graphing correctly.

Calculator-Based Lab (CBL) activities can be helpful in reducing certain kinds of graphing errors, notably scaling errors.[40] Given that the data from these experiments were real-life data, they were messy and therefore required students to explore various window parameters. It may also be useful to place greater emphasis on the common behavior patterns of functions, including asking students to state the domain, range, intercepts, and asymptotes and to justify thinking by using general properties of the various functions. Also, *Math Flyer* is an app that works with graphing functions by altering parameters. It can help students see how graphs transform dynamically when scales are changed. *Math Flyer* is explored more fully in the next section, **Transformations of Functions**.

TRANSFORMATIONS OF FUNCTIONS

Common Core Standards: 8.F.2, 5, F.IF.4, 7, 8, 9

It is not always immediately apparent how altering a parameter of a function will change the behavior of a function (or how it won't). It is important for students to recognize that the graphs of $f(x) = 3x + 2$ and $g(x) = -2x + 4$ share a property of "straightness" that is not shared by $h(x) = x^2$, for example, or that $k(x) = x^2 + 4$ has the same fundamental shape as h. Without such recognition, students may expect that changing a in ax has the same effect as changing it in $x^2 + ax$.

Using technology, students are able to develop a deeper understanding of the various components of functions. Function transformations can be approached by systematically graphing families of functions from equations and changing the parameters. This can help students see relationships and make generalizations about them. In the same sort of way, other technologies such as the Function Probe[41] or *Linear Model* app (see Figure 5.8) allow students to begin by examining the graph and then focusing on what happens to the tables and equations when the function is transformed. A critical aspect of understanding functions is connecting concepts across representations, but research indicates that students have significant difficulties moving among the different forms.[42] The *Linear Model*

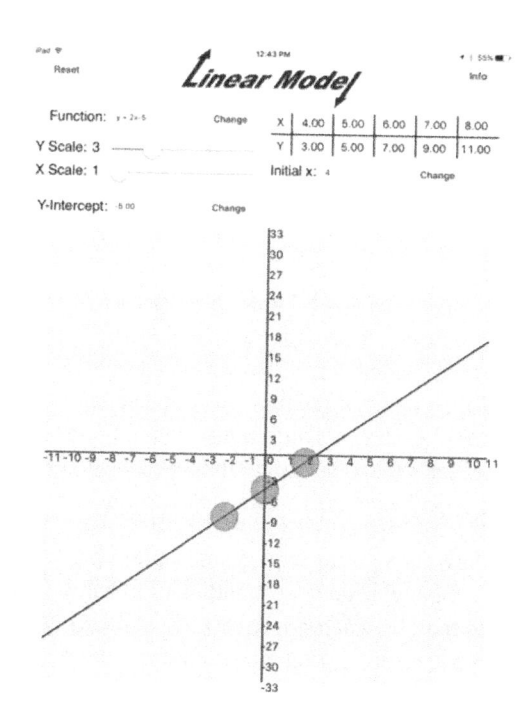

FIGURE 5.8. Linear Model App.

app, in particular, shows a table, graph, and equation changing simultaneously as a change is made in one of the forms. Also, connecting verbal or written descriptions as another representation would further students' conceptual development with functions.

In one study, students were given the equations of $y = x + b$ and $y = mx$ set up in a standard spreadsheet program. They could then change the parameter and observe the result in the graph and a table of values. They were given targets to hit and had to change the parameters to do so. This achieved two goals: first, students gained a more intuitive understanding of the role the parameters play. Second, students were able to see the same function presented in three different ways simultaneously, all giving the same information about the function but provided differently, encouraging "children to develop flexibility and facility with moving among the multiple representations of a function, and to recognize equivalent forms of the same functions."[43]

In another study, a 16-year old student, Ron, explored transformation of absolute value, quadratic, and step functions using Function Probe[44]. In each case, he worked from the graph of the base function and selected the function family based on the shape. He then made predictions about the effect on the tabular values and the values of the coefficients in the algebraic equations, struggling somewhat with the effect of sign on horizontal and vertical transformations.[45]

Teachers

Teachers can help students begin to generalize the properties and behaviors of individual functions and families of functions and how the various changes they make influence the representation. For example, if they wanted to create an "upward" motion, how would they accomplish that on a linear, quadratic, absolute value, etc. function? Similarly, if they wanted to alter the steepness, how would they do so? *Math Flyer* (see Figure 5.9) is an app that allows for easy transformations of functions. It has a built-in library of standard functions with sliders associated with parameters of the functions. Moving the sliders results in real-time changes to the graphs of the functions. The app provides instant and automatic feedback regarding students' intuitions and predictions about the behavior of functions and eliminates the tedium of graphing each and every function. However, it might not be appropriate if the students have not first acquired this skill.

- Apps that allow students to explore multiple representations simultaneously can help them gain an understanding of transformations of functions. This approach may be teacher-intensive, as students can very quickly explore many changes, prompting many questions at once.
- There is mixed research on the value of graphing calculators for helping students understand multiple representations of a function.[46]
- One researcher suggests these five approaches to learning functions:[47]

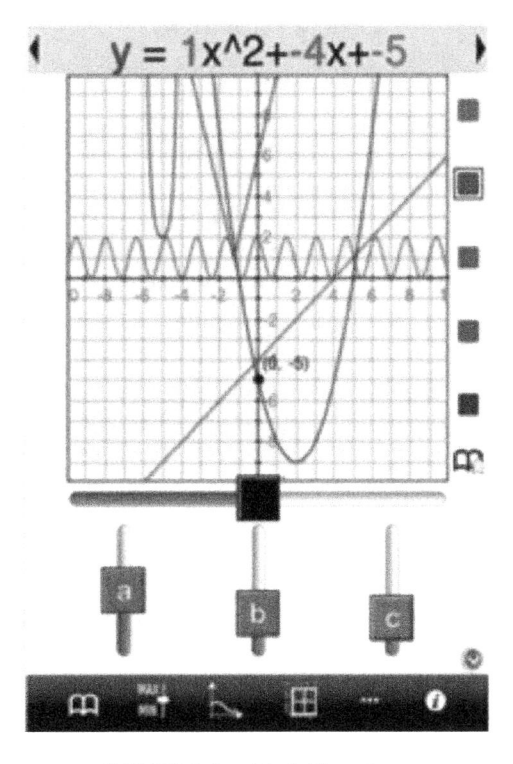

FIGURE 5.9. Math Flyer App.

- *Substitution to a Template:* Use the algebraic form [such as $y = A f(Bx + C) + D$] as a template to identify the impact of each parameter on the function (i.e., A is the magnitude of the vertical stretch factor, 1/B is the magnitude of the horizontal stretch factor, -C/B is the magnitude of the horizontal translation, and D is the magnitude of the vertical translation).
- *Function Building:* Start with the identity function: $y = x$. Build on that function with a series of actions. For instance, a vertical stretch creates $y = mx$. A vertical translation produces $y = mx + b$.
- *Symmetric Actions:* Examine the parallel structure between the vertical and horizontal transformations.
- *Visual Curve Matching:* Start with visual representations of functions, such as fitting a curve to some data. Examine the translations visually.
- *Scaling the Axes:* Transform the function by changing one or both of the axes and explore the results.

LINEAR FUNCTIONS

Common Core Standards: 8.F.3, 4, 5, F-LE.1, 2, F.BF.1, F.IF.4, 5, 7

Linear functions comprise one of the simplest and yet most important classes of functions for students to understand, but because the nature of linear functions involves proportional reasoning, they can be somewhat difficult for students. Their power and utility are apparent throughout mathematics and the sciences. We frequently make linear approximations to nonlinear functions, for example, and the savings in computational complexity can be substantial.

In one study, students were asked to explore two dynamic physical models and explain their thinking. Students were also given pictures, tables, and equations to represent the situations. A specific protocol for the interview was not provided since the interviewer followed the lead of the student's thinking with her questions. Included are some transcribed interactions between the student and the interviewer, as well as samples of the student's written work.

One of the models could be a gear elevating system. As shown in the illustration in Figure 5.10, there are gears with varying diameters and a weighted object at the end of the cord. By turning the crank one turn clockwise, the object, if on the 3-gear with a 3-inch circumference, would rise 3 inches. A counter-clockwise turn would lower the object 3 inches. The 4-gear has a 4-inch circumference and would raise or lower the object 4 inches with each complete turn of the crank. These can be modeled with elementary linear functions, where the height raised is a function of the number of turns.

The second model was the Etch-a-Sketch™, where one complete clockwise turn of the left knob creates a 3 cm horizontal segment moving from left to right, while a counter-clockwise turn moves from right to left. One complete clockwise turn of the right knob creates a 3 cm vertical segment from a lower position to an

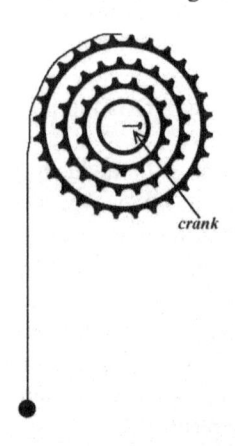

FIGURE 5.10. Illustration of the gear elevating system. Based on Hines, Klanderman & Khoury (2001, p. 364)

upper position, a counter-clockwise turn moves from an upper position to a lower position. If the knobs are turned simultaneously, a diagonal line is produced.

In Table 5.13, the interviewer found that Brad's thinking, using the gear elevating system and its corresponding tables, was limited in that he viewed each turn as a repeatable action rather than a generalized process.[48] That is, Brad needed to turn the handle 8 times before he could know what the 9th turn would yield. The focus was on how the heights changed, rather than on the relationship between turns and height. The interviewer moved to using the Etch-a-Sketch as a graph of the movement of the object in the gear elevating system. Initially, Brad treated the graph as though its slope were 1: a "unit" change in the horizontal position resulted in a "unit" change in vertical position, but a unit for horizontal motion was 1in, while a unit for vertical motion was 2in (using a 2-gear). This led to the conversation in Table 5.14.

The purpose of the study was to understand the processes by which students create, make sense of, and link the various representations of functions (e.g., tables, graphs, equations). The findings, while limited to the thinking of one "typical" eighth-grade student, provide teachers with a potential instructional sequence or sense of a function learning trajectory. This particular instructional sequence is enhanced with dynamic models used to support viewing functions as the covariation between two related variables and developing understanding of proportional relationships (see ***Analysis of Change: Ratio and Proportional Reasoning***).

The research offers a well-developed discussion of various understandings of functions while offering a potential developmental sequence or progression for deepening understanding of functions:

1. Begin with qualitative descriptions that are later quantified,
2. Move from input-output as individual actions toward a general pattern,
3. Discuss covariation among variables as a generalized process as opposed to an iterative process, and
4. Discuss the way a ratio can become a rate.[49]

TABLE 5.13. First Interview for Gears

Brad:	Height equals the number of cranks, that's it [writes $h = c$] ...
Interviewer:	If you know the number of cranks, how do you get the height?
Brad:	One crank equals ... [writes 1c = 3 in., h × 3, 1c = 3 in. in height] ... 1c = 3 in. = h, 2c = 6 in. = h, 3c = 9 in. = h [writes three equations].
Interviewer:	What does c mean?
Brad:	The number of cranks.
Interviewer:	Why did you put a 3 in the third equation?
Brad:	So you'd know how many cranks.

Based on Hines (2002)

TABLE 5.14. Second Interview for Gears[93]

Interviewer:	Would the line on the Etch-a-Sketch™ screen go up faster on the 3-gear?
Brad:	[Moves both knobs on the Etch-a-Sketch™ faster to create a graph similar to that representing the function on the 2-gear.]
Interviewer:	The line looks about the same. But the 3-gear is going up faster.
Brad:	[Turns one knob clockwise and one counter-clockwise.]
Interviewer:	Are you telling me the object on the gear goes down?
Brad:	Wait! The lines would be the same on the Etch-a-Sketch™.
Interviewer:	Even though this is going up faster?
Brad:	Yes.
Interviewer:	What if I wanted to show both relationships on the screen? How should I turn the height knob? [The right knob on the Etch-a-Sketch™—to show the action of the 3-gear.]
Brad:	It'll just go higher.
Interviewer:	If you saw that, would you be able to tell which one belongs with which gear?
Brad:	Oh yeah, this one goes up faster. It has to be on the 3-gear.
Interviewer:	Even though the number of cranks [on the gears] stays the same, this one goes up faster?
Brad:	Yeah, I want to try the 10-gear. It goes up 10 on each rotation. It should go to 20 in two rotations.

Based on Hines (2002)

Teachers

- Linear functions are in many ways a natural starting point for an exploration of functions. Students can explore parameter changes that have clear geometric interpretations, and computations are simple enough that they do not obscure the idea of function being taught.
- The gear elevating system provides a context for exploring functions and the associated variables in a non-symbolic way. Exploring with this model can support understanding regarding how the change in one variable influences the change in the other variable (see **Covariation** for more discussion of this approach).
- Physical models, such as a gear elevating system or Etch-A-Sketch™, can provide a context for exploring functions to help students understand how change in one variable influences change in the other variable.
- Students may struggle to understand physical models as generalized processes rather than individual actions.

GENERALIZING TO FUNCTIONS OF MULTI-VARIABLES

Common Core Standard: F.IF.2

Students' understanding of functions can benefit from consideration of functions of more than one independent variable. Seeing the generalization (a larger context) can clarify the idea of function just as the concept of "three" is, in some sense, easier to grasp than the concept of "one." Students are exposed to multivariable functions in the form of formulas (V = l x w x h) but do not usually consider these in their roles as functions. One stumbling block for students in generalizing to multivariable functions is that the input variables can change in two dimensions instead of just one: there are infinitely many choices of direction from a given point instead of only two (left or right). In addition, students must revisit graphing points, this time in a three-dimensional space rather than just in the plane.

Consider the problems in Table 5.15.[50] Problem 18 helps determine whether students can operate in three dimensions by translating a point, recognizing actions on a point in (a, b), and generalizing the results into a geometrical object (a plane) (c). Problem 19 can help teachers explore whether students can coordinate their understandings of the Cartesian plane, real numbers, and three-dimensional space. It also requires them to convert from a symbolic or verbal representation to a geometric one. These steps can help prepare students to consider functions of two variables. The fourth problem requires the coordination of three-dimensional thinking with the two-dimensional thinking required for a Cartesian plane as well as graphical representation and verbal descriptions. This question is difficult for students because it involves visualizing the intersection between surfaces. Con-

TABLE 5.15. Problems Involving Multiple Variables

Problem 17
a. Represent, in three-dimensional space, the point that is obtained when the ordered pair (1, 2) is assigned a height $z=3$.
b. Represent points (0, −2, 2) and (−3, 2, −2).
c. Find the coordinates (x, y, z) of point A.

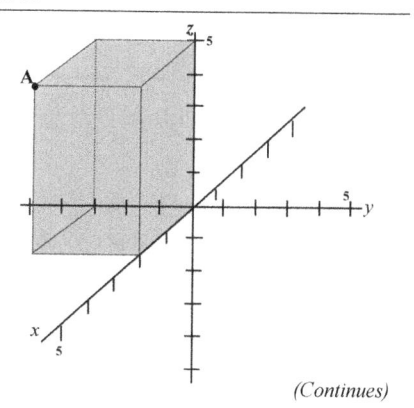

(Continues)

TABLE 5.15. Continued

Problem 18

In this problem, positive x is east, negative x is west, positive y is north, negative y is south, positive z is up, negative z is down.

a. Starting at point $(2, -1, 4)$ move the point 4 units west, 3 north, and 2 down. At what point does it end?

b. How would you move point $(-3, 4, -1)$ to end at point $(2, -2, 2)$?

c. If you start at point $(3, -2, -4)$ and move freely in the directions east–west and up–down, give an equation that describes the set of all points that may be reached.

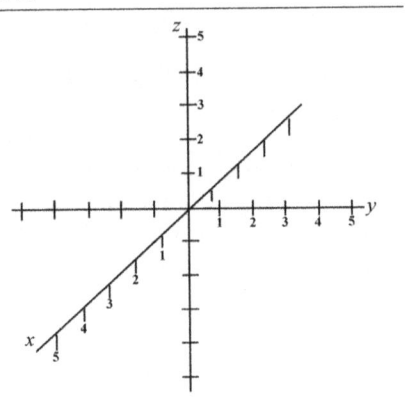

Problem 19

a. Draw the set of points in a plane that satisfies the equation $x^2 + y^2 = 1$.

b. Draw in three dimensions the set of points that are the result of assigning a height $z = 3$ to the points in a horizontal plane satisfying the equation $x^2 + y^2 = 1$.

c. Draw in three dimensions the set of points that may be reached by starting with points in a horizontal plane that satisfy the equation $x^2 + y^2 = 1$ and assigning them an arbitrary height z.

Problem 20

a. Consider the grid for Problem 2 above:

b. Draw a two-dimensional plane the points on the surface that satisfy $z=1$.

c. Draw a two-dimensional plane the points on the surface that satisfy $x=1$.

d. Draw a two-dimensional plane the points on the surface that satisfy $y=1$.

Problem 21

a. A rental car company charges 100 shekels for a day and an additional 5 shekels per kilometer. The company would like to have a clear description of the price that any client may have to pay when returning a rented car. Suggest such a description.

b. You have won a 1000 shekels coupon from the rental car company. Provide a detailed description of all your options to spend the exact amount for renting and driving a car using the maximum of your winnings.

---

structions such as the one this problem asks for are needed when analyzing graphs of two-variable functions. Finally, Problem 21 provides an approach to considering functions of two variables.[51] It may be more suitable for students who are ready to move beyond graphing points in three dimensions, or teachers may find it more instructive to allow students to develop those ideas independently through the problem, depending on the readiness of the students.

Teachers

- Multivariable functions can expand students' notions of functions and help them understand the big picture of what a function is.
- Real-world problems are messy, and real phenomena usually depend on many variables.
- Students will need some preparation to consider multivariable situations (e.g., an introduction to graphing in three dimensions).

Real-world problems can help motivate many ideas in mathematics, and multivariable functions are no exception. The rental car example in Problem 5 is easy for students to grasp and begin thinking about; there is no barrier to immediately discussing multiple input variables, as there might be with a more abstract example.

Interpreting two-dimensional renderings of three-dimensional figures takes practice. Having manipulatives with which to build graphs such as that of the function describing the rental car options can help students make the transition from two-dimensional graphs to three-dimensional graphs. Once students have an understanding of graphing in three dimensions, visualization software can help make the graphs more precise and more malleable (i.e., rotatable). When students can move fluidly between algebraic and graphical representations with multiple variables, they will have developed a richer understanding of function.

PROMOTING GENERALIZATION

Common Core Standards: 7.EE.1, 7.EE.2, F.IF.3, A-SSE.1, A-SSE.2, A-SSE.3

A main goal in studying patterns and functions is to promote generalizations. The ability to generalize is what separates success from failure in algebra (see also **Variables & Expressions: Letter Used as a General Number**). Generalization is reasoning that leads to a mathematical rule about relationships or properties.[52] It is moving from a focus on individual cases to articulating the "patterns, procedures, structures, and the relations across and among them."[53] Students can express generalizations in four ways: (a) arithmetic, (b) algebraic, (c) graphical/pictorial, and (d) verbal.[54] In order to generalize, they need to understand the notion of variable and the contingency relationship between two variables, and be able to move between representations.[55]

In upper elementary grades, students begin to generalize relationships between two quantities and can express these relationships using words and "non-standard" equations. As they move into middle and high school, they explore more complex relationships and can more readily generate equations or inequalities to correspond with the relationships. An important role for teachers is to ask questions that help students make conjectures that lead them to develop generalizations.

One hazard for the new teacher in particular is the fact that "What's the next entry in my pattern?" problems are not well defined. There can be multiple rules or ways of reasoning that justify different answers for the next number in the pattern. Thus, when asking students to extend patterns, it is essential that the student explain his or her reasoning, and the teacher must be prepared to evaluate the legitimacy of the explanation and equivalent forms of responses (see **Patterns and Generalizations**, **Justifying Generalizations**). For example, a pattern beginning 1, 2, 3 could continue with 4 (counting), 5 (a Fibonacci-like rule), 6 (the sum of all previous numbers), 7 (the product of all previous numbers plus 1), etc.

Three stages are necessary when generalizing: seeing a pattern, verbally articulating what the student sees, and then recording the pattern that the student recognizes symbolically.[56] In interviews with students, researchers found that students may be able to present their thinking using words and gestures before they can accurately utilize algebraic expressions.[57] However, students often struggle to use the signs and symbols of algebra correctly when asked to present generalized thinking. When looking at patterns, students may arrive at a "rule" through naïve induction (trial and error) or through generalization. The pattern that they see is limited. For instance, for the pattern in Figure 5.11 used in one study, students said "times two plus one" or "times two plus three" and may have only checked one or two cases to confirm their strategy. One group of students symbolized their strategy by writing $n \times 2(+3)$.

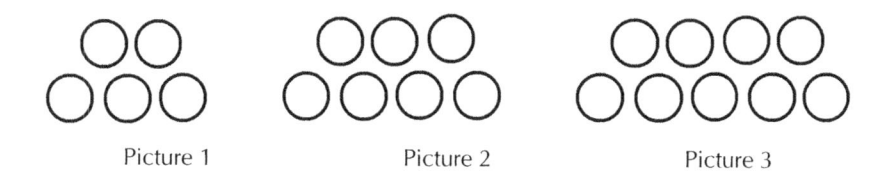

```
  OO      OOO     OOOO
 OOO     OOOO    OOOOO
Picture 1  Picture 2   Picture 3
```

FIGURE 5.11. The sequence of pictures given to students in a Grade 8 class. Radford (2010)

In contrast, one student, Mel, reasoned that "the top line always has one more circle than the number of the picture and the bottom line always has two circles more than the number of the picture" and recorded the formula $(n + 1) + (n + 2)$.[58] This is an example of a generalization because Mel noticed some common features across the three examples that he could project into the future. Mel reasoned what he needed to add on to the picture to create the next picture in the pattern.

Another study examined the types of rules (i.e., recursive or explicit) students generate for given problem situations and the connections they make between the rule and the situation and between the types of rules they generate (see Figure 5.12). "Recursive rules involve recognizing and using the change from term-to-term in the dependent variable.... Explicit rules use index-to-term reasoning that relates the independent variable to the dependent variable(s), allowing for the immediate calculation of any output value."[59] Of the four target students in the study, two rarely understood the connection between the recursive and explicit rules, while two used the recursive relationship to build the explicit rule. In both cases, the generalization came from considering a subdivision of the picture. Mel saw the nth picture as two rows in which were $n + 1$ and $n + 2$ circles, which led him

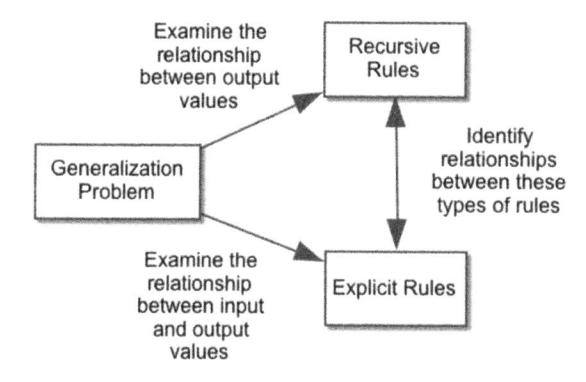

FIGURE 5.12. Conceptual model for generalizing numeric situations. Lannin, Barker, & Townsend (2006, p. 303)

278 • HOW STUDENTS THINK WHEN DOING ALGEBRA

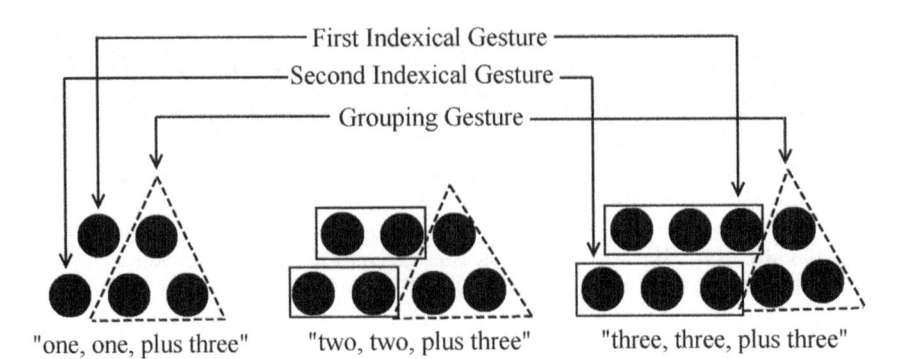

FIGURE 5.13. Mimi identifies a triangle in each picture.
Radford (2010, p. 50)

to an explicit rule—using the number of the picture in his thinking about the rule. Others saw it as a previous picture with two additional circles, giving a recursive rule. For instance, Mimi saw it as a previous picture with an additional triangle of circles (see Figure 5.13)—not using the picture number in her thinking to get a generalized equation of $2n + 3$. Similarly, Doug saw two being added to Picture 2 in Figure 5.14 as it grew. He noticed what was in common and how it changed from one picture to the next. The two of them reasoned recursively.

Students often reason recursively and have a difficult time transitioning to developing explicit rules and recognizing the generalizability of their solutions. That is, they focus on specific instances instead of realizing that their rule is generalizable across instances. Whether students see the generalizability can be influenced by the kind of tasks they are given and by how well they understand the relevant mathematical operations and how they are related.

Learning to use algebraic symbols knowledgeably requires students to release their physical and mental attachment to concrete objects, figures, and tables when generalizing. It also requires students to move away from specific numbers associated with those physical realities in order to see a general rule.[60] Once they are able to move from those mental limitations to algebraic notation, it is just as

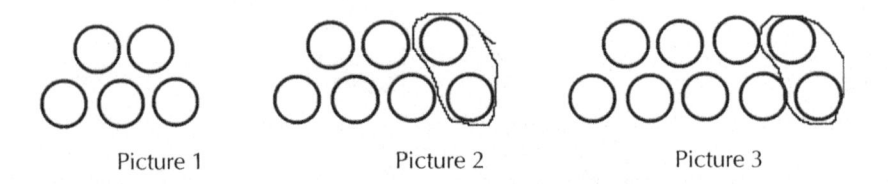

FIGURE 5.14. Doug emphasizes the last two circles.

important to be able to return to those representations and reconnect the general with the particular. Then students can see the physical and numeric in the generated rule, giving a more complete and complex understanding of the pattern.

Teachers

- People naturally want to generalize; we seek patterns in everything to help us understand the world. However, hasty generalizations are often quite wrong, so students must be taught to articulate the reasoning for their generalizations so that it can be explored.
- Generalizations can appear in a recursive form or an explicit form. Both have their uses, and students should try to generalize in both ways.

Teachers can influence reasoning through class discussions centered on developing efficient strategies, thus moving students from recursive to explicit rules.[61] Students can work in small groups on some of the following exercises:

- An arithmetic investigation (continuing the pattern on the basis of some given information, including answering questions about specific figures such as the 10th, 25th and 100th);
- An expression of generalization in natural language; and
- An algebraic expression for the nth figure.[62]

These are some questions that press learners to clarify what constitutes a pattern:

- Would the following _____ fit into the given sequence? If yes, how can you tell and where would it appear? If not, why not?

Questions that encourage learners to consider what is staying the same, what is changing, and how is it changing:

- What are the variables in this situation? What quantities are changing?
- How are the variables related? As one variable increases, what happens to the other variable?

Questions that prompt additional representations and relationships among them:

- How can you represent the relationship using words, concrete objects, pictures, tables, graphs, or symbols?
- What connections do you see across the representations?

Questions that encourage generalization:

- What does the 10th, 25th, or 100th figure of the pattern look like?
- What does the nth figure of the pattern look like?

- How would you find the total number of tiles (cubes, dots, toothpicks, etc.) in the pattern?

Teachers can also use tools that support generalization, such as technology. Appropriate technology allows students to move quickly and easily among representations, to see connections among representations, and to transform representations in order to more deeply understand how the components of functions influence how they look (see *Transforming Functions*).

JUSTIFYING GENERALIZATION

Common Core Standards: 7EE.1, 7.EE.2, A-SSE.1, A-SSE.2, A-SSE.3

"Engaging teachers in discussions focused on the details of students' competencies in justifying and proving may provide a basis for enhancing both teachers' own understandings of proof and their perspectives regarding proof in school mathematics."[63] Accordingly, consider the two problems in Table 5.16. Both problems are about proportional relationships, but they are presented in very different ways. Each has two phases: finding the relationship between the variables and justifying the generalization. These problems can be useful as formative assessments of the two phases. In the Connected Gears problem, students will need to understand that the gears mesh; that is, "more teeth" corresponds to a larger radius. The goal is to work with students to induce "a habit of mind whereby one naturally questions and conjectures in order to establish a generalization."[64] For teachers, it is important to go beyond simply having students recognize patterns, and to help them "attend to those that are algebraically useful or generalizable" and be able to argue on their behalf.[65]

One researcher used these two problems to identify five ways to generalize and justify or develop algebraic proof (see Table 5.17).[66] Students practice generalization by considering linear relationships and identifying them as such. In the cases of the Connected Gears and the Frog Walking problems, students examined ratios and how ratios stayed the same even when the numbers changed. In this process,

TABLE 5.16. Problems on Proportional Relationships

Problem 22: Connected Gears

You have 2 gears on your table, one with 8 teeth and one with 12 teeth. Answer the following questions:

1. If you turn the small gear a certain number of times, does the big gear turn more revolutions, fewer, or the same amount? How can you tell?

2. Devise a way to keep track of how many revolutions the small gear makes. Devise a way to keep track of the revolutions the big gear makes. How can you keep track of both at the same time?

3. How many times will the small gear turn if the big gear turns 64 times? How many times will the big gear turn if the small gear turns 192 times?

Problem 23: Frog Walking

The table shows some of the distances and times that Frog traveled. Is he going the same speed the whole time, or is he speeding up or slowing down? How can you tell?	Distance	Time
	3.75	1.5
	7.5	3
	12	4.8
	15	6
	40	16

Both the 'Connected Gears' and 'Frog Walking' problems are from Ellis (2007, p. 206)

TABLE 5.17. Student Approaches to Generalization Tasks

Authoritarian:

External Conviction
The main source of conviction is a statement made by a teacher or appearing in a text.

Student: [Looking at a table of data] Since the y-values increase by the same number each time, this will be a straight line.
Interviewer: Why?
Student: That's what Ms. R told us.

Symbolic:

External Conviction
Students view and manipulate symbols without reference to any functional or quantitative reference. (Given three connected gears that rotate 6, 4, and 3 times, respectively, students decide that another triple could be 12, 8, and 6 rotations.)

Interviewer: Why is that valid?
Student: There's a pattern in all of them. So if you do one thing to the small one, you have to do it to the middle one and the big one to keep the ratios the same.
Interviewer: Why does that work?
Student: It's kind of like changing fractions from 1/2 to 3/6. It's the same thing, just in different form.

Inductive:

Empirical
Students ascertain and persuade by quantitatively evaluating (directly measuring, calculating, substituting specific numbers into expressions, etc.) a conjecture in one or more specific cases.

Student 1: All of these pairs must be the same speed because cross-multiplying gives the same answer each time.
Interviewer: How do you know that means they're the same speed?
Student 1: Because 27 times 5 and 7 and 1/2 times 18 equals the same thing.
Student 2: I tried it for all the pairs in the table and it works every time.

Perceptual:

Empirical
Conviction lies in perceptual observations by means of rudimentary mental images [The student examined a $y = mx$ table of ordered pairs: (2, 9); (5, 22.5); (12, 54); (16, 72)].

Student: It couldn't be a straight line.
Interviewer: Why not?
Student: If you made your graph, it doesn't look like it'd be a straight line [sketching a graph that appears curved]: 18's up here, 9's like right here, 2 and 22.5 that's like right there, and then 16, 72 would be all the way up there.

Transformational:

Deductive
Transformations involve goal-oriented operations on mental objects and the ability to anticipate the results of those operations. Students validate conjectures by deduction. [A student explains why he thinks $(3/4)m = b$ should describe the relationship between two gears with 12 and 16 teeth.]

Student: Since there's 3/4 of the teeth on the small one, the big one always has 1/4 teeth to make up every turn. Making it, the big one turns 3/4 of a turn every time the small one turns once. And so, say it went through 12 teeth on the small gear and 12 on the big gear. That's only 3/4 of a turn for the big gear, while it's a full turn for the small gear.

students were 1) searching and identifying common features across cases, 2) extending reasoning beyond the original example, or 3) making statements derived from broader results from particular cases. Although correct algebraic generalizations and deductive forms of proof are a critical instructional goal, students' incorrect, non-deductive generalizations may serve as an important bridge toward this goal, which has important implications for teaching. These generalizations give insight into how the students are thinking and provide the teacher a way to highlight the misconception with a suitable example or other form of instruction. Generalizing and justifying are often intertwined and mutually influence each other as students increase their sophistication of explaining.

Researchers hypothesize that there is a learning trajectory for justification (see Table 5.18) that shows how justifying is "likely to proceed from inductive toward deductive and toward greater generality."[67] This trajectory can be a useful tool for teacher and student reflection regarding the sophistication of their justification approach.

Research examining the implementation of this framework indicated that students have an overwhelming reliance (70% of responses from 350 middle school students) "on the use of examples as a means of demonstrating and/or verifying the truth of a statement (Level 1/Level 2)" and don't recognize the limitations of

TABLE 5.18. Justification Level[94]

Level 0 No justification	Student is ignorant of the need for—or existence of—justification or proof.
Level 1 Appeal to an external authority	Student refers to the correctness stated by some other individual or reference material.
Level 2 Empirical evidence	Student justifies through the correctness of particular examples. Student is aware of the idea of a proof but considers checking a few cases as sufficient.
Level 3 Strategic evidence	Student is aware that checking a few cases is not sufficient but is satisfied that checking extreme cases or random cases is proof.
Level 4 Generic example	Student expresses deductive justification in a particular instance. Student believes that use of a generic example forms a proof for a class of objects.
Level 5 Limited deductive justification	Student is aware of the need for a general argument, is able to understand the generation of such an argument, and is able to produce such arguments in a limited number of familiar contexts.
Level 6 Deductive justification	Student is aware of the need for a general argument, is able to understand the generation of such an argument (including more formal arguments), and is able to produce such arguments in a variety of contexts (both familiar and unfamiliar). Student achieves validity through a deductive argument that is independent of particular instances.

that approach.[68] Through questioning and encouraging justification, teachers can help students advance in the sophistication of their arguments.

Lastly, an important part of understanding justification and proof is disproof. Textbooks in algebra generally do not address the use of counter-examples well.[69] What does it take to prove something is not true? While several examples do not prove something, if there is one instance in which a statement is not true, then a mathematical statement can be refuted. This is confusing for many students: they "are not convinced by a counterexample and view it as an exception that does not contradict the statement in question."[70] Students generally have difficulty coming up with a counterexample— either giving "an example that does not satisfy the necessary conditions or an example that is impossible."[71]

Teachers

- Students often do not have difficulty with seeing a pattern, but need help moving from proof by example to proof by relationship.
- Instead of allowing students to "prove" their rule works by showing that it results in the right answers for a few cases, have students explain each component in their rule and how it represents a feature of the situation as it changes across the pattern. Students need to see the limits and inadequacy of a justification that relies on empirical evidence or tests of particular numbers. A comprehensive understanding of how the unknowns operate in a generated formula or rule includes not only a sense of generic examples as well as specific, but also the connection between the two.
- Simply asking students to generalize and justify can go a long way toward having them succeed at it. In other words, tasks/problems should include portions that encourage justification and generalization, given that much of the difficulty students face stems from lack of familiarity and experience with justifying and generalizing.
- Patterns can be legitimately extended in infinitely many ways. Thus, justifying the extension the student chooses is the only way he or she can convincingly argue that the extension is not just a random guess.

Students begin by searching for patterns in the numbers. It is only after they are asked to explain their thinking that they move from specific patterns to general principles. Teachers should ask their students to (a) identify an appropriate domain for their rules and (b) justify their rules.[72] This can help students stay connected to the model and also employ more sophisticated strategies than "guess-and-check."

While proof, as justification, occurs mostly in geometry classes, it has an important role to play in other classes, such as algebra. Consider incorporating justification early into instruction and emphasizing problems that allow for justification. Teachers should focus on quantities and the language of quantitative relationships, rather than just number patterns. Students often attend to number patterns alone, even within the context of a quantitatively rich problem, so teachers might need to intervene to draw students' attention back toward the quantitative referents. Ask students to shift from pattern descriptions to phenomenon descriptions. Over time, these become habits of reasoning that students bring to a problem situation. The Connected Gears and Frog Walking problems are good examples of these kinds of problems. The problems supported each other in that students found commonalities between them that helped them to generalize the idea of a linear relationship.

SEQUENCES AND PICTORIAL GROWTH PATTERNS

Common Core Standards: F.IF.3, F.BF.1, 2

Number sequences are, for many students, the most natural place to begin attempting to generalize. Students begin looking for patterns in elementary school with shapes, colors, letters, etc. such as "ABCABCABC." By middle school, they are familiar with numbers and number patterns, and they are accustomed to looking for structure in them. As mentioned previously, it is important that the teacher ask for justification when students continue a pattern, and be prepared to evaluate the student's justification. Consider Problem 24:

Look at the following sequence of numbers: 2, 5, 8, 11, 14, 17 …
Write an equation that would generate this pattern of values.

The easiest approach to Problem 24 is to recognize the repeated addition—that each term is adding three to the last term, saying "start at two and keep adding 3." However, in one study, the majority of 9- to 13-year-old students tended to generalize "an erroneous direct proportion method, that is, determining the n-th element as the n-th multiple of the difference" In other words, the 20th term for Problem 24 would be incorrectly identified as 20 * 3.[73] In order to construct the accurate formula for the sequence of $y = 3x - 1$, students need to recognize the relationship between the numbers as well as the position of each number in the sequence. Researchers suggest that "big numbers may bridge the gap between concrete small numbers and abstract algebraic symbolism."[74] In Problem 24, for instance, a teacher might ask whether 147 would be a number in the pattern and extend to generalization from there.

When working with number sequences, researchers have found that "computer spreadsheets facilitate investigation of mathematical patterns and functions."[75] "Spreadsheet technology can help students make connections between their informal ideas and the formal representations."[76] A spreadsheet can be a useful tool "not only for developing generalization but also for expressing it in algebraic terms."[77] The main idea is for students to write a formula that produces the given number sequence, as in Figure 5.15 for Problem 24. There are multiple benefits to the process, including getting a deeper understanding of the potential of a variable, as a cell, to represent a broad range of quantities, including a formula. In addition, other benefits include having to create a rule/formula, being able to reason flexibly and recursively, and connecting symbolic representations to graphical representations.[78]

Pictorial Growth Patterns

Pictorial growth patterns are a common vehicle used to explore patterns and functions. Students look for relationships from one figure to the next regarding what is staying the same and what is changing, laying the foundation for generalizations about larger figures and any figures in the pattern. This representation can be accessible for students and allows them opportunities to explore and ana-

	n	3x-1
1		
2	1	fx ~ 3× (A2 ▾)-1
3	2	5
4	3	8
5	4	11
6	5	14
7	6	17
8	7	20
9	8	23
10	9	26

FIGURE 5.15. Spreadsheet Solution to Problem 24.
Ferrini-Mundy, Lappan, & Phillips (1997)

lyze relationships among the various representations of the function (e.g., tables, graphs, and verbal and symbolic rules).

Consider Problem 25: Pool Tiles.

Tat Ming is designing square swimming pools. Each pool has a square center that is the area of the water. Tat Ming uses gray tiles to represent the water. Around each pool is a border of white tiles. Figure 5.16 gives examples of the three smallest square pools that he can design with gray tiles for the interior and white tiles for the border.

- Create a table that shows the number of gray tiles and white tiles in the first six pools. Also list the pool area and the area of the border.
- Make a graph that shows the number of gray tiles in each square pool. On the same coordinate graph show the number of white tiles in each square pool.

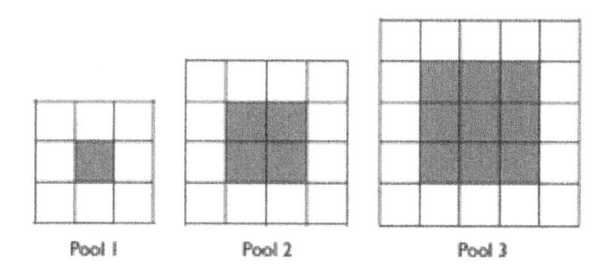

Pool 1   Pool 2   Pool 3

FIGURE 5.16. Examples for Problem 25: Pool Tiles.
Ferrini-Mundy, Lappan, & Phillips (1997)

Students engaged with this problem in different ways. Even though the sixth-grade student in the interview (Table 5.19) could express the generalization about Tat Ming's Pool verbally, she could not yet express both generalizations symbolically. She could express the pool's area as n^2 where n is the pool number, but she could only describe the pool's border verbally. However, the ability to verbally describe a relationship appears to be an important first step toward algebraically describing the relationship.[79] The verbal description is, in fact, an algebraic description without algebraic notation. This would suggest that with the right question or with a little more time, this sixth grader could come up with symbolic expressions for both relationships in this task.

An important role for teachers is to help students recognize patterns in different representations (picture [Pool Tile problem], words [Table 5.19], table [Table 5.20], and graph [Figure 5.17]). In addition, students need to understand how the pattern is connected across the representations. For instance, how do the pictorial patterns of white and gray tiles in the Pool Tile problem connect to the graph and table?

TABLE 5.19. Interview for Problem 25: Pool Tiles

Interviewer:	As the area of the pool increases how does the number of white tiles change? How does the number of gray tiles change? How does this relationship show up in the table and in the graph?
Student:	The number of white tiles in the border increases by 4 each time because you are adding one to each side of the pool as you build the next biggest deck. The number of gray tiles add three, then five, then seven, so it should go 9, then 11. The difference between each of those is two... The [graph] is steady for the pool deck, just adding 4 each time. The pool jumps up [motions arm upward quickly] much faster. The jumps between pools are bigger each time.
Interviewer:	Use your graph to find the number of gray tiles in the seventh square.
Student:	There are 49 gray tiles. I know that 7 times 7 is 49. It should also be 13 larger but that is not as easy.
Interviewer:	Can there ever be a border for a square pool with exactly twenty-five white tiles? Explain why or why not.
Student:	No, because it would be in there [points to a place in the graph where it would show up] and it is not. Unless you could have like point five or something. If you were allowed to have something other than integer blocks then you could probably do it.
Interviewer:	Find the number of gray (or white) tiles in the 10th pool. The 25th pool. The 100th pool.
Student:	There are (10 × 10) or 100 gray tiles and (10 × 4 + 4) or 44 white tiles. There are (25 × 25) or 625 gray tiles and (25 × 4 + 4) or 104 white tiles. There are (100 × 100) or 10,000 gray tiles and (100 × 4 + 4) or 404 white tiles. [She saw the pools as squares and found the areas simply by "squaring the pool number or n-squared." She saw the borders as consecutive additions of 4 and said there were "as many of them as the pool number you were on plus one more 4."]

TABLE 5.20. Table for Problem 25: Pool Tiles

Pool Number	Pool Area	Pool Border
1	1	8
2	4	12
3	9	16
4	16	20
5	25	24
6	36	28

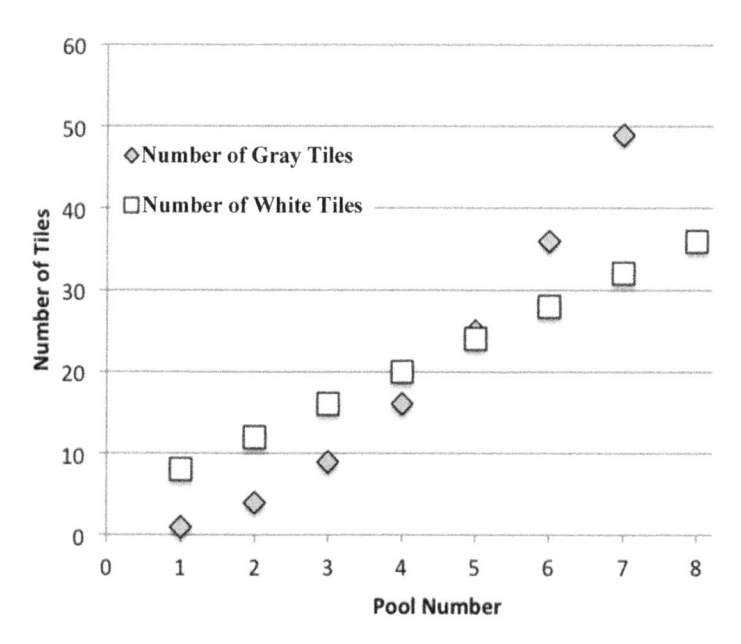

FIGURE 5.17. Graph of data to compare two patterns.
Ferrini-Mundy, Lappan, & Phillips (1997)

Image 1   Image 2   Image 3

FIGURE 5.18. Problem 26 Marcia's Tiling Squares.

Problem 26: Marcia's Tiling Squares Problem[80]

Marcia is using black and white square tiles to make patterns. Marcia wants to know how many white tiles and black tiles there will be in the tenth pattern, but she does not want to draw all the patterns and count the squares.

It is important that teachers recognize the many different ways students can perceive a pattern. In a study with 22 ninth-grade students, twenty-three different strategies were used to solve Marcia's Tiling Squares problem (Figure 5.18), although not all of them were successful. Almost half of the students who worked on the task used visual strategies such as grouping, proportioning, chunking, or examining visual finite differences. For instance, a visual grouping strategy for the black tiles might count each "arm" of the pattern and then multiply to get the total, as Student 1 did (Table 5.21). A strategy for counting the white tiles could be identifying four groups of three white tiles forming an L shape around the center black cross with an additional four center white squares on each side.

The other strategies were primarily numerical, such as systematic trial and refinement, extending a table, and numerical use of finite differences in a table. Students who focused primarily on the numbers in a pattern tended to search for properties by considering procedural rules for combining, obtaining numbers, and making calculations. However, some had little understanding of what coefficients in a formula represented in the relationship. Students may have used variables simply as placeholders with no meaning except to generate a sequence of num-

TABLE 5.21. Interviews for Problem 26: Marcia's Tiling Squares

Student 1:	[Black tiles] I looked at pattern 3 and I saw the three pattern, three tiles, that are on each side so I thought I looked at the pattern two and it just added one so I multiplied four times four with all the sides and just added one in the middle [for pattern 4].
Student 2:	You can see here, like, it's three, three, three, three, plus twelve, and four, and the same here [referring to the next pattern]. The thing is you just add four more and if you are doing a table you just add 8, that's 16, 24, 32, 40.

bers. Students who used more visual methods may have focused on perceptual strategies, seeing the functional relationship among the numbers and diagrams. In that case, variables were seen not only as placeholders, but also as representing the relationship. Students who used both are "representationally fluent" because they are able to understand how sequences of numbers can consist of both properties and relationships.[81]

These types of pictorial patterns can also be very useful in helping students see the equivalence of different structures of expressions. For instance, in looking at the white tile pattern in Figure 5.18, students might see the three white tiles at the end of each arm staying consistent as 4(3). Then add the inside L of the arm as 4(1), 4(3), and 4(5) tiles across the three different images or 4(1), 4(1 + 2), 4(1 + 4). This could result in the expression $4(3) + 4[1 + 2(n - 1)]$. Another student might see the four corners of the figure as 4(3), 4(5), and 4(7) or as 4(1 + 2), 4(1 + 4), and 4(1 + 6), plus the 4 single white tiles in the middle of each arm consistently across the three images. This might result in the expression $4(1 + 2n) + 4(1)$. Another student might see a U of five tiles around the top black tile and the same at the bottom or 2(5) and add the three white tiles on either side or 2(3). Following that visual grouping leads to 2(7) + 2(5)—or the three white tiles on each side plus two additional—for the next image and 2(9) + 2(7) for the third image. This would lead to the expression $2(3 + 2n) + 2(1 + 2n)$. Another student might simply look at the total number of tiles (16, 24, and 32) and recognize that they are multiples of 8, leading to an expression of $8(n + 1)$. There are many other potential ways of chunking these tiles and resulting expressions. These four different approaches (see Table 5.22) are all equivalent, the first three simplifying to the $8(n + 1)$ expression. Students often have difficulty understanding how expressions can be equivalent, but when viewing tile patterns, they can recognize that they are all simply different ways of counting the tiles.

Researchers found that students typically used one of four methods to solve ladder types of problem in Table 5.23:[82]

1. *Counting Method:* Counting from a drawing.
2. *Difference Method:* Multiplying the number of rungs by 3, the common difference. For example, assuming repeated addition of 3 or 3n.
3. *Whole Object Method:* Taking a multiple of the number of matches required for a smaller ladder and multiplying times the needed ladders.
4. *Linear Method:* Using a pattern that recognizes both multiplication and addition are involved and that the order of operation matters. For example, $M(n) = an + b$.

TABLE 5.22. Equivalent Expressions for Problem 26: Marcia's Tiling Squares

$4(1 + 2n) + 4(1)$	$4(3) + 4[1 + 2(n - 1)]$
$2(3 + 2n) + 2(1 + 2n)$	$8(n + 1)$

TABLE 5.23. Problem 27: Ladder

With 8 matches, I can make a ladder with two rungs like this:

With 11 matches, I can make this ladder with 3 rungs:

Problem	Response
28. How many matches would you need to make a ladder with 20 rungs?	20 rungs* (answers typically in 60s) (16%, 32%, 22%, 49%, 50%)
29. How many matches are needed for a ladder with 1000 rungs?	1000 rungs* (answers range from 300–3000) (8%, 13%, 13%, 24%, 24%)

*500 4th-, 5th-, 6th-, 7th-, and 8th-grade students, respectively, in Stacey (1989) and Bourke & Stacey (1988).

These approaches show an increasing level of understanding of the problem. The simplest, counting, is effective, but does not show that the student is attempting to generalize. The Difference Method recognizes that there is a regular increase in the number of matches needed as the number of rungs increases, but fails to account for other parts of the ladder. The Whole Object Method has a similar issue, but looks at the ladder as a whole rather than at the constituent parts (rungs, in this case). The Linear Method takes into account both the regularity of the pattern and the fact that the ladder has some parts not accounted for by the increases.

A student's efforts to "link visual and symbolic representations is important and yet it seems that the majority of students are failing to achieve this goal."[83] Research suggests that it is not effective to "introduce students to algebraic conventions by moving directly from patterns into algebra notation."[84] Instead, it might be useful to have students verbally describe the situation first[85] and/or have students create tables of data to help them see generalizations.[86] Researchers found that "students who gave a correct verbal description were more likely than other students to write a correct algebraic rule."[87]

Teachers

- Mathematics educators agree that pictorial growth patterns provide a context for supporting algebraic thinking—and, in particular, the generalization of relationships between two variables.
- Physically building the models prompts observations about what stays the same and what changes as you move from one figure of the pattern to the next. Consider having students construct the pattern, and be sure to watch

when students translate a diagram into a numeric representation since inappropriate generalizations are easy to make.[88]
- "Backwards questions," in which students attempt to find *earlier* figures in the pattern, can provide opportunities for the problem solver to analyze a situation from a different starting point or make a connection among representations or tools in different orders. Such questions can encourage thinking from multiple perspectives.[89]

ENDNOTES

1 p. 251, citing the work of Hershkowitz & Schwarz, 1999 as well as Heid, 1996
2 Lloyd & Wilson, 1998, p. 270
3 Lloyd & Wilson, 1998, p. 270. Based on Stein, Baxter, & Leinhardt, 1990
4 Hatisaru & Erbas, 2010; Hollar & Norwood, 1999
5 Davidenko, 1997
6 Elia, Panaoura, Eracleous, & Gagatsis, 2007, p. 539 based on Vinner, 1992
7 Norman, 1992
8 Dubinsky & Wilson, 2013, p. 85 based on Carlson, 1998; Clement, 2001; Sierpinska, 1992
9 Even, 1990
10 Breidenbach, Dubinsky, Hawks, & Nichols, 1992
11 Even, 1990
12 Markovitz, Eylon, & Bruckheimer, 1986
13 Sajka, 2003
14 Adapted from Vinner & Dreyfus, 1989, problems #3-7.
15 Vinner & Dreyfus, 1989, pp 359–360.
16 Vinner & Dreyfus, 1989,
17 The trajectory for functions is defined at http://turnonccmath.net.
18 Kaput, 1999, p. 134.
19 Even, 1993
20 Markovitz, Eylon, & Bruckheimer, 1986
21 Elia, Panaoura, Eracleous, & Gagatsis, 2007, p. 538 based on Even, 1998
22 Even, 1993
23 Even, 1993, p. 105
24 Jones, 2006
25 Clement, 2001, p. 746
26 Even, 1993, p. 95
27 Even, 1993, p. 106
28 Even, 1993
29 Davis & McGowen, 2002
30 Davis & McGowen, 2002, p. 5
31 Davis & McGowen, 2002, p. 4
32 Falcade, Laborde, & Mariotti, 2007
33 Knuth, 2000
34 Yerushalmy & Shternberg, 2001, p. 259
35 Yerushalmy & Shternberg, 2001, p. 256
36 Yerushalmy & Shternberg, 2001, p. 256
37 Yerushalmy & Shternberg, 2001, p. 260
38 Knuth, 2000, p. 506
39 Yerushalmy & Shternberg, 2001
40 Pilipczuk, 2008.

41 Confrey, 1991
42 Bloom, Comber, & Cross, 1986; Goldenberg, 1988; Kalchman, 1998; Markovits, Eylon, & Bruckheimer, 1986; Schoenfeld, Smith, & Arcavi, 1990
43 Kalchman, 1998, p. 11
44 Borba & Confrey, 1996
45 Borba & Confrey, 1996, p.333
46 Gray & Thomas, 2001
47 See Confrey, 1994 for a more detailed description of these approaches. We left one approach out, as it was complex to explain in this space.
48 Hines, 2002
49 Thompson, 1994
50 Problems #1-5 are from Trigueros & Martínez-Planell, 2010.
51 Yerushalmy & Shternberg, 2001
52 Carpenter & Franke, 2001; Ellis, 2007; English & Warren, 1995; Lee, 1996
53 Kaput, 1999, p. 137
54 Cañadas, Castro, & Castro, 2009
55 Kalchman, 1998
56 Mason, Pimm, Graham, & Gower, 1985
57 Radford, 2010
58 Radford, 2010
59 Lannin, Barker, & Townsend, 2006, p. 300
60 Radford, 2000
61 Radford, 2000
62 Radford, 2000.
63 Knuth, Slaughter, Choppin, & Sutherland, 2002, p. 1699
64 Blanton & Kaput, 2002, p. 25
65 Ellis, 2007, p. 195 based on Blanton & Kaput, 2002; English & Warren, 1995; Lee, 1996; Lee & Wheeler, 1987; Orton & Orton, 1994; Stacey, 1989
66 Ellis, 2007
67 Simon & Blume, 1996, p. 9
68 Knuth, Slaughter, Choppin, & Sutherland, 2002, p. 1694
69 Zaslavsky & Ron, 1998
70 Lin, Yang, & Chen, 2004, p. 231. Based on Galbraith, 1981 and Harel & Sowder, 1998
71 Lin, Yang, & Chen, 2004, p. 232. Based on Zaslavsky & Ron, 1998
72 Lannin 2005
73 Zazkis, Loljedahl, & Chernoff, 2008, p. 132 based on Stacey, 1989 and Zazkis & Liljedahl, 2002
74 Zazkis, Liljedahl, & Chernoff, 2008, p. 140
75 Dugdale, 1998, p. 203 based on Abramovich, 1995; Cornell & Siegfried, 1991; Dugdale, 1994; Hoeffner, Kendall, Stellenwerf, Thames, & Williams, 1990; Masalski, 1990; Maxim & Verhey, 1991; Neuwirth, 1995; Sutherland & Rojano (1993)
76 Lannin, 2005, p. 236
77 Zazkis, Liljedahl, & Chernoff, 2008, p. 132 based on Ainley, Bills & Wilson, 2005; Bills, Ainley & Wilson, 2006; Lannin, 2005
78 Abramovich, 2000; Drier, 2001, Healy & Hoyles, 1999; Lannin, 2005
79 MacGregor and Stacey 1993, Quinlan 2001
80 Becker & Rivera, 2005
81 Becker & Rivera, 2005
82 Stacey, 1989
83 Warren, 1992, p. 253
84 Quinlan, 2001, p. 426
85 MacGregor & Stacey, 1993

86 Quinlan, 2001
87 MacGregor & Stacey, 1993, p. 1-186
88 Billings, 2008
89 Billings, 2008
90 Knuth, 2000
91 Pilipczuk, 2008, p. 672
92 Pilipczuk, 2008, p. 673
93 Based on Hines, 2002
94 Based on Lannin (2005), p. 236, adapted from Simon & Blume (1996) and Waring (2000)

REFERENCES

Abramovich, S. (1995). Technology-motivated teaching of advanced topics in discrete mathematics. *Journal of Computers in Mathematics and Science Teaching, 14*(3), 391–418.

Abramovich, S. (2000). Mathematical concepts as emerging tools in computing applications. *Journal of Computers in Mathematics and Science Teaching, 19,* 21–46.

Ainley, J., Bills, L., & Wilson, K. (2005). Designing spreadsheet-based tasks for purposeful algebra. *International Journal of Computers for Mathematical Learning, 10*(3), 191–215.

Becker, J. R., & Rivera, F. (2005). *Generalization strategies of beginning high school algebra students.* Paper presented at the 29th Conference of the International Group for the Psychology of Mathematics Education, Melbourne, Australia.

Billings, E. M. H. (2008). Exploring generalization through pictorial growth patterns. In C. Greenes & R. Rubenstein (Eds.), *Algebra and algebraic thinking in school mathematics* (pp. 279–293). Reston, VA: National Council of Teachers of Mathematics.

Bills, L., Ainley, J., & Wilson, K. (2006). Modes of algebraic communication—Moving between natural language, spreadsheet formulae and standard notation. *For the Learning of Mathematics, 26*(1), 41–46.

Blanton, M., & Kaput, J. (2002). *Developing elementary teachers' algebra "eyes and ears": Understanding characteristics of professional development that promote generative and self-sustaining change in teacher practice.* Paper presented at the annual meeting of the American Educational Research Association, New Orleans, LA.

Bloom, L., Comber, G., & Cross, J. (1986). A microcomputer approach to exponential functions and their derivatives. *Mathematics in School, 15*(5), 30–32.

Borba, M. C., & Confrey, J. (1996). A student's construction of transformations of functions in a multiple representational environment. *Educational Studies in Mathematics, 31*(3), 319–337.

Breidenbach, D., Dubinsky, E., Hawks, J., & Nichols, D. (1992). Development of the process conception of function. *Educational Studies in Mathematics, 23*(3), 247–285.

Brown, J. P. (2007). *Early notions of functions in a Technology-Rich Teaching and Learning Environment (TRTLE).* Paper presented at the 30th annual conference of the Mathematics Education Research Group of Australasia.

Cañadas, M., Castro, E., & Castro, E. (2009). *Graphical representation and generalization in sequences problems.* Paper presented at the Congress of the European Society for Research in Mathematics Education.

Carlson, M. P. (1998). A cross-sectional investigation of the development of the function concept. In A. H. Schoenfeld, J. Kaput, & E. Dubinsky (Eds.), *Research in col-*

legiate mathematics education III (pp. 114–162). Washington, DC: Mathematical Association of America.

Carpenter, T., & Franke, M. (2001). Developing algebraic reasoning in the elementary school: Generalization and proof. In K. S. H. Chick, J. Vincent, & J. Vincent (Ed.), *Proceedings of the 12th ICMI Study Conference: The future of the teaching and learning of algebra* (pp. 155–162). Melbourne, Australia: The University of Melbourne.

Clement, L. (2001). What do students really know about functions? *Mathematics Teacher*, 94(9), 745–748.

Confrey, J. (1991). The concept of exponential functions: A student's perspective. In L. Steffe (Ed.), *Epistemological foundations of mathematical experience. Recent research in psychology.* (pp. 124–159). New York, NY: Springer Verlag.

Confrey, J. (1994). *Six approaches to transformation of functions using multi-representational software.* Paper presented at the XVIII Psychology of Mathematics Education, Lisbon: Lisbon University.

Cornell, R. H., & Siegfried, E. (1991). Incorporating recursion and functions in the secondary school mathematics curriculum. In M. J. Kenney & C. R. Hirsch (Eds.), *Discrete mathematics across the curriculum, K–12* (pp. 149–157). Reston, VA: National Council of Teachers of Mathematics.

Davidenko, S. (1997). Building the concept of function from students' everyday activities. *Mathematics Teacher*, 90, 144–149.

Davis, G. E., & McGowan, M. A. (2002). *Function machines & flexible algebraic thought. Annual meeting of the International Group for the Psychology of Mathematics Education.* University of East Anglia: Norwick, UK.

Drier, H. S. (2001). Teaching and learning mathematics with interactive spreadsheets. *School Science and Mathematics*, 101(4), 170–179.

Dubinsky, E., & Wilson, R. T. (2013). High school students' understanding of the function concept. *Journal of Mathematical Behavior*, 32(1), 83–101.

Dugdale, S. (1994). K–12 teachers' use of a spreadsheet for mathematical modeling and problem solving. *Journal of Computers in Mathematics and Science Teaching*, 13(1), 43–68.

Dugdale, S. (1998). A spreadsheet investigation of sequences and series for middle grades through precalculus. *Journal of Computers in Mathematics and Science Teaching*, 17, 203–222.

Elia, I., Panaoura, A., Eracleous, A., & Gagatsis, A. (2007). Relations between secondary pupils' conceptions about functions and problem solving in different representations. *International Journal of Science and Mathematics Education*, 5(3), 533–556.

Ellis, A. B. (2007). Connections between generalizing and justifying: Students' reasoning with linear relationships. *Journal for Research in Mathematics Education*, 38(3), 194–229.

English, L., & Warren, E. (1995). General reasoning processes and elementary algebraic understanding: Implications for instruction. *Focus on Learning Problems in Mathematics*, 17(4), 1–19.

Even, R. (1990). Subject matter knowledge for teaching the case of functions. *Educational Studies in Mathematics*, 21(6), 521–544.

Even, R. (1993). Subject-matter knowledge and pedagogical content knowledge: Prospective secondary teachers and the function concept. *Journal for Research in Mathematics Education, 24*(2), 94–116.

Even, R. (1998). Factors involved in linking representations of functions. *The Journal of Mathematical Behavior, 17*(1), 105–121.

Falcade, R., Laborde, C., & Mariotti, M. (2007). Approaching functions: Cabri tools as instruments of semiotic mediation. *Educational Studies in Mathematics, 66*(3), 317–333.

Ferrini-Mundy, J., Lappan, G., & Phillips, E. (1997). Experiences with patterning. In B. Moses (Ed.), *Algebraic thinking, grades k–12: readings from NCTM's school-based journals and other publications* (pp. 112–119). Reston, VA: National Council of Teachers of Mathematics.

Galbraith, P. L. (1981). Aspects of proving: A clinical investigation. *Educational Studies in Mathematics, 12*, 1–28.

Goldenberg, E. P. (1988). Mathematics, metaphors, and human factors: Mathematical, technical, and pedagogical challenges in the educational use of graphical representation of functions. *Journal of Mathematical Behavior, 7*, 135–173.

Gray, R., & Thomas, M. (2001). *Quadratic equation representations and graphic calculators: Procedural and conceptual interactions.* Paper presented at the 24th Annual Mathematics Education Research Group of Australasia Conference, Sydney, Australia.

Harel, G., & Sowder, L. (1998). Students' proof schemes: Results from exploratory studies. In A. H. Schoenfeld, J. Kaput, & E. Dubinsky (Eds.), *Research in collegiate mathematics education 3* (pp. 234–283). Washington, DC: Mathematical Association of America.

Hatisaru, V., & Erbas, A. K. (2010). Students' perceptions of the concept of function: The case of Turkish students attending vocational high school on industry. *Procedia— Social and Behavioral Sciences, 2*(2), 3921–3925.

Healy, L., & Hoyles, C. (1999). Visual and symbolic reasoning in mathematics. Making connections with computers. *Mathematical Thinking and Learning, 1*, 58–84.

Heid, M. K. (1996). A technology-intensive functional approach to the emergence of algebraic thinking. In N. Bednarz, C. Kieran, & L. Lee (Eds.) *Approaches to algebra: Perspectives for research and teaching* (pp. 239–56). Dordrecht, Netherlands: Kluwer Academic Press.

Hershkowitz, R., & Schwarz, B. (1999). The emergent perspective in rich learning environments: Some roles of tools and activities in the construction of socio-mathematical norms. *Educational Studies in Mathematics, 39*(1–3), 149–66.

Hines, E. (2002). Developing the concept of linear function: One student's experiences with dynamic physical models. *Journal of Mathematical Behavior, 20*(3), 337–361.

Hines, E., Klanderman, D. B., & Khoury, H. A. (2001). The tabular mode: not just another way to represent a function. *School Science and Mathematics, 101*(7), 362–371.

Hoeffner, K., Kendall, M., Stellenwerf, C., Thames, P., & Williams, P. (1990). Teaching mathematics with technology: Problem solving with a spreadsheet. *Arithmetic Teacher, 38*(3), 52–56.

Hollar, J. C., & Norwood, K. (1999). The effects of a graphing-approach intermediate algebra curriculum on students' understanding of function. *Journal for Research in Mathematics Education, 30*(2), 220–226.

Jones, M. (2006). Demystifying functions: The historical and pedagogical difficulties of the concept of function. *Rose-Hulman Undergraduate Math Journal, 7*(2), 1–20.

Kalchman, M. (1998). *Developing children's intuitive understanding of linear and non-linear functions in the middle grades.* Paper presented at the Annual Meeting of the American Educational Research Association, San Diego.

Kaput, J. (1999). Teaching and learning a new algebra with understanding. In E. Fennema, & T. Romberg (Eds.), *Mathematics classrooms that promote understanding* (pp. 133–155). Mahwah, NJ: Erlbaum.

Knuth, E. (2000). Student understanding of the Cartesian connection: An exploratory study. *Journal for Research in Mathematics Education, 31*(4), 500–508.

Knuth, E., Slaughter, M., Choppin, J., & Sutherland, J. (2002). *Mapping the conceptual terrain of middle school students' competencies in justifying and proving.* Paper presented at the 24th Annual Meeting of the North American Chapter of the International Group for the Psychology of Mathematics Education. Athens, GA.

Lannin, J. K. (2005). Generalization and justification: The challenge of introducing algebraic reasoning through patterning activities. *Mathematical Thinking & Learning, 7*(3), 231–258.

Lannin, J. K., Barker, D. D., & Townsend, B. E. (2006). Recursive and explicit rules: How can we build student algebraic understanding? *Journal of Mathematical Behavior, 25*(4), 299–317.

Lee, L. (1996). An initiation into algebraic culture through generalization activities. In N. Bednarz, C. Kieran, & L. Lee (Eds.), *Approaches to algebra* (pp. 87–106). Dordrecht, The Netherlands: Kluwer.

Lee, L., & Wheeler, D. (1987). *Algebraic thinking in high school students: Their conceptions of generalization and justification* (Research Report). Montreal, Canada: Concordia University, Department of Mathematics.

Lin, F. L., Yang, K. L., & Chen, C. Y. (2004). The features and relationships of reasoning, proving and understanding proof in number patterns. *International Journal of Science and Mathematics Education, 2*(2), 227–256.

Lloyd, G. M., & Wilson, M. (1998). Supporting innovation: The impact of a teacher's conceptions of functions on his implementation of a reform curriculum. *Journal for Research in Mathematics Education, 29*(3), 248–274.

MacGregor, M., & Stacey, K. (1993). Seeing a pattern and writing a rule. In I. Hirabayashi, N. Nohda, K. Shigematsu, & F. Lin (Eds.), *Proceedings of 17th conference of the International Group for the Study of the Psychology of Mathematics Education* (Vol. 1, pp. 181–188). Tsukuba, Japan: Program Committee.

Markovits, Z., Eylon, B., & Bruckheimer, M. (1986). Functions today and yesterday. *For the Learning of Mathematics, 6*(2), 18–28.

Masalski, W. J. (1990). *How to use the spreadsheet as a tool in the secondary school mathematics classroom.* Reston, VA: National Council of Teachers of Mathematics.

Mason, J., Pimm D., Graham, A., & Gower, N. (1985). *Routes to/Roots of Algebra.* Milton Keynes, UK: Open University.

Maxim, B. R., & Verhey, R. F. (1991). Using spreadsheets to introduce recursion and difference equations in high school mathematics. In J. Kenney & C.R. Hirsch (Eds.), *Discrete mathematics across the curriculum, K–12* (pp. 158–165). Reston, VA: National Council of Teachers of Mathematics.

Neuwirth, E. (1995). Spreadsheet structures as a model for proving combinatorial identities. *Journal of Computers in Mathematics and Science Teaching, 14*(3), 419–434.

Norman, A. (1992). Teachers' mathematical knowledge of the concept of function. In G. Harel & E. Dubinsky (Eds.), *The concept of function. Aspects of epistemology and pedagogy* (Vol. 25, MAA notes, pp. 215–232). Washington, DC: Mathematical Association of America.

Orton, A., & Orton, J. (1994). Students' perception and use of pattern and generalization. In J. P. da Ponto & J. F. Matos (Eds.), *Proceedings of the 18th International Conference for the Psychology of Mathematics Education* (Vol. III, pp. 407–414). Lisbon, Portugal: PME Program Committee.

Pilipczuk, C. (2008). Graphing functions: Resolving students' misconceptions by using messy data and calculator-based laboratory activities. *Mathematics Teacher, 101*(9), 670–676.

Quinlan, C. (2001). *From geometric patterns to symbolic algebra is too hard for many*. Paper presented at the 24th Annual Mathematics Education Research Group of Australasia Conference, Sydney, Australia.

Radford, L. (2010). Layers of generality and types of generalization in pattern activities. *PNA, 4*(2), 37–62.

Sajka, M. (2003). A secondary school student's understanding of the concept of function—A case study. *Educational Studies in Mathematics, 53*(3), 229–254.

Schoenfeld, A., Smith, J., & Arcavi, A. (1990). The microgenetic analysis of one student's evolving understanding of a complex subject matter domain. In R. Glaser (Ed.), *Advances in instructional psychology*. Hillsdale, NJ: Lawrence Erlbaum Associates.

Sierpinska, A. (1992). On understanding the notion of function. In G. Harel & E. Dubinsky (Eds.), *The concept of functions: Aspects of epistemology and pedagogy* (pp. 25–28). Washington, DC: The Mathematical Association of America.

Simon, M. A., & Blume, G. W. (1996). Justification in the mathematics classroom: A study of prospective elementary teachers. *Journal of Mathematical Behavior, 15,* 3–31.

Stacey, K. (1989). Finding and using patterns in linear generalising problems. *Educational Studies in Mathematics, 20*(2), 147–164.

Stein, M. K., Baxter, J. A., & Leinhardt, G. (1990). Subject matter knowledge and elementary instruction: A case from functions and graphing. *American Educational Research Journal, 27,* 639–663.

Sutherland, R., & Rojano, T. (1993). A spreadsheet approach to solving algebra problems. *Journal of Mathematical Behavior, 12,* 353–383.

Thompson, P. W. (1994). The development of the concept of speed and its relationship to concepts of rate. In G. Harel & J. Confrey (Eds.), *The development of multiplicative reasoning in the learning of mathematics* (pp. 181–234). Albany, NY: SUNY Press.

Trigueros, M., & Martínez-Planell, R. (2010). Geometrical representations in the learning of two-variable functions. *Educational Studies in Mathematics, 73*(1), 3–19.

Vinner, S. (1992). The function concept as a prototype for problems in mathematics learning. In G. Harel & E. Dubinsky (Eds.), *The concept of function: Aspects of epistemology and pedagogy* (pp. 195–214). Washington, DC: Mathematical Association of America.

Vinner, S., & Dreyfus, T. (1989). Images and definitions for the concept of function. *Journal for Research in Mathematics Education, 20*(4), 356–366.

Waring, S. (2000). *Can you prove it? Developing concepts of proof in primary and secondary schools.* Leicester, UK: The Mathematical Association.
Warren, E. (1992). Algebra: Beyond manipulating symbols. In A. Baturo & T. Cooper (Eds.), *New directions in algebra education* (pp. 252–258). Brisbane: Queensland University of Technology, Centre for Mathematics and Science Education.
Yerushalmy, M., & Shternberg, B. (2001). Charting a visual course to the concept of function. In A. A. Cuoco & F. R. Curcio (Eds.), *The roles of representation in school mathematics: 2001 Yearbook* (pp. 251–268). Reston, VA: National Council of Teachers of Mathematics.
Zaslavsky, O., & Ron, G. (1998). Students' understanding of the role of counter-examples. *PME, 22*(4), 225–232.
Zazkis, R., & Liljedahl, P. (2002). Generalization of patterns: The tension between algebraic thinking and algebraic notation. *Educational Studies in Mathematics, 49*(3), 379–402.
Zazkis, R., Liljedahl, P., & Chernoff, E. (2008). The role of examples in forming and refuting generalizations. *ZDM, 40*(1), 131–141.

CHAPTER 6

MODELING AND WORD PROBLEMS

> I am never content until I have constructed a
> mechanical model of the subject I am studying.
> If I succeed in making one, I understand. Otherwise, I do not.
> —*Lord Kelvin, 1884*

INTRODUCTION

The phrase "word problems" evokes a strong response from most adults. In a survey, algebra teachers cited word problems as "the most serious deficiency of incoming students."[1] Often found at the end of the exercises in a textbook, problems that involve reading and context have a reputation for being difficult. Most traditional textbook exercise sets start with concepts devoid of context and then end with word problems that are perceived to be the challenges. However, in the same survey, algebra teachers also cited "motivation" as the single biggest challenge they face as teachers.[2] Word problems and modeling have the potential to make mathematics meaningful and therefore motivational when used effectively.

In spite of the reputation of word problems, there is data to indicate that when students are presented problems in a context that is understandable to them, they actually perform quite well.[3] Particularly when solving simple algebra word problems when a variable is mentioned just once, the context often helps students make sense of the problem. Struggles with story problems are often more about comprehension of the task and not the generation of a solution.[4] The difficulty students often have with word problems is how to represent the context symbolically.[5]

Teachers' beliefs about the potential difficulty of a problem for their students do not always align with the actual ability of students to solve tasks.[6] Teachers can misread what aspects of a problem are challenging for students and their level of understanding. Based on their interpretation of students' development in algebra, teachers often change how they have students interact with word problems.[7] For instance, teachers' beliefs about the value of students' informal methods influence the shape of their instructional practice. They may believe that only formal methods are worthwhile. However, students who use informal and formal problem-solving methods in combination can perform better than those who simply use formal methods.[8] Teachers tend to believe that students engage in cycles of learning formal, symbolic mathematics and then applying that understanding to word problems in arithmetic and algebra. Teachers start with the symbolic forms because they believe they are simplest. However, some research indicates that starting with words and students' informal strategies and then moving to symbolic forms may be more beneficial.[9] "Instruction that bridges formal algebra instruction to previously grounded representations helps students learn processes such as algebraic modeling of verbally presented relations."[10] Instruction that integrates students' informal methods is particularly helpful when students struggle with manipulation of symbols. Of course, these informal strategies have limitations, particularly as students engage in word problems that have more complex structures.

Researchers suggest that the process of engaging with word problems can be divided into the comprehension phase and the solution phase.[11] First, students must understand the text of the story, as well as any other forms of data such as charts or diagrams, and "create corresponding internal representations of the quantitative and situation-based relationships expressed in that text."[12] The comprehension phase includes the transformation of the quantitative relationships into a symbolic representation and is usually the source of most problem-solving difficulties and errors.[13] Lastly, students work toward a solution. Solving word problems is not typically linear, but rather a back-and-forth process of comprehending part of the task and working to represent that understanding. A first step in the transformation process is **_Defining Variables to Represent Generalizations_**[1] related to the problem context. The research on difficulties with variables is extensive and warrants

---

[1] Note: **_bold and italicized_** print indicates a reference to an entry in the book

an entire chapter in this book (See *Variables and Expressions*). In this chapter, we note that one such misconception stems from generalizing contexts into variables and confusing letters that represent qualities or quantities.

The next natural step for mathematics students is to ***Translate Word Problems into Algebraic Sentences***, or change the English context into algebraic symbols. In elementary school, student intuition about the translation is often accurate. However, as their learning progresses, some students become dependent on specific examples that they can replicate in similar situations. In this chapter we discuss specific tasks that can be used to address common student errors with translation. Another strategy supported by research is paying close attention to the features of algebraic expressions and how they impact an expression's value.

A significant aspect of making decisions about the algebraic expression that best represents a given situation is ***Choosing Operations*** that best fit the task. Many tasks in current algebra curricula follow similar patterns and students have come to recognize these patterns. Without really understanding their actions, students can make decisions that work, even when they do not understand why. For example, given a context where one number is bigger than another, they might choose to use division since we rarely have reasons to divide a small number by one that is large. They might also try taking both numbers and doing all four operations and then just make a decision about which answer looks best. These strategies are not mathematically mature, but they often produce correct answers in the carefully controlled contexts we find in mathematics classes.

After the word problem has been translated into an expression, students now need to proceed with a solution. They must ***Choose Between Algebraic and Arithmetic Solution Strategies***, regardless of whether ***Drawing Models as an Intermediate Step in Solving a Word Problem*** would be helpful, or whether ***Linear Versus Non-linear Modeling*** would be most appropriate for the context. These decisions should be made with mathematical justification. However, the decisions can sometimes stem from something as simple as copying whatever was demonstrated in class recently. Teachers can challenge this behavior by assigning specific tasks and by demonstrating problem-solving strategies that work across contexts. Specific illustrations of this follow in the entries in this chapter.

We would be remiss if we did not acknowledge the ***Issues of Language in a Task***—the challenges that language presents to students solving word problems. Mathematics teachers have much to learn from the research done in the area of English Language Learners. The strategies and techniques supported by the research can be applied to word problems. We want to be sure that students who understand the mathematics are not held back by language constraints. The way we write tasks and the way we assess student learning can address this issue straight on.

Finally, there is one key choice for algebra teachers to make: should I start with learning equations and symbols and then move to word problems (the traditional method of teaching algebra) or should I start with story problems and develop students' verbal skills in algebra before focusing on symbolic equations? Research indicates that students have greater success when teachers start with stories and move to equations.[14] "Contexts may provide accessibility or scaffolding for students, with concrete and familiar situations providing a bridge between what the students know and the abstract mathematics they are trying to learn."[15]

However, the word problems in algebra textbooks are typically stereotyped and oversimplified and make it difficult for students to apply their real-world knowledge. Doing these types of problems may actually harm students' ability to learn problem solving.[16] Engaging students in real-world problems with complex and messy situations in which students must determine what data is important and which mathematical tools are most appropriate is the most effective way to develop students' "mathematical power," in the words of the National Council of Mathematics Teachers. The practice of tackling authentic problems can develop critical citizenship skills and foster creativity, self-reliance, persistence, and confidence.[17] Perhaps most importantly for algebra teachers, these kinds of problems can be a gateway for mathematics learning by allowing students to use their mathematical intuition and prior knowledge, and enabling them to "ground abstract concepts that might be otherwise difficult to grasp."[18] Dan Meyer's work in this area is notable and his TED talk on the subject is worth viewing (https://www.ted.com/talks/dan_meyer_math_curriculum_makeover?language=en).[19]

DEFINING VARIABLES TO REPRESENT GENERALIZATIONS

Common Core Standard: 6.EE.1

As described in the introduction to this book, one of the primary values of algebra is the ability to generalize using letters as symbols for unknown quantities and using arithmetic operations to find those unknowns.[20] A key task in modeling is to have meaning associated with the variables that are tied to the context. Meaning, in the definition of variables, comes from three primary sources: the algebraic structure, the context of the problem, and the exterior of the problem context.[21] Students' understanding of algebraic structure or "syntax" (e.g., equivalent equations or other representations) helps them "rearrange or recombine relations and other algebraic entities, thereby deriving new relations that may be useful in solving new kinds of problems or helpful in furthering our understanding of the original relations."[22] The meaning of the variable is influenced by the structure in which it resides. The context of the problem provides the tangible meaning of the letter, such as the number of buses needed for a school field trip—it is a real-world problem. The exterior of the problem context is what the student brings to the definition of the variable—"the meaning constructed by the individual learner."[23] What experience with field trips and buses does the student have? This meaning can include "metaphors, gestures, and bodily actions" that have some impact on an individual student's interpretation.[24]

Students can often accurately describe relationships in word problems, but have difficulty representing the problem situation (see also ***Variables & Expressions: Letter Used as General Number*** and ***Patterns & Functions: Promoting and Justifying Generalization.***).[25] They are often "imprecise, inconsistent, paradoxical, and over-associative" in their definition of variables.[26] For instance, in one study, 74.1% of first-year engineering students had difficulty defining variables in the following problems.[27]

1. In a classroom there are 2 chairs beside each table. If there are n chairs, how many chairs and tables are there altogether?
2. On a math test, the number of students who pass is 3 times the number of students who fail. If the number of students who pass is p, how many students are in this class?
3. n coins are to be divided between Jack and Jill. Jill must first get k coins. The rest of the coins are divided so that for every x coins that Jill gets, Jack gets z coins. How many coins does Jack get?

In spite of the fact that the questions contained all the variables needed to respond, 42% of students introduced variables that were the first letters of the objects. In Problem 1, for example, 8.1% of students defined the variables as "t = tables and c = chairs" and then came up with a formula such as $A = (2c + 1t)$, using the variables as adjectives rather than quantities. This type of error has been confirmed in other research.[28]

When students attempt to translate words to symbols, they must assign variables to the quantities of interest. There are some common student errors when students use variables to model word problems:[29]

- The use of algebraic letters as abbreviated words (e.g., a means "apple," not "number of apples")(see also *Variables & Expressions: Letter Used as an Object*).
- Attempts to translate directly from key words to mathematical symbols, from left to right, without concern for meaning (e.g., "There are six times as many cats as dogs" is translated incorrectly as $6 \times c = d$).
- Use of the "equals" sign to indicate that what is on the left is loosely associated with what is on the right (e.g., $20p = t$ could mean "there are 20 pupils for every teacher" in a student's mind).
- The misleading influence of mental pictures (e.g., groups of 20 pupils and an individual teacher seen in the mind's eye, and represented on paper as $20p + t$, $20p = t$ or $20p:t$).

In Table 6.1, researchers asked secondary students to pick variable names that would help indicate their relationship. Students could discuss the relationship accurately but were unable to use the "algebraic code" to represent the relationship.[30] The responses demonstrate that students struggled to understand how to convey the relationship between the two numbers by defining the appropriate variable. In another study, when given a word problem, 87% of 245 college students did not define the variable, or named the variable in an imprecise way, such as "*b* is for books."[31]

In the interview for Problem 5 (see Table 6.2), a student articulates that she did not understand what the variable she chose represented.[32] The student acknowledges that she didn't understand what the letter represented and was still imprecise in defining letter *P*, even after prompting by the interviewer. Due to their lack of understanding of how a letter might be defined for a word problem, students can arrive at incorrect conclusions or fail to understand what their answer is telling them about the problem.[33] This failure likely stems from students' attempts to have the variable represent both quantity and quality. The letter *P* in Problem 1 cues the student that the letter represents "pounds," or the weight of nuts (qual-

TABLE 6.1. Problem 4: Variables and Odd Numbers

Problem	Responses
By means of a good choice of names to designate two subsequent odd numbers, show that their sum is a multiple of 4.	• $x + y$ • $2h + 1 + 2k + 1$ • $2h + 1 + 2k + 1 + 2$ • instead of $2h + 1 + 2h + 3$

Source: Arzarello, Bazzini, & Chiappini (1994)

TABLE 6.2. Problem 5: Pecans and Cashews: Student Interview on Defining Variable

A person went to the store and bought pecans and cashews. He bought a total of 100 nuts. The number of pounds of pecans he bought was the same as the number of pounds of cashews. Eight pecans weigh 1 pound and 12 cashews weigh 1 pound. How many pounds of pecans did he buy?	
Interviewer:	You multiplied [the 5] times 8 and got 40. Then you divided 40 by 8 to get 5 again.
Student:	Yeah, so I probably could have stopped but I didn't recognize that that was the answer.
Interviewer:	Why do you think that ... you didn't recognize the answer?
Student:	Because probably not until I looked back did I realize that the P I was trying to find was pounds. If I realized it first, then I would have said P is pounds and P is 5 then 5 pounds of pecans is what he bought. But not until you asked me what was P, what was P standing for, then I realized that it had to be pounds.

Source: Rosnick (1982, p. 6)

ity), and it stands for the number of pounds (quantity) in the problem equation. To assist students in overcoming this confusion, teachers may want to provide many opportunities for students to define variables for simple word problems, translate the words into multi-variable equations, and then explain how the quantities in the problem statement connect with the variables in the equation. For an in-depth review of this topic, see ***Variables & Expressions: Representation—What Can Variables Stand For?, Letter as an Object,*** and ***Letter Used as an Unknown Number.***

Another common type of problem in algebra is generalizing from a pattern. Researchers have found three specific strategies that students use when they try to define a variable for the general term for a pattern, such as in Figure 6.1.[34] First, n is considered as a place holder to be taken by numbers, with the expression for the pattern determined by trial-and-error. The formula is seen as a procedural mechanism. For example, with the pattern in Figure 6.1, students found the formula $n \times 2 + 1$ (where "n" stands for the number of the figure). In Interview 1 (see Table 6.3), the student knew only that the formula works to get the right number of toothpicks for each figure. The rest of the interview indicates that students really struggled with the meaning of the variable and its purpose in the equation. The second strategy recognizes the recursive nature of the problem. Students saw that the number of toothpicks in one figure was two more than the previous figure (see Table 6.3, Interview 2). However, students did not equate what they noticed numerically with a general form. They saw the general pattern, but only through each instance of recursion. Their definition of the variable was "all that is unknown," as is apparent with Student 5's final comment. In the third strategy, students saw the connection between the visual pattern and the general form (see Table 6.3, Interview 3). They understood how the symbolic expression reproduced the shape of the figures of the pattern.

308 • HOW STUDENTS THINK WHEN DOING ALGEBRA

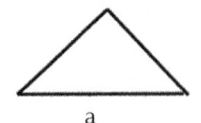

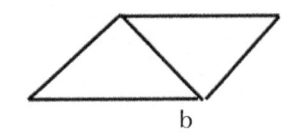

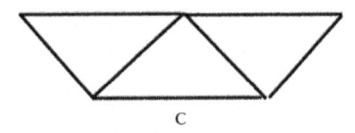

FIGURE 6.1. Toothpick Pattern For Naming a Generalized Variable.

Good problem solvers usually have "a glimpse of a possible path, and put it implicitly in their first trials of naming; usually they are able to incorporate the relationships among the elements and prefigure transformations apt to reach the solution."[35] As students define the variable for the problem, they keep in mind how they will use the variable to achieve a solution. Novices, on the other hand, tend to name variables more randomly and superficially, not anticipating their role in the algebraic sentence.

TABLE 6.3. Interviews for Toothpick Pattern

Interview 1	
Interviewer:	How did you get your formula?
Student 2:	"n," after this, bracket, n times 2, minus 1.
Student 3:	Equals "n."
Student 2:	You don't have to write equals "n." Do we?
Student 3:	Yes. You have to write it.
Student 2:	Just ... we don't need "n."
Student 1:	You need a formula.
Student 2:	OK, "n" bracket "n" times 2, minus 1. [And she writes $n = (n \times 2 - 1)$.]
Interview 2	
Student 4:	One plus two, two plus three, three plus four, four plus five, ...
Student 5:	[Interrupting] Five plus six. Oh! [realizing that S1's idea works] O.K.
Student 6:	It's always the next one. One plus two, two plus three (...), three plus four ....
Student 4:	How do you say it [in algebraic symbols]... the figure plus the next figure?
Student 5:	n plus n?
Interview 3	
Student 7:	The base of the triangle is the one thing that is constant. Otherwise, you add two sides to the left, then two sides to the right and back to the left. So, it is always the base, one, plus two times the figure number.
Student 8:	So one plus twice the figure number will equal the total toothpicks [writes $1 + 2n = t$].

Based on Radford (2000, pp. 3–4)

Teachers

- Students can often accurately discuss what is going on in a word problem, but may not be able to represent that understanding algebraically. Help them make the connection between their words and the meaning of the variable(s) used.
- Help students understand the difference between qualitative information in a problem and quantitative information that can be represented by a variable.
- Ask students to justify each variable they choose and how it relates to the problem. "Students need to make the link between the representation and the problem context."[36]
- Help students be precise in the language they use to define a variable.
- Provide students with opportunities to define variables as representing an object. Contrast that with tasks where variables represent an unknown number or a range of numbers.
- In pattern-recognition problems, encourage students to tie aspects of their formulas to aspects of the visual.
- The use of tables may help students understand the role of variables in generalizations of patterns.[37]

TRANSLATING WORD PROBLEMS INTO ALGEBRAIC SENTENCES

Common Core Standards: 6.EE.1, 2, 6, 9, 8.F.4, CED.A.1, CED.A.2

Research indicates that students have "informal strategies that allow them to effectively deal with algebraic word problems."[38] They bring "rich, verbal resources" and experiences that influence their interpretation of word problems.[39] Even as early as fourth grade, students are capable of developing algorithms to solve equations, and of understanding the use of variables.[40] Sixth-grade students are generally capable of solving problems in specific cases and can develop equations to represent the problem in nonstandard (but correct) forms and generalize the problem situation.[41] However, students have multiple issues transitioning from informal problem solution methods to using more formal algebraic processes and developing symbolic representations of the problem.[42] It is not their ability to perform algorithms, but their ability to represent a given word problem that determines students' success.[43]

Often, teachers try to support students working on word problems with strategies in which specific words are attached to mathematical operations. For example, teachers might say "more than" means addition or "of" means multiplication, etc. However, this strategy can lead to simplistic approaches to word-problem solving that negatively impact students' perseverance in making sense of problems.[44] Researchers find that students who employ this "direct translation strategy" of selecting numbers and key words from the problem and basing their solution plan on them are much less successful.[45] Students are much more successful when they "construct a meaningful representation of the situation" rather than dissect words in isolation.[46] Consider the following problems:[47]

Susan collected 6 rocks, which were 4 more than Jan collected. How many rocks did Jan collect?

Elliott ran 6 times as far as Andrew. Elliott ran 4 miles. How far did Andrew run?

Using the phrases "more than" as an indication to add in Problem 6 and "times" as an indication to multiply in Problem 7 leads to incorrect solutions. A key-word approach focuses students' attention "primarily on the numbers (or *values*) in the situation and not on the quantities and the relationships between quantities."[48] In contrast, successful word-problem solvers exhibit "comprehension of relevant textual information, the capacity to visualize the data, the capacity to recognize the deep structure of the problem, the capacity to correctly sequence their solution activities, and the capacity and willingness to evaluate the procedure used to solve the problem."[49]

There is no doubt that word problems can be challenging for students. However, it is important for teachers to deconstruct what aspect is causing the struggle. Consider the problems in Table 6.4, in which students were asked to translate the words into algebraic sentences.[50] The results are shared in Table 6.5. It is interest-

TABLE 6.4. Word Problems

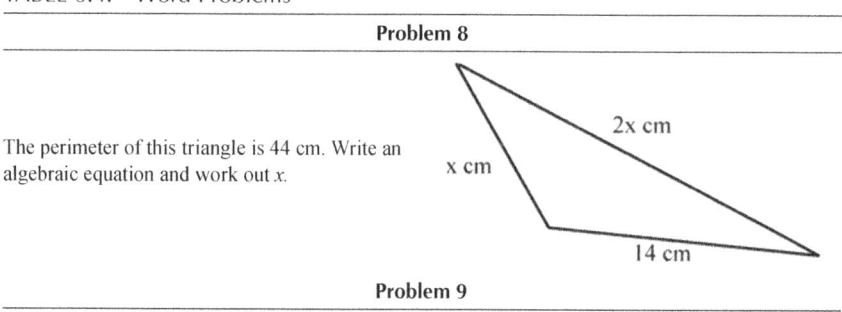

Problem 8

The perimeter of this triangle is 44 cm. Write an algebraic equation and work out x.

(Triangle with sides labeled x cm, $2x$ cm, and 14 cm)

Problem 9

Some money is shared between Mark and Jan so that Mark gets $5 more than Jan gets. Jan gets $$x$. Use algebra to write Mark's amount. The money to be shared is $47. Use algebra to work out how much Jan and Mark would get.

Problem 10

I think of a number, multiply it by 8, subtract 3, and then divide by 3. The result is twice the number I first thought of. What was the number? Write an equation and solve it.

---

ing to note that more students had difficulty writing the equation than answering the first two problems. Only about one third of students were able to create equations for each problem, while twice that amount were able to get accurate answers for the first two problems, typically using trial and error. In Problem 10, the most abstract of the problems, only 17% were able to get an accurate solution.

For many students, the challenge with word problems lies in the translation between English and mathematical notation and then vice versa, especially when the task is not in a familiar context.[51] "Common solution error patterns are directly related to the linguistic sophistication possessed by the solver."[52] Researchers have found that students' level of literacy is a predictor of their performance on algebraic word problems.[53] The difficulty of the word problem is often correlated with text comprehension factors such as ambiguous words. For instance, words such as "some" and "altogether" were often misinterpreted by students. When researchers used the same problem structure but made simple changes in wording, such as using the words "more than" instead of "less than," students' success significantly increased.[54] The "errors" students made with word problems were often correct interpretations or solutions to miscomprehended problems.[55] Generally,

TABLE 6.5. Percent of 524 10th-Grade Students with Correct Responses

Problem 8		Problem 9		Problem 10	
Equation	Answer	Equation	Answer	Equation	Answer
38%	63%	30%	73%	30%	17%

Source: Stacey & MacGregor (2000)

word problems can be easier for students to solve when the "problem situation activates real-world knowledge that aids students in arriving at correct solutions."[56]

Because mathematics is such a dense language, every symbol matters; yet most students are not accustomed to reading text carefully. For example, the difference between $f'(x) = 6x$ and $f(x) = 6x$ is enormous. While students might be able to solve a problem conceptually, they may still struggle with symbolically representing the same problem. This leads to difficulties when they are faced with a problem they cannot solve by simple reasoning and need algebraic symbols to represent aspects of the problem. "Unsuccessful problem solvers base their solution plan on numbers and key words that they select from the problem (the direct translation strategy), whereas successful problem solvers construct a model of the situation described in the problem and base their solution plan on this model (the problem-model strategy)."[57] Students use the direct translation strategy as a shortcut that bypasses any need to develop meaning for the problem. They look for words that suggest mathematical operations (e.g., "more" means addition and "less" means subtraction)[58] so they can "compute first and think later."[59] In contrast, students using the problem-model strategy create a mental model of the relationships in the problem and are more likely to correctly solve the problem as a result.

The tasks in Table 6.6 were given to ten sixth-grade students.[60] In the tasks, students could typically compute specific cases for each of the problems with little difficulty. In a number of instances, they were even able to develop reasonable symbolic representations of each problem. However, very few of the students were able to use the equations they wrote to explain the general case, or to use the equation to solve a randomly chosen specific case. This demonstrates that students come to early algebra instruction with significant prior knowledge, and underscores the need for teachers to build on this knowledge. These tasks are also a good resource for teachers to use in their classes. The researchers in this study found that, as students computed multiple cases over the same context (e.g., The Border Problem for different dimensions), algebraic generalization became easier.

It is apparent that students become highly dependent on examples and strategies found in their textbooks. One research study compared errors in problem solving between Japanese and Thai secondary school students and found they were able to solve problems that were similar to those found in their textbooks, but unable to create models for problems with features unlike those in their respective texts.[61] It appeared that they memorized processes rather than understood the concepts underlying the problems. In contrast, researchers found that students who learned in a problem-based environment were more flexible and better able to apply their mathematical understanding to translate word problems into mathematical notation.[62] They had opportunities to explore their thinking first and then look for mathematical notation to represent that thinking. When teachers gradually introduced algebraic notation after students informally represented their

TABLE 6.6. Word Problem Translation Tasks

Task	Results
Problem 11: Refund Problem (direct variation/proportionality)	
In some states, a deposit is charged on aluminum pop cans and is refunded when the cans are returned. In New York, the deposit is 5 cents a can. a) What would be the refund for returning 6 (or 10 or 12) cans? b) Describe how the storeowner would figure the amount of refund for any number of returned cans. c) Let R represent the amount of refund and let C represent the number of cans returned; write an equation for the amount of refund. d) Can you use your equation to find out how many cans would have to be returned to get a refund of $3.00? How much refund would you get for 100 cans?	• All 10 students were able to compute specific cases correctly. • 9 students were able to describe the functional relation. • 7 students were able to represent the problem symbolically (writing an equation). • However, only 2 students were able to use the equation to solve the problem.
Problem 12: Hours and Wage Problem (linear relationship):	
Mary's basic wage is $20 per week. She is also paid another $2 for each hour of overtime she works. a) What would her total wage be if she worked 4 hours of overtime in 1 week? 10 hours of overtime? b) Describe the relationship between the number of hours of overtime Mary works and her total wage. c) If H stands for the number of hours of overtime Mary works and if W stands for her total wage, write an equation for finding Mary's total wage. d) Can you use your equation to find out how much overtime Mary would have to work to earn a total wage of $50? One week Mary earned $36. A coworker of Mary's had worked 1 hour less overtime than Mary had worked. What would her coworker's wage be for the week?	• All 10 students were able to compute specific cases correctly. • 8 students were able to describe the functional relation, 1 student described the relation recursively. • 6 students were able to represent the problem symbolically (writing an equation). • However, only 3 students were able to use the equation to solve the problem.
Problem 13: Border Problem (linear relationship, geometric context, visuals provided for the student)	
Here is a 10-by-10 grid. How many squares are in the border? a) Here is a 5-by-5 grid. How many squares are in the border of this grid? We don't have a 100-by-100 grid, but how could you figure out how many squares would be in the border? b) Describe how to figure out the number of squares in the border of any size grid (N by N). c) Let N represent the number of squares along one edge of a grid and let B represent the number of squares in the border. Write an equation for finding the number of squares in the border. d) Can you use your equation to find the size of a grid with a border that contains 76 squares?	• 9 students were able to compute specific cases correctly. • 9 students were able to describe the functional relation. • 7 students were able to represent the problem symbolically (writing an equation). • However, only 2 students were able to use the equation to solve the problem.

(*continues*)

TABLE 6.6. Continued

Task	Results
Problem 14: Concert Hall Problem (arithmetic sequence)	
The first row of a concert hall has 10 seats. Each row thereafter has 2 more seats than the row in front of it. a) How many seats are in Row 10? (If student had difficulty, ask how many seats in Rows 2, 3, and 4.) b) The ticket manager needs to know how many seats are in each row. If she knows the number of the row, explain how she can figure out how many seats are in that row. c) Let R represent the number of the row and let S represent the number of seats in that row. Can you give an equation for finding the number of seats? d) How many seats are in Row 21? If the last row has 100 seats, how many rows are in the concert hall?	• All 10 students were able to compute specific cases correctly. • 2 students were able to describe the functional relation. 5 students described the relation recursively. • Only 2 students were able to represent the problem symbolically (writing an equation).
Problem 15: Paper Folding Problem (exponential)	
Fold this piece of paper in half and then open it up. How many regions were made? a) Fold the paper in half twice. How many regions? How many regions will be made with three such folds? b) Describe how to find the number of regions for any number of such folds. c) Write an equation for finding the number of regions if you know the number of folds. Let R represent the number of regions and let F represent the number of folds. d) Suppose we have a magical piece of paper that can be folded indefinitely. If you fold the paper in half 10 times, how many regions are formed?	• All 10 students were able to compute specific cases correctly. • 3 students were able to describe the functional relation and the rest of the students described the relation recursively. • 3 students were able to represent the problem symbolically (writing an equation).

Source: Swafford & Langrall (2000)

thinking, the notation helped students record what they understood and helped them to structure their thinking.[63]

Students come to algebra with significant prior knowledge that allows them to solve complex tasks for specific cases. When algebraic notation is introduced in a way that connects to this knowledge, even young students can be empowered to solve challenging problems. Students will come to depend on examples presented in curricula, so it is critical that we select rich, problem-based texts that allow students to apply their intuition while solving problems.

Teachers

- Don't take shortcuts to answers by giving students key-word strategies for solving word problems. Find ways to support students in making meaning of the problem instead.

- Tasks such as those in Table 6.6 can allow students to build on their intuitive knowledge of a problem's context prior to establishing notation to represent a problem.
- Students are often able to compute specific examples correctly. Give students opportunities to connect their algebraic representation to the specific examples in order to confirm their thinking.
- Emphasize the important role that each symbol and each letter play in an expression. Give students experiences that allow them to see the impact of changing small parts of expressions.

CHOOSING OPERATIONS

Common Core Standards: 6.EE.6, 7, 7.EE.A.2, 7.EE.B.3, CED.A.1

When translating word problems into operational notation, students often struggle to identify the correct operation(s) to use (e.g., addition, subtraction, multiplication, or division). Students often use non-mathematical strategies to select an operation (or operations) to use when solving a word problem.[64] Even their strategies for choosing correct operations can be limited and superficial. Consider Problem 16:

> A bag of snack food contains 4 vitamins and weighs 228 grams. How many grams of snack food are in 6 bags altogether?

Researchers examining students' responses to this problem developed a categorization of the strategies typically used:[65]

1. *Random guess.* Students make a guess at the operation to be used without considering all reasonable options. In Problem 16, a student might come to the answer 24, because $4 \times 6 = 24$, and that is a calculation that is easy to complete.
2. *Guess based on recent classroom material.* Students make a guess on the operation to be used, based on which operation was covered or discussed most recently in class. Again, the student might say that the answer is 24 because, in a recent example, the teacher demonstrated a solution that involved multiplying two numbers.
3. *The numbers determine the operation.* Students determine the operation based on which operation seems most likely to be used on two numbers in the problem. In interviews about Problem 16, a student indicated that the solution was probably to divide 228 by 6 because "…there's a big number and little number."[66]
4. *Try all and choose the best.* Students try all four operations with the two numbers in the problem and choose the answer that seems most reasonable or is simplest to compute. This works especially well when the solutions involve a single step and the student has some sense of what a reasonable answer might be.
5. *Look for key words or phrases.* Students determine an operation based on key words or phrases. For example, some students are taught that "altogether" means the numbers should be added. In the case of Problem 16, addition would lead to an incorrect answer.
6. *Add/multiply for larger, subtract/divide for smaller.* Students reason about whether or not the answer should be larger or smaller than the numbers given in the problem. If the answer should be larger, the student tries addition and multiplication and then chooses the more reasonable

answer. If the answer should be smaller, the student tries subtraction and division and then chooses the more reasonable answer.
7. *Meaning-based.* Students understand what to do based on the meaning of the problem. Teachers desire this solution, but students need to be pressed to justify their thinking. Justification is a key component to any solution, in order to know if it is meaning-based, rather than blind luck.

The first six strategies are mathematically immature, yet students use them because (1) they work often enough, (2) teachers tend not to push students to use a more sophisticated approach, (3) teachers may not question students about why they used a particular strategy, and (4) these are the only strategies students have at their disposal. However, students who have a well-developed mathematical understanding of the operations will select operations based on which fit the story best. Noteworthy in this research was the fact that students who were able to draw a picture to represent Problem 16 were more likely to use a meaning-based solution strategy [see **Drawing Models as an Intermediate Step**]. The goal is for all students to choose operations based on their understanding of both the problem situation and the meanings of the operations.

It is important to acknowledge that "correct answers are not a safe indicator of good thinking" and teachers should require students to justify their choice of an operation, in addition to requiring that students give an answer to the problem [67] Teachers could have students consider questions like "How did you know to [divide] here?" or "What words made you think addition was the appropriate operation to use?"—on individual written work, in small-group discussion, and in whole-class discussion. Teachers can then emphasize the benefits of choosing an operation based on understanding the meaning of the operation. Shortcomings of limited or unsophisticated strategies can be highlighted in class discussion after having students solve problems that involve extraneous information, fractions or decimals, or multi-step solutions.

Teachers

- Try to identify when students are implementing one of the six "mathematically immature" strategies.
- Use questioning to press for justification of operation choice. Ask students to explain more than what they did—ask why they did it.
- Draw explicit connections between what the problem is describing and the correct operation.

CHOOSING BETWEEN ALGEBRAIC AND ARITHMETIC SOLUTION STRATEGIES

Common Core Standards: 6.EE.6, 7, 9, 7.EE.3, 4, CED.A.1, 2, 3

Most word problems presented in standard algebra curricula can be solved either through algebraic or arithmetic methods. Moreover, a given word problem is usually more efficiently solved with one or the other type of solution. When solving word problems, some students automatically apply algebraic methods that are unnecessarily cumbersome. Other students refuse to use algebraic methods (relying solely on arithmetic calculations) because they do not see the utility of using algebra, or they are unable to create an equation that makes sense to them in the context of the problem. Research indicates that many students are unable to think flexibly about options for problem solving.[68] Typically, when students are in an algebra class, they are asked to solve all word problems algebraically to practice algebra problem-solving skills. However, this does not afford students the opportunity to learn how to flexibly apply either arithmetic strategies or algebraic strategies. Students need practice in determining whether an algebraic, arithmetic, or possibly a geometric strategy would be most appropriate for a given task.

If a problem can be solved most efficiently through an arithmetic solution strategy, then it does not make sense to force students to use algebraic approaches. Researchers find that students often use the informal strategies *guess-and-check* and *unwind* effectively to solve many word problems. The unwind strategy is a powerful problem-solving tool in algebra, particularly for "start unknown" word problems. It consists of students working backward, inverting operations from the given result value, as in Problem 17 of Table 6.7. A more formal approach in which the student translated the words into an algebraic sentence is also shown. In one study, researchers found that the unwind approach was most common, with 40% of students using this type of strategy.[69] In order to come to appreciate the value of algebra, students need to encounter problems that are not easily solvable using arithmetic calculations with these two strategies. Research indicates that

TABLE 6.7. Problem 17: Solution Strategies

You have 175 songs downloaded onto your iPod from Limewire and iTunes. You download 4 more every week. By the time you have 275 songs, how many weeks will have passed?"

unwind strategy	Equation-solving strategy
275 − 175 ――― 100 25 40⟌100	$275 = 4x + 175$ $-175 \quad\quad -175$ ――――――― $100 = 4x$ $\frac{100}{4} = \frac{4x}{4}$ $25 = x$

Source: Walkington, Sherman, & Petrosino (2012)

using students' informal approaches as a starting point for conversations about word problems can have significant advantages for their performance.[70] However, teachers should be aware of how the "unwinding strategy breaks down when an unknown quantity has multiple occurrences. Likewise, the guess-and-test strategy can be inefficient, highly demanding of cognitive resources, and limited to finding numerical answers that are likely to be guessed."[71]

If students are only asked to solve problems with algebra that could just as easily be solved with arithmetic calculations or by guess-and-check, they will likely not see the point in using algebra. For example, the task in Table 6.8 is a typical word problem. There would be little benefit for a student to choose algebraic methods to solve Problem 18, as it is more easily understood and solved using an arithmetic strategy. The algebraic approach has a "distance from the problem" in that it is more difficult for students to see how "$120 + B + 2B = 345$" is directly related to the dynamics and context of the problem.[72] In contrast, each action of the arithmetic approach can be directly tied to the problem. Furthermore, the algebraic solution offers more opportunities for error because of the different number and type of manipulations involved.

A word problem that translates into an equation in which only one variable appears (i.e., $2x + 3 = 7$) is much more closely related to arithmetic approaches, since "every stage involves only calculating with known numbers."[73] The algebraic nature of a word problem is really engaged when variables appear on both sides of the equal sign or a second variable is involved. It is with these types of

TABLE 6.8.   Problem 18: Arithmetic and Algebraic Approaches

An elementary school with 345 pupils has a sports day. The pupils can choose between rollerblading, swimming, or a bicycle ride. Twice as many pupils choose rollerblading as bicycling, and there are 30 fewer pupils who choose swimming than rollerblading. One-hundred-twenty pupils want to go swimming. How many choose rollerblading and bicycling?[115]

Example of Arithmetic Solution:

The number of students who chose rollerblading is $120 + 30 = 150$.

The number who chose bicycling is $150/2 = 75$.

Example of Algebraic Solution:

Let R and B be the numbers of students who chose rollerblading and bicycling, respectively.

$120 + B + R = 345$

$2B = R$

$R - 30 = 120$

Substitute $2B$ for R in the first equation.

$120 + B + 2B = 345$

$=> 3B = 225$

$=> B = 75$

problems that the equation can become somewhat detached from the context for students.[74]

Teachers should recognize that formulating an equation is not an intuitive way for many students to represent a problem. To help bridge the gap between arithmetic and algebraic solution strategies, "supports need to be put in place for students to make connections between formal algebraic representation, informal arithmetic-based reasoning, and situational knowledge."[75] When working on word problems, "Teachers view themselves as responsible for assuring that students have learned the canonical method for solving linear equations."[76] They don't tend to be open to non-standard approaches that might be more efficient or easier to comprehend.[77] Teachers should not discourage creative uses of arithmetic to solve problems, but rather engage students in discussing when algebraic approaches are helpful and more efficient.

Teachers

- Most students choose arithmetic over algebraic solution strategies, as they do not see the utility of algebra. Even when the problem is too difficult to be solved arithmetically, they are more likely to choose some form of guess-and-check rather than algebraic methods.
- Use simple problems early in instruction to illustrate that arithmetic and algebraic strategies yield the same results.
- To help students transition to algebraic strategies, teachers must use problems that cannot be solved with simple arithmetic strategies.

DRAWING MODELS AS AN INTERMEDIATE STEP IN SOLVING A WORD PROBLEM

Common Core Standards: 6.RP.3, 6.EE.B.5, 6, 7; 7.EE.B.4; CED.A.1

Models in mathematics can range from simple equations (such as total cost as a product of unit price and number bought, $C = p*n$), gestures,[78] drawings,[79] and using a geometric shape such as a cylinder to model a coin[80] to elaborate flow charts or diagrams that communicate mathematical relationships. Students often struggle with word problems because words themselves are abstract. When students draw visual models as an intermediate step in the process of translating a word problem into a symbolic representation, they can make concrete connections between the words and the symbols. However, some researchers question the effectiveness of diagrams as problem-solving tools.[81] There are two primary potential purposes for drawing models: 1) communicating mathematics, and 2) a way for students to "develop and examine their own ideas."[82] In this entry, we focus on the latter—the value of student-constructed pictorial representations as a means to help them think about mathematical relationships in word problems.

Researchers suggest there are five potential advantages to students drawing diagrams to help them model algebraic word problems:[83]

1. Diagrams can relieve working memory by acting as a place where students can chunk pieces of information together.[84]
2. Diagrams create a useful assessment opportunity for teachers, as they can see visually how students are thinking about a problem.[85]
3. Diagrams help students display information holistically, thereby potentially helping students perceive implicit information within a problem.[86]
4. Diagrams require students to reorganize information, which may help them see new relationships.[87]
5. Diagrams provide a visual alternative to words.[88]

One educator suggests that there are five steps that students need to work through in order to solve a word problem successfully:[89]

1. Reading the problem,
2. Comprehending what was read,
3. Transforming the words into a mathematical strategy,
4. Applying a mathematical procedure, and
5. Writing the answer.

Most errors that students make occur in the first three steps—during the process of interpreting the words—before they even begin to solve the problem. In order to help students understand word problems, teachers often focus on key words such as "more" and "times." This strategy is useful in the short term, but limited in the long term because key words don't help students understand the problem situation (i.e., what is happening in the problem). Key words can also be misleading because the same word may mean different things in different situations. Consider the following two examples:

1. There are 7 boys and 21 girls in a class. How many more girls than boys are there?
2. There are 21 girls in a class. There are 3 times as many girls as boys. How many boys are in the class?

In the first problem, if students focus on the word "more," they may add when they actually need to subtract. In the second problem, if students focus on the word "times," they may multiply when they actually need to divide.

Students who draw models of word problems may overcome many of the difficulties associated with translating a word problem directly into algebraic symbols.[90] A number of studies have shown that diagrams are helpful, from arithmetic word problems to algebraic word problems.[91] In fact, some research indicates that students can do better by drawing their own diagrams than students who were given a diagram.[92] For example, significantly more students doing Problem 20 (Table 6.9) recognized that you could not cut 10-cm-tall triangles with 5-cm-wide metal as they drew their own diagram. "By actively engaging with the diagram students are, in fact, engaging in the process of meaning construction."[93]

There are many different kinds of drawing models students may use. The *model method*, used in Singapore, which is one of the top-performing countries on international tests, involves representing key information of the word problem using rectangles. Table 6.10 includes examples of a model for solving part-whole problems, (additive) comparison problems, and multiplication and division problems.[94] After each problem set, there is an example of a rectangle that could be

TABLE 6.9. Problems with Sheet Metal

Problem 19	Problem 20
A metal sheet of dimensions 20 cm by 5 cm was purchased to cut out metal disks with a diameter of 5 cm. How many disks can be obtained from the sheet?	A metal sheet of dimensions 20 cm by 5 cm was purchased to cut out metal triangles having dimensions of base 10 cm and height 10 cm. How many triangles can be obtained from the sheet?
Percent Correct	Percent Correct
Given diagram / Drew diagram	Given diagram / Drew diagram
70%[a] / 80%	14% / 55%

[a]Percent of 88 university students answering correctly in Mudaly (2012)

TABLE 6.10. The Model Method

Problems	Representations
Part-Whole Problems	
21. On Saturday, 1050 people went on a cruise. On Sunday, 1608 people went on a cruise. How many people went on the cruise over the 2 days?	$a + b = x$
22. There were 2659 visitors at Orchid Gardens. 447 of them were adults and the rest were children. How many children visited Orchid Gardens?	$x + a = b$
Comparison Problems (Additive)	
23. Dunearn Primary School has 280 pupils. Sunshine Primary School has 89 pupils more than Dunearn Primary. Excellent Primary has 62 pupils more than Dunearn Primary. How many pupils are there altogether?	$a + b = x$
24. A cow weighs 150 kg more than a dog. A goat weighs 130 kg less than the cow. Altogether the three animals weigh 410 kg. What is the weight of the cow?	$x + b = c$
Multiplication and Division Problems	
25. Bala took 24 pictures. David took 3 times as many pictures as Bala. How many pictures did the two boys take in all?	
26. Mary and John have $48 altogether. John has three times as much money as Mary. How much money does Mary have?	$a + b = x$
27. If the volume of water in container A is 1/4 the volume of water in container B, and the total volume of water in both containers is 250 litres, find the volume of water in container A.	

Swee Fong & Lee (2009)

used to represent the task. In each case, the first rectangle represents the arithmetic word problem, while the second rectangle represents the algebraic word problem.

The model method is based on a three-phase problem-solving process.[95] The approach has teachers demonstrate all three problem-solving phases for students before asking them to use the model method on their own.

Phase 1:   Text Phase (T) Students read the text.
Phase 2:   Structural Phase (S) Students represent the text in the structure of the model. Students alternate between text and model to check that the model accurately depicts the text.
Phase 3:   Procedural-Symbolic Phase (P) Students use the model to plan and develop a sequence of logical equations that will lead to a solution of the problem.

Consider Problem 21 in Table 6.10. First, the teacher reads the text with students, confirming that students understand the words and vocabulary. Next, the class builds a rectangular model. In this case, the problem can be represented by the part-whole arithmetic model (see Figure 6.2)

Finally, the model is used to determine which equation and procedures should be used to generate the answer. In this case: $1050 + 1608 = x$.[96]

Models alone do not make a problem directly understandable. Students need instruction on features of models and how their design represents a context.[97] "Representational competency," or the ability to understand and create representations of mathematical ideas, is a critical component to participation in the community of a classroom. Teachers need to help their students understand the features of a representation and how it supports the solution of a task.[98] Student-created models can be effective tools to bridge the gap between the task and the solution, and can aid students in understanding a task and illuminate areas where they struggle. Research indicates that computer-based, interactive models may also be effective in helping students reason through algebraic word problems.[99]

Teachers

- Use the 3-phase Model Method to facilitate the translation of a task from words to symbols.

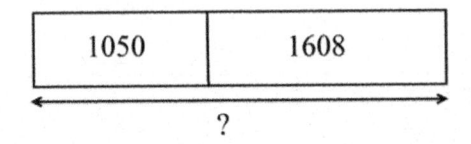

FIGURE 6.2.   Part-Whole Arithmetic Model.

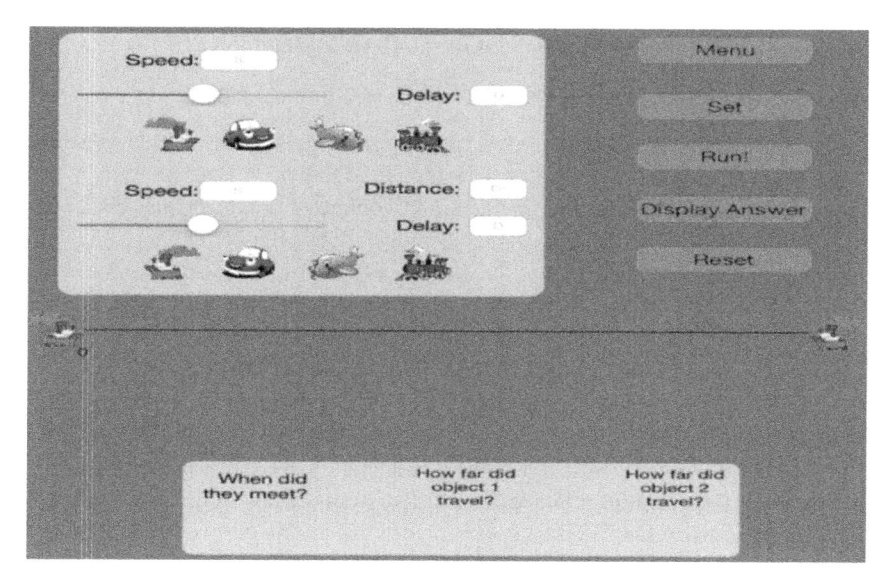

FIGURE 6.3. Distance Rate Time Solver App

- At each phase of the process, students should compare the text, the rectangle, and the equation to confirm that each is correct.
- To encourage understanding through models, students could compare correct models to incorrect (or partially correct) models, discuss multiple correct models to illuminate different strategies, or compare their symbolic representations against models.
- When students are struggling with word problems and a particular representation, research indicates that a teacher-supported switch to a new representation may facilitate improved understanding.[100]
- The Center for Algebraic Thinking app *Distance Rate Time Solver* can help students model DRT types of problems (see Figure 6.3).

LINEAR VERSUS NON-LINEAR MODELING

Common Core Standards: 7.G.1, HSF.LE.A.1, 2; HSF.LE.B.5

Students have a well-known proclivity for assuming linearity in relationships, even when linearity is not appropriate for the situation.[101] For example, when students are asked how the surface area and volume of a cube change when the side lengths are doubled, many neglect the multiplying effect of having a shape in multiple dimensions and say that the surface area and volume also double. This type of thinking is evident in Problem 28 of Table 6.11, in which most students responded with linear thinking. In a follow-up interview (see Table 6.12), the students explained how they understood the problem. In each case, students fail to understand how changes to multiple dimensions result in a non-linear relationship.

At the heart of this confusion is many students' sense that everything is in a proportional relationship.[102] For instance, there is an illusion of proportionality in the following problems:

1. It takes 15 minutes to dry 1 shirt outside on a clothesline. How long will it take to dry 3 shirts outside?
2. John's best time to run 100 meters is 17 seconds. How long will it take him to run 1 kilometer?

Students typically respond to the first problem by saying you need to triple the drying time for the 3 shirts and to the second problem by saying it will take him 10 times as long.[103] In each case, they neglect to make sense of the context and tend to overgeneralize and jump to a proportional relationship. Researchers have documented this tendency with the problems in Table 6.13. Over 90% of 120 twelve- to thirteen-year old students got the proportional problem correct while approximately 5% got the non-proportional problem correct.[104]

A common approach in algebra courses is for the teacher to present a sample word problem and then offer a solution path for the students to follow. Research-

TABLE 6.11. Problem 28: Non-Linear Word Problem

Problem	Response
Michael is a publicity painter. In the last few days, he had to paint a logo on several store windows. Yesterday, he made a drawing of a 16 in. high logo on the window of a print shop. He needed 6 oz of paint. Now he is asked to make an enlarged version of the same logo on a business window. This copy should be 64 in. high. Approximately how much paint will Michael need to do this?	Since the new logo is four times as tall as the original, Michael will need 24 oz of paint, which is four times the original amount. (38 of 40 twelve- to sixteen-year-old students gave this type of answer as their first choice. None of the students included a drawing in their solution.)

Adapted from De Bock, Van Dooren, Janssens, & Verschaffel (2002)

TABLE 6.12. Interview for Problem 28

Interviewer:	How do you know your answer is correct?
Student 1:	The height is multiplied by 4, so the amount of paint needed will be multiplied by 4.
Interviewer:	How can you be sure your answer is correct?
Student 1:	It's an easy problem. I just used the three numbers and the formula, so it must be correct.
Student 2:	It's logical, the logo becomes four times bigger.
Student 3:	It's 4 times bigger, not only the height but also the width. You can see it on the drawing. The whole thing is enlarged by factor 4, so you will need 4 times as much paint.

ers suggest that an alternative approach when trying to help students understand the dynamics of non-linear relationships is to present a task with no particular solution path.105 As students have difficulty with the problem, then a series of interventions are introduced one phase at a time until the student succeeds. Each phase is explained in Table 6.14, using Problem 28 from Table 6.11 to illustrate it.

Much of a traditional algebra curriculum focuses on linear relationships. When presented with non-linear tasks, it is not surprising that students rely on previous linear strategies to solve the problems. Using the phases illustrated above may aid students in recognizing their errors. Not surprisingly, students better understand their error in over-generalizing linearity as they are shown more phases. The usefulness to teachers is that these phases are cumulative and progress from lesser to greater levels of cognitive support and direct instruction.

TABLE 6.13. Problem 29: Proportional Versus Non-Proportional

Proportional Problem	Non-Proportional Problem
Farmer Gus needs approximately 4 days to dig a ditch around a square pasture with a side of 100 m. How many days would he need to dig a ditch around a square pasture with a side of 300 m?	Farmer Carl needs approximately 8 hours to manure a square piece of land with a side of 200 m. How many hours would he need to manure a square piece of land with a side of 600 m?

TABLE 6.14. Potential Phases of Intervention

Phase 1: Solving the Word Problem

Give students the task without any additional suggestions or solution strategies:

Michael is a publicity painter. In the last few days, he had to paint a logo on several store windows. Yesterday, he made a drawing of a 16 in. high logo on the window of a print shop. He needed 6 oz of paint. Now he is asked to make an enlarged version of the same logo on a business window. This copy should be 64 in. high. Approximately how much paint will Michael need to do this?

Phase 2: Cognitive Conflict

Use a form of cognitive conflict in which the student is told two different answers that other students have given, and the frequency with which they were given:

Last period I gave this task to the class. 43% said it was 24 oz. 43% said it was 96 oz. 14% had other answers. Who was right?

Phase 3: Argument with Peer

Use an argument from a fictitious peer. Pretend you are another student and argue with the student, mostly asking questions:

One student told me that if the logo becomes four times as high while keeping the same shape, not only its height is multiplied by 4, but also the width is multiplied by 4, so that you have to multiply the amount of paint by 16.

$$16 \times 6 \text{ oz} = 96 \text{ oz}$$

Phase 4: Argument with Peer 2

Give a solution of a fictitious peer, in which rectangles are drawn around the logo. This gives a visual clue in the form of a rectangle that the student may recognize.

Phase 5: Direct Link to Content

Link the solution explicitly to a calculation of areas:

Can you calculate the area of the two rectangles?

How much larger is the area of the large rectangle, compared to the small one?

How much larger is the area of the large logo, compared to the small one?

How much more paint do you need to paint the large logo?

Adapted from De Bock, Van Dooren, Janssens, & Verschaffel (2002)

Teachers

- Students tend to believe that all word problems in algebra are linear relationships. It is important for teachers to provide tasks that dispel this belief.
- Students should be given a chance to solve problems without a provided solution strategy.
- If errors present, there are ways to challenge student thinking without dictating a strategy. By progressively introducing the challenges in the five phases listed in Table 6.14, teachers can help students adjust their thinking while encouraging true understanding.

ISSUES OF LANGUAGE IN A TASK

Common Core Standards: 6.EE.2, 9; 7.EE.3, 4, 8.EE.7, CED.A.1

The way that word problems are written, including information given explicitly and word order, can have a major impact on the difficulty level for students, even if the mathematical content is the same. "Text complexity and comprehension failures are central to the difficulty of word problems."[106] Reasoning about the situation may play a more critical role in success on word problems than the algebra.[107] This is especially true for English Language Learners. Grammatical structure and word choice influence students' access to the ideas and dynamics of a problem. For example, one study presented students with the task in Table 6.15.[108] A follow-up interview revealed the misconception in Table 6.16.

Students may translate the "students and professors" into $6s = 6p$ because the problem as stated says there are "as many students as professors." Such an interpretation may be more likely to occur when working with ELL students. Indeed, when working with ELL students, language proficiency is a major indicator of performance in mathematics.[109]

Next, consider Problem 31 in Table 6.17. Even when students understand the mathematics represented in the word problems, they may be unable to see the relationship between the mathematics and statements like "how many more ___ are there than ___?" Without the unknown quantities made explicit, students may struggle to understand the problem. Additionally, students may be hesitant to do operations or make inferences about unknown quantities.

Word choice can also significantly impact students' understanding of the problem intended by the author. This is particularly true for ELL students because of their limited English vocabulary. Consider Problem 32 in Table 6.18, in which a student misunderstands the word "initial," which impedes his ability to do the problem accurately.

TABLE 6.15. Problem 30: Students and Professors

Problem	Response
Write an equation to represent the following sentence: "There are six times as many students as professors."	Student: $6s = 6p$

TABLE 6.16. Interview of Problem 30

Interviewer:	Please explain how you came to that equation.
Student:	"As many students as professors" means that there has to be an equal number, so $s = p$. The "six times" in front means that each side should be multiplied by 6.

TABLE 6.17. Problem 31: Marbles

Problem	Response
1) John had 5 marbles. He got some more. Now he has 8. How many did he get? 2) John had some marbles. He got 5 more. Now he has 8. How many did he get?	In a study, 61% of kindergarten students correctly solved Question 1, while only 9% correctly solved Question 2.[116]

Source: Reed (1999)

The example in Table 6.18 is part of a larger issue around vocabulary use and development. Even when ELL students display understanding of decontextualized vocabulary and understanding of mathematics, they do not necessarily display competence in translating narrative problems to algebraic representations. Studies reveal that syntax variables, such as the problem length (measured by total number of words), the level of vocabulary, the number of sentences, and the degree of separation of data within a problem are reasonably good predictors of performance among age groups.[110] Lower levels of language proficiency compound these issues.

One strategy to assist students in solving these tasks is drawing pictures or building models for the problem (See ***Drawing Models as an Intermediate Step in Solving a Word Problem***). A chart that is developed prior to writing an equation may be useful. For Problem 30, we might write:

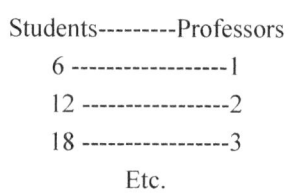

$$\begin{array}{rcl} \text{Students} & \text{---------} & \text{Professors} \\ 6 & \text{------------------} & 1 \\ 12 & \text{----------------} & 2 \\ 18 & \text{----------------} & 3 \\ & \text{Etc.} & \end{array}$$

Students should then translate the problem statement into an algebraic sentence. Following this, they may substitute their numbers from the chart into the equation

TABLE 6.18. Problem 32: Word Choice in Word Problems

Some rental cars have mobile phones installed. In one car, the cost of making a call from the mobile telephone is $1.25 per minute with an initial fee of $2.50. If a call cost a total of twenty dollars, how many minutes did the call last?
Interviewer: OK, um, do you know what this word here means?
Student: No.
Interviewer: OK, "initial?" OK. And what about the 1.25? How come you didn't use that one?
Student: The 1.25 is, like, the . . . per minute, how much it costs per minute [long pause]. Per minute, but I think that "initial" is plus tax, or the whole thing together.

Source: Walkington, Sherman, & Petrosino (2012)

TABLE 6.19. Examples of Reworded Tasks

Original Problem Wording	Reworded Problem
33. Bob got 2 cookies. Now he has 5 cookies. How many cookies did Bob have in the beginning?	Bob had some cookies. He got two more cookies. Now he has 5 cookies. How many cookies did Bob have in the beginning?
34. Tom and Ann have 9 candies altogether. Tom has 3 candies. How many candies does Ann have?	Tom and Ann have 9 candies altogether. Three of those candies belong to Tom. The rest belong to Ann. How many candies does Ann have?
35. Ann has 6 puppies. Sue has 3 puppies. How many puppies does Ann have more than Sue?	There are 6 children but there are only 3 chairs. How many children won't get a chair?

to confirm the accuracy of the sentence. Discussions about the truth (or not) of the number sentences will build a sense of usefulness of the algebraic representation. Another strategy is to reverse the process: given a mathematical sentence (such as $s = 6p$), write a sentence in English that describes it.

Rewording problems may also make them easier to solve. Table 6.19 has ways to reword problems that make it significantly easier for students to solve.[111] While the problems are elementary focused, there are insights for algebra teachers. First,

TABLE 6.20. Problem 36: Dieting Problem

\multicolumn{2}{l	}{Two people who have been on diets are talking:}

Two people who have been on diets are talking:
Dieter A: "I lost 1/8 of my weight—I lost 19 pounds."
Dieter B: "I lost 1/6 of my weight, and now you weigh 2 pounds less than I do."
What was Dieter B's original weight?

Teacher:	Before you tell me what operations you performed, can anyone in the class tell me what this problem was about?
Ellen:	There are two dieters who lost weight.
Teacher:	Can you tell me anything else about the problem?
Maria:	After the diet, Dieter A weighed less than Dieter B.
Teacher:	And how do you know that?
Jose:	Because Dieter B said that Dieter A now weighs 2 pounds less than Dieter B.
Teacher:	What other relationships do you notice?
Anna:	Dieter B lost a greater fraction of her original weight than Dieter A.
Teacher:	And how do you know?
Anna:	Because Dieter B lost 1/6 of her weight, but Dieter A lost only 1/8 of her weight.
Teacher:	Ok, so we have this problem about two dieters who have lost some weight, and we know something about who lost more weight, and we know something about who lost a greater fraction of their original weight.

Source: Clement & Bernhard (2005, pp. 363–364)

TABLE 6.21. Problem 37: Positive Influence of Situational Understanding

You have 80 toys and lose 6 every day. After how many days will you have 8 toys?	
Interviewer:	Why wouldn't you want to get a decimal for the answer to this one?
Student:	Because you can't lose half an object. Because if you have a toy and you lost half of it, it doesn't make any sense.

the more explicit the wording of the problem, the easier it is for students to solve. Given that we want students to solve complex problems as well, it is good for teachers to help students deconstruct word problems to clarify the language used and facilitate students making explicit what is implicit. Consider the dialogue in Table 6.20. In this conversation, the teacher helps the students deconstruct the relationships between the dieters' weights and focus on the changes in quantities before working toward a solution.

One question for algebra teachers is whether writing story problems based on individual students' interests and experiences can have a positive effect (see Problem 37). The typical intent of personalizing word problems is to increase motivation and tap into students' prior knowledge as a means to enhancing their access to the mathematics. Some studies found that personalized problems did enhance learning with students.[112] Understanding and seeing relevance in the problem helped students have access to the mathematics and solutions. In Problem 38 (see Table 6.21), for instance, the student was able to make sense of the solution because of the context. However, other studies found that personalization is not always effective and may distract students' attention, particularly with students who have lower interest in math.[113] In Problem 39 (see Table 6.22), for example, the student became distracted by her own experience in math classes, in which about half of students get A's and B's and the other half do not.

TABLE 6.22. Problem 38: Negative Influence of Situational Understanding

The number of students getting A or B in algebra class is given by the equation $y = .25x$ where x is the total number of students taking algebra. If 40 students earned an A or a B in algebra last year, how many total students were enrolled?	
Student:	80 students were enrolled. [pause] [mumbling]
Interviewer:	So how did you get 80 for that one?
Student:	You just times the umm 40 students times 2, cause there's always a half that doesn't get like the full stuff done, like umm . . . pretty much, there's so many students and then, it divides umm how many students get an A or a B, and the other students don't get an A or B. So I guess it divides how many A's or B's I have.

Problem 39: Personalized Story Problem[114]

You are playing your favorite battle game on the Xbox 360. When you started playing today, there were 80 enemies left. You stop an average of 6 enemies every minute.
1. How many enemies are left after 10 minutes?
2. How many enemies are left after 7 minutes?
3. Write an algebra rule that represents this situation using symbols.

If there are only 8 enemies left, how long have you been playing today?

When translating text-based tasks into algebraic expressions and equations, all students can struggle to make sense of the language. Students who are English Language Learners face additional challenges when they have limited vocabulary. Yet, they also bring rich experiences that influence the way that they interpret and understand story problems. It is critical for teachers to properly assess each student's abilities with regard to language and identify specific supports for areas that may hinder students' understanding of word problems.

Teachers

- Misconceptions about the English language may compound difficulties that students have with translating mathematical tasks into new representations.
- While these language issues are especially pertinent for ELL students, they are also relevant for all students.
- Rewording problems or discussing what is implicit in the problem may help students comprehend the text and move toward the mathematical representation.

ENDNOTES

1. Walkington, Sherman, & Petrosino, 2012, p. 174 based on Loveless, Fennel, Williams, Ball, Banfield, 2008
2. Loveless, Fennel, Williams, Ball, Banfield, 2008, p. 9–xiii
3. Baranes, Perry, & Stigler, 1989; Carraher, Carraher, & Schliemann, 1987; Koedinger & Nathan, 2004; Walkington, Sherman, & Petrosino, 2012
4. Koedinger & Nathan, 2004
5. Koedinger, Alibali & Nathan, 2008
6. Nathan & Koedinger, 2000
7. Nathan & Koedinger, 2000
8. Koedinger & Tabachneck, 1994
9. Koedinger & Anderson, 1997; Nathan, Kintsch, & Young, 1992
10. Nathan & Koedinger, 2000, p. 186, based on Koedinger & Anderson, 1997 and Nathan, Kintsch, & Young, 1992
11. Koedinger & Nathan, 2004 based on Cummins, Kintsch, Reusser, & Weimer, 1988; Hall, Kibler, Wenger, & Truxaw, 1989; Lewis & Mayer, 1987, Mayer, 1982
12. Koedinger & Nathan, 2004, 131
13. Koedinger & Nathan, 2004 based on Cummins, Kintsch, Reusser, & Weimer, 1988 and Lewis & Mayer, 1987
14. Koedinger & Nathan, 2004; Nathan & Koedinger, 2000a, 2000b; Nathan & Petrosino, 2003; Nathan, Long, & Alibali, 2002
15. Walkington, Sherman, & Petrosino, 2012, p. 174 based on Boaler, 1994
16. Baranes, Perry, & Stigler, 1989; Boaler, 1994; Cooper & Harries, 2009; Greer, 1997; Inoue, 2005; Kazemi, 2002; Palm, 2008; Reusser & Stebler, 1997; Roth, 1996
17. Walkington, Sherman, & Petrosino, 2012, based on Blum & Niss, 1991 and Durik & Harackiewicz, 2007
18. Walkington, Sherman, & Petrosino, 2012, p. 175
19. Also consider Dan Meyer's website that encourages the use of unstructured problems and provides numerous examples (http://blog.mrmeyer.com).
20. Kieran, 1992; Usiskin, 1988, 1997
21. Kieran, 2007 based on Radford, 2004
22. Booth, 1989, p. 57
23. Kieran, 2007, p. 711 based on Noss and Hoyles, 1996
24. Radford, 2004
25. Swafford & Langrall, 2000
26. Rosnick, 1982, p. 4
27. Hubbard, 2004
28. Stacey & MacGregor, 1997
29. Stacey & MacGregor, 1993
30. Arzarello, Bazzini, & Chiappini, 1994
31. Rosnick, 1982
32. Rosnick, 1982, p. 6
33. Rosnick, 1982
34. Radford, 2000
35. Arzarello, Bazzini, & Chiappini, 1994, p. 42
36. Swafford & Langrall, 2000, p. 109
37. Swafford & Langrall, 2000
38. Johanning, 2004, p. 372, based on Bednarz & Janvier, 1996; Hall, Kibler, Wenger, & Truxaw, 1989a; Kieran, Boileau, & Garancon, 1996; Nathan & Koedinger, 2000; Rojano, 1996
39. Walkington, Sherman, & Petrosino, 2012, p. 181

40 Brizuela & Schliemann, 2003
41 Swafford & Langrall, 2000
42 Johanning, 2004
43 De Corte, Verschaffel, & De Win, 1985; Lee, Ng, & Ng, 2009; Lewis, 1989
44 Clement & Bernhard, 2005, Sowder, 1988
45 Hegarty, Mayer, & Monk, 1995
46 Carpenter, Fennema, Franke, Levei, & Empson, 1999 and Hegarty, Mayer, & Monk, 1995, p. 29
47 Clement & Bernhard, 2005, p. 360
48 Clement & Bernhard, 2005, p. 361
49 From Jonassen, 2003, p. 269, based on Lucangeli, Tressoldi, and Cendron, 1998
50 Stacey & MacGregor, 2000
51 Koedinger & Nathan, 2004; Walkington, Sherman, & Petrosino, 2012
52 Cummins, Kintsch, Reusser, & Weimer, 1988, p. 435
53 Lee, Ng, Ng, & Lim, 2004
54 Cummins, Kintsch, Reusser, & Weimer, 1988
55 Cummins, Kintsch, Reusser, & Weimer, 1988; Inoue, 2005
56 Koedinger & Nathan, 2004, p. 135, based on Carraher, Carraher, & Schlieman, 1987 and Baranes, Pery, & Stigler, 1989
57 Hegarty, Mayer, & Monk, 1995, p. 18
58 Briars & Larkin, 1984; Littlefield & Rieser, 1993
59 Stigler, Lee, & Stevenson, 1990, p. 15
60 Swafford & Langrall, 2000
61 Ya-amphan & Bishop, 2004
62 Brizuela, Carraher, & Schliemann, 2000; Brizuela & Schliemann, 2003
63 Brizuela, Carraher, & Schliemann, 2000
64 Johanning, 2004
65 Adapted from Sowder, 1988
66 Sowder, 1988, p. 228
67 Sowder, 1988, p. 227
68 Stacey & MacGregor, 2000
69 Koedinger & MacLaren, 1997; Walkington, Sherman, & Petrosino, 2012
70 Nathan & Koedinger, 2000
71 Nathan & Koedinger, 2000, p. 187, based on Tabachneck, Koedinger, & Nathan, 1995
72 Bednarz, Radford, Janvier, & Lepage, 1992, p. 1-71
73 Stacey & MacGregor, 2000, p. 150, based on Filloy & Rojano, 1989
74 Based on Stacey & MacGregor, 2000
75 Walkington, Sherman, & Petrosino, 2012, p. 174
76 Buchbinder & Chazan, 2013, p. 5
77 Buchbinder & Chazan, 2013
78 Schwartz & Black, 1996
79 Diezmann, 1995; Mudaly, 2012
80 National Governors Association Center for Best Practices, Council of Chief State School Officers, 2010
81 Simon, 1986
82 Cox & Brna, 1995, p. 4
83 Diezmann, 1995
84 van Essen & Hamaker, 1990
85 Kersch & McDonald, 1991; Shigematsu & Sowder, 1994
86 English & Halford, 1995; Larkin & Simon, 1987
87 Larkin & Simon, 1987
88 Mayer & Gallini, 1990
89 Newman, 1977

90   Schoenfeld, 1985; Swee Fong & Lee, 2009
91   Cox, 1996; Mudaly, 2012; Proudfit, 1980; Singley, Anderson, Gevins, & Hoffman, 1989; Swee Fong & Lee, 2009; van Essen & Hamaker, 1990
92   Grossen & Carnine, 1990; Mudaly, 2012
93   Mudaly, 2012, p. 26
94   Swee Fong & Lee, 2009
95   Swee Fong & Lee, 2009
96   For a more detailed explanation of the model method see Hong, Mei, & Lim, 2009 and Swee Fong & Lee, 2009
97   Rau, 2016
98   Booth & Koedinger, 2012
99   Cox, 1996; Nathan, Kintsch, & Young, 1992; Moyer-Packenham, 2005; Moyer-Packenham, Salkind, & Bolyard, 2008; Suh & Moyer, 2007; Zbiek, Heid, Blume, & Dick, 2007
100 Cox, 1996
101 De Bock, Van Dooren, Janssens, & Verschaffel, 2002; De Bock, Verschaffel, & Janssens, 1998; Freudenthal, 1983; Van Dooren, De Bock, & Depaepe, 2003
102 Gagatsis, 1998; Greer, 1993; Nesher,1996; Verschaffelet, De Corte, Lasure, 1994, 2000; Wyndhamn and Saljo, 1997
103 De Bock, Van Dooren, Janssens, & Verschaffel, 2002
104 De Bock, Van Dooren, Janssens, & Verschaffel, 2002; De Bock, Verschaffel & Janssens, 1998
105 De Bock, Van Dooren, Janssens, & Verschaffel, 2002
106 Nathan, Kintsch, & Young, 1992, p. 330, based on Carpenter, Corbitt, Kepner, Lindquist, & Reys, 1980; Cummins, Kintsch, Reusser, & Weimer, 1988; Kintsch & Greeno, 1985; Lewis & Mayer, 1987
107 Hall, Kibler, Wenger, & Truxaw, 1989b; Nathan, Kintsch, & Young, 1992
108 Mestre & Gerace, 1986
109 Mestre & Gerace, 1986
110 Mestre & Gerace, 1986
111 From Reed, 1999 based on De Corte, Verschaffel, & De Win, 1985 and Hudson, 1983
112 Anand & Ross, 1987; Cordova & Lepper, 1996; Davis-Dorsey, Ross, & Morrison, 1991; Walkington, Sherman, & Petrosino, 2012
113 Cummins, Kintsch, Reusser, & Weimer, 1988; Renninger, Ewen, & Lasher, 2002; Vicente, Orrantia, & Verschaffel, 2007
114 Adapted from Walkington, Sherman, & Petrosino, 2012
115 Van Dooren, Verschaffel & Onghena, 2003
116 Reed, 1999

REFERENCES

Anand, P., & Ross, S. (1987). Using computer-assisted instruction to personalize arithmetic materials for elementary school children. *Journal of Educational Psychology, 79*(1), 72–78.

Arzarello, F., Bazzini, L., & Chiappini, G. (1994). *The process of naming in algebraic problem solving.* Paper presented at the 18th Annual Conference of the International Group for the Psychology of Mathematics Education (PME), Lisbon, Portugal.

Baranes, R., Pery, M., & Stigler, J. W. (1989). Activation of real-world knowledge in the solutions of word problems. *Cognition and Instruction, 6,* 287–318.

Bednarz, N., & Janvier, B. (1996). Emergence and development of algebra as a problem-solving tool: Continuities and discontinuities with arithmetic. In N. Bednarz, C. Kieran, & L. Lee (Eds.), *Approaches to algebra: Perspectives for research and teaching* (pp. 115–136). Boston, MA: Kluwer Academic Publishers.

Bednarz, N., Radford, L., Janvier, B., & Lepage, A. (1992). *Arithmetical and algebraic thinking in problem solving*. Paper presented at the 16th International Conference for the Psychology of Mathematics Education, Durham, NH.

Blum, W., & Niss, M. (1991). Applied mathematical problem solving, modeling, applications, and links to other subjects: State, trends & issues. *Educational Studies in Mathematics, 22*, 37–68.

Boaler, J. (1994). When do girls prefer football to fashion? An analysis of female underachievement in relation to 'realistic' mathematics contexts. *British Education Research Journal, 20*(5), 551–564.

Booth, J. L., & Koedinger, K. R. (2012). Are diagrams always helpful tools? Developmental and individual differences in the effect of presentation format on student problem solving. *British Journal of Educational Psychology, 82*(3), 492–511.

Booth, L. R. (1989). A question of structure. In S. Wagner & C. Kieran (Eds.), *Research issues in the learning and teaching of algebra* (pp. 57–59). Reston, VA: NCTM.

Briars, D. J., & Larkin, J. H. (1984). An integrated model of skill in solving elementary word problems. *Cognition and Instruction, 1*, 245–296.

Brizuela, B. M., & Schliemann, A. D. (2003). *Fourth graders solving equations*. Medford, MA: Tufts University.

Brizuela, B., Carraher, D., & Schliemann, A. (2000). *Mathematical notation to support and further reasoning ("to help me think of something")*. Chicago, IL: NCTM Research Pre-session.

Carpenter, T. P., Corbitt, M. K., Kepner, H. S., Lindquist, M. M., & Reys, R. E. (1980). Solving verbal problems: Results and implications for national assessment. *Arithmetic Teacher, 28*, 8–12.

Carpenter, T. P., Fennema, E., Franke, M. L., Levi, L., & Empson, S. B. (1999). *Children's mathematics: Cognitively guided instruction*. Portsmouth, NH: Heinemann

Carraher, T. N., Carraher, D. W., & Schliemann, A. D. (1987). Written and oral mathematics. *Journal for Research in Mathematics Education, 18*, 83–97.

Clement, L., & Bernhard, J. (2005). A problem solving alternative to using key words. *Mathematics Teaching in the Middle Grades, 10*, 47–75.

Cooper, B., & Harries, T. (2009). Realistic contexts, mathematics assessment, and social class. In B. Greer, L. Verschaffel, W. Van Dooren, & S. Mukhopadhyay (Eds.), *Word and worlds: Modelling verbal descriptions of situations*. Rotterdam, the Netherlands: Sense Publishers.

Cordova, D., & Lepper, M. (1996). Intrinsic motivation and the process of learning: Beneficial effects of contextualization, personalization, and choice. *Journal of Educational Psychology, 88*(4), 715–730.

Cox, R. (1996). *Analytical reasoning with multiple external representations* (Unpublished doctoral dissertation). The University of Edinburgh, Scotland.

Cox, R., & Brna, P. (1995). Supporting the use of external representations in problem solving: The need for flexible learning environments. *Journal of Artificial Intelligence in Education, 6*(2/3), 239–302.

Cummins, D., Kintsch, W., Reusser, K., & Weimer, R. (1988). The role of understanding in solving word problems. *Cognitive Psychology, 20*, 439–462.

Davis-Dorsey, J., Ross, S., & Morrison, G. (1991). The role of rewording and context personalization in the solving of mathematical word problems. *Journal of Educational Psychology, 83*(1), 61–68.

De Bock, D., Van Dooren, W., Janssens, D., & Verschaffel, L. (2002). Improper use of linear reasoning: An in-depth study of the nature and the irresistibility of secondary school students' errors. *Educational Studies in Mathematics, 50*(3), 311–334.

De Bock, D., Verschaffel, L., & Janssens, D. (1998). The predominance of the linear model in secondary school students' solutions of word problems involving length and area of similar plane figures. *Educational Studies in Mathematics, 35*(1), 65–83.

De Corte, E., Verschaffel, L., & De Win, L. (1985). Influence of rewording verbal problems on children's problem representations and solutions. *Journal of Educational Psychology, 77*, 460–470.

Diezmann, C. M. (1995). Evaluating the effectiveness of the strategy 'Draw a diagram' as a cognitive tool for problem solving. In B. Atweh & S. Flavel (Eds.), *Proceedings of the 18th Annual Conference of Mathematics Education Research Group of Australasia* (pp. 223–228). Darwin, Australia: MERGA.

Durik, A., & Harackiewicz, J. (2007). Different strokes for different folks: How individual interest moderates effects of situational factors on task interest. *Journal of Educational Psychology, 99*(3), 597–610.

English, L. D., & Halford, G. S. (1995). *Mathematics education: Models and processes.* Mahwah, NJ: Lawrence Erlbaum.

Filloy, B., & Rojano, T. (1989). Solving equations: The transition from arithmetic to algebra. *For the Learning of Mathematics: An International Journal of Mathematics Education, 9*(2), 19–25.

Freudenthal, H. (1983). *Didactical phenomenology of mathematical structures*. Dordrecht: D. Reidel.

Gagatsis, A. (1998). Solving methods in problems of proportion by Greek students in secondary education, ages 13–16. *Scientia Paedagogica Experimentalis, 35*(1), 241–262.

Greer, B. (1993). The mathematical modeling perspective on word problems. *Journal of Mathematical Behaviour, 12*, 239–250.

Greer, B. (1997). Modelling reality in mathematics classrooms: The case of word problems. *Learning and Instruction, 7*(4), 293–307.

Grossen, G., & Carnine, D. (1990). Diagramming a logic strategy: Effects on difficult problem types and transfer. *Learning Disability Quarterly, 13*, 168–182.

Hall, R., Kibler, D., Wenger, E., & Truxaw, C. (1989a). Exploring the Episodic Structure of Algebra Story Problem Solving. *Cognition and Instruction, 6*(3), 223–283.

Hall, R., Kibler, D., Wenger, E., & Truxaw, C. (1989b). The situativity of knowing, learning, and research. *American Psychologist, 53*, 5–26.

Hegarty, M., Mayer, R. E., & Monk, C. A. (1995). Comprehension of arithmetic word problems: A comparison of successful and unsuccessful problem solvers. *Journal of Educational Psychology, 87*(1), 18–32.

Hong, K. T., Mei, Y. S., & Lim, J. (2009). *The Singapore model method for learning mathematics*. Singapore: EPB Pan Pacific.

Hubbard, R. (2004). *An investigation into the modelling of word problems*. Paper presented at the 27th annual conference of the Mathematics Education Research Group of Australasia.

Hudson, T. (1983). Correspondences and numerical differences between disjoint sets. *Child Development, 54*, 84–90.

Inoue, N. (2005). The realistic reasons behind unrealistic solutions: The role of interpretive activity in word problem solving. *Learning and Instruction, 15*(1), 69–83.

Johanning, D. I. (2004). Supporting the development of algebraic thinking in middle school: A closer look at students' informal strategies. *Journal of Mathematical Behavior, 23*, 371–388.

Jonassen, D. H. (2003). Designing research-based instruction for story problems. *Educational Psychology Review, 15*(3), 267–296.

Kazemi, E. (2002). Exploring test performance in mathematics: The questions children's answers raise. *Journal of Mathematical Behavior, 21*(2), 203–224.

Kersch, M. E., & McDonald, J. (1991). How do I solve thee? Let me count the ways. *Arithmetic Teacher, 39*(2), 38–41.

Kieran, C. (1992). The learning and teaching of school algebra. In D. A. Grouws (Ed.), *Handbook of research on mathematics teaching and learning* (pp. 390–419). New York, NY: Macmillan.

Kieran, C. (2007). Learning and teaching algebra at the middle school through college levels. In F. Lester (Ed.), *Second Handbook of Research on Mathematics Teaching and Learning* (pp. 707–762): Information Age Publishing.

Kintsch, W., & Greeno, J. G. (1985). Understanding and solving word arithmetic problems. *Psychological Review, 92*, 109–129.

Koedinger, K. R., Alibali, M., & Nathan, M. J. (2008). Trade-offs between grounded and abstract representations: Evidence from algebra problem solving. *Cognitive Science. 32*(2), 366–397.

Koedinger, K. R., & Anderson, J. A. (1997). Illustrating principled design: The early evolution of a cognitive tutor for algebra symbolization. *Interactive Learning Environments, 5*, 161–180.

Koedinger, K. R., & MacLaren, B. (1997). *Implicit strategies and errors in an improved model of early algebra problem solving.* Paper presented at the 19th Annual Conference of the Cognitive Science Society, Stanford University, CA.

Koedinger, K., & Nathan, M. J. (2004). The real story behind story problems: Effects of representations on quantitative reasoning. *The Journal of the Learning Sciences, 13*(2), 129–164.

Koedinger, K. R., & Tabachneck, H. J. M. (1994). *Two strategies are better than one: Multiple strategy use in word problem solving.* Paper presented at the annual meeting of the American Education Research Association, New Orleans, LA.

Larkin, J. H., & Simon, H. A. (1987). Why a diagram is (sometimes) worth ten thousand words. *Cognitive Science, 11*, 65–99.

Lee, K., Ng, E. L., & Ng, S. F. (2009). The contributions of working memory and executive functioning to problem representation and solution generation in algebraic word problems. *Journal of Educational Psychology, 101*(2), 373–387.

Lee, K., Ng, S.-F., Ng, E.-L., & Lim, Z.-Y. (2004). Working memory and literacy as predictors of performance on algebraic word problems. *Journal of Experimental Child Psychology, 89*(2), 140–158.

Lewis, A. B. (1989). Training students to represent arithmetic word problems. *Journal of Educational Psychology, 81*, 521–531.

Lewis, A. B., & Mayer, R. E. (1987). Students' misconceptions of relational statements in arithmetic word problems. *Journal of Educational Psychology, 79,* 363–371.

Littlefield, J., & Rieser, J. J. (1993). Semantic features of similarity and children's strategies for identification of relevant information in mathematical story problems. *Cognition & Instruction, 11,* 133–188.

Loveless, T., Fennel, F., Williams, V., Ball, D., & Banfield, M. (2008). Chapter 9: Report of the Subcommittee on the National Survey of Algebra I Teachers. In *Foundations for success: Report of the National Mathematics Advisory Panel.* Retrieved from http://www2.ed.gov/about/bdscomm/list/mathpanel/report/nsat.pdf

Lucangeli, D., Tressoldi, P. E., & Cendron, M. (1998). Cognitive and metacognitive abilities involved in the solution of mathematical word problems: Validation of a comprehensive model. *Contemporary Educational Psychology, 23,* 257–275.

Mayer, R. E. (1982). Memory for algebra story problems. *Journal of Educational Psychology, 74,* 199–216.

Mayer, R. E., & Gallini, J. K. (1990). When is an illustration worth ten thousand words? *Journal of Educational Psychology, 82*(4), 715–726.

Mestre, J., & Gerace, W. (1986). The interplay of linguistic factors in mathematical translation tasks. *Focus on Learning Problems in Mathematics, 8*(1), 59–72.

Moyer-Packenham, P. S. (2005). Using virtual manipulatives to investigate patterns and generate rules in algebra. *Teaching Children Mathematics, 11*(8), 437–443.

Moyer-Packenham, P. S., Salkind, G., & Bolyard, J. J. (2008). Virtual manipulatives used by K–8 teachers for mathematics instruction: Considering mathematical, cognitive, and pedagogical fidelity. *Contemporary Issues in Technology and Teacher Education, 8*(3), 202–218

Mudaly, V. (2012). Diagrams in mathematics: To draw or not to draw? *Perspectives in Education, 30*(2), 22–31.

Nathan, M. J., & Koedinger, K. R. (2000). Teachers' and researchers' beliefs about the development of algebraic reasoning. *Journal for Research in Mathematics Education, 31*(2), 168–190.

Nathan, M. J., Kintsch, W., & Young, E. (1992). A theory of algebra word problem comprehension and its implications for the design of learning environments. *Cognition & Instruction, 9*(4), 329–389.

Nathan, M., & Koedinger, K. (2000a). An investigation of teachers' beliefs of students' algebra development. *Cognition and Instruction, 18,* 209–238.

Nathan, M., & Koedinger, K. (2000b). Teacher and researchers' beliefs about the development of algebraic understanding. *Journal for Research in Mathematics Education, 31*(2), 168–190.

Nathan, M., & Petrosino, A. (2003). Expert blind spot among pre-service teachers. *American Educational Research Journal, 40*(4), 905–928.

Nathan, M., Long, S., & Alibali, M. (2002). The symbol precedence view of mathematical development: A corpus analysis of the rhetorical structure of textbooks. *Discourse Processes, 33*(1), 1–21.

National Governors Association Center for Best Practices & Council of Chief State School Officers (2010). *Common core state standards for mathematics.* Washington, DC: National Governors Association Center for Best Practices, Council of Chief State School Officers.

Nesher, P. (1996). *School stereotype word problems and the open nature of applications* (pp. 335–343). Selected Lectures from the 8th International Congress on Mathematical Education, Sevilla, Spain.

Newman, M. A. (1977). An analysis of sixth-grade pupils' errors on written mathematical tasks. *Victorian Institute for Educational Research Bulletin, 39*, 31–43.

Noss, R., & Hoyles, C. (1996). *Windows on mathematical meaning.* Dordrecht, The Netherlands: Kluwer.

Palm, T. (2008). Impact of authenticity on sense making in word problem solving. *Educational Studies in Mathematics, 67*, 37–58.

Proudfit, L. A. (1980). *The examination of problem-solving processes by fifth-grade children and its effect on problem-solving performance* (Order No. 8105942). Available from ProQuest Dissertations & Theses Global. (303024786). Retrieved from http://proxy.lib.pacificu.edu:2048/login?url=http://search.proquest.com/docview/303024786?accountid=13047.

Radford, L. (2000). *Students' processes of symbolizing in algebra: A semiotic analysis of the production of signs in generalizing tasks.* Paper presented at the the 24th Conference of the International Group for the Psychology of Mathematics Education, Hiroshima, Japan.

Radford, L. (2004). *Syntax and meaning.* Paper presented at the 28th International Conference of the International Group for the Psychology of Mathematics Education, Bergen, Norway.

Rau, M. A. (2017). Conditions for the effectiveness of multiple visual representations in enhancing STEM learning. *Educational Psychology Review, 29*(4), 717–761.

Reed, S. K. (1999). *Word problems: Research and curriculum reform.* Mahwah, NJ: Lawrence Erlbaum.

Renninger, K., Ewen, L., & Lasher, A. (2002). Individual interest as context in expository text and mathematical word problems. *Learning and Instruction, 12*(4), 467–490.

Reusser, K., & Stebler, R. (1997). Every word problem has a solution: The social rationality of mathematical modeling in schools. *Learning and Instruction, 7*(4), 309–327.

Rojano, T. (1996). Problem solving in a spreadsheet environment. In N. Bednarz, C. Kieran, & L. Lee (Eds.), *Approaches to algebra: Perspectives for research and teaching* (pp. 137–145). Boston, MA: Kluwer Academic Publishers.

Rosnick, P. (1982). *Students' symbolization process in algebra.* University of Massachusetts, Amherst: Cognitive Development Project, Department of Physics and Astronomy.

Roth, W. (1996). Where is the context in contextual word problems: Mathematical practices and products in grade 8 students' answers to story problems. *Cognition and Instruction, 14*(4), 487–527.

Schoenfeld, A. (1985). *Mathematical problem solving.* Orlando: Academic Press.

Schwartz, D. L., & Black, J. B. (1996). Shuttling between depictive models and abstract rules: Induction and fallback. *Cognitive Science, 20*(4), 457–497.

Shigematsu, K., & Sowder, L. (1994). Drawings for story problems: Practices in Japan and the United States. *Arithmetic Teacher, 41*(9), 544–547.

Simon, M. (1986). Components of effective use of diagrams in math problem solving. *Proceedings of the North American Chapter of the International Group for the Psychology of Mathematics Education* (pp. 1–7). East Lansing, MI: PME.

Singapore Model Method for Learning Mathematics. Singapore: EPB Panpac Education.

Singley, M. K., Anderson, J .R., Gevins, J .S., & Hoffman, D. (1989). The algebra word problem tutor. In D. Bierman, J. Breuker, & J. Sandberg (Eds.), *Artificial intelligence and education.* Proceedings of the 4th International Conference on AI and Education (pp. 267–275), Amsterdam, Netherlands.

Sowder, L. (1988). Children's solutions of story problems. *The Journal of Mathematical Behavior, 7,* 227–238.

Stacey, K., & MacGregor , M. E. (1997). Ideas about symbolism that students bring to algebra. *The Mathematics Teacher, 90*(2), 110–113.

Stacey, K., & MacGregor, M. (2000). Learning the algebraic method of solving problems. *Journal of Mathematical Behavior, 18*(2), 149–167.

Stacey, K. M., & MacGregor, M. (1993). Origins of students' errors in writing equations. In A. B. T. Cooper (Ed.), *New directions in algebra education,* (pp. 205–212). Brisbane, Australia: Queensland University of Technology.

Stigler, J. W., Lee, S-Y., & Stevenson, H. W. (1990). *Mathematical knowledge of Japanese, Chinese, and American elementary school children.* Reston, VA: National Council of Teachers of Mathematics.

Suh, J., & Moyer, P. S. (2007). Developing students' representational fluency using virtual and physical algebra balances. *The Journal of Computers in Mathematics and Science Teaching, 26*(2), 155–173.

Swafford, J. O., & Langrall, C.W. (2000). Grade 6 students' preinstructional use of equations to describe and represent problem situations. *Journal for Research in Mathematics Education, 31*(1), 89–112.

Swee Fong, N., & Lee, K. (2009). The model method: Singapore children's tool for representing and solving algebraic word problems. *Journal for Research in Mathematics Education, 40*(3), 282–313.

Tabachneck, H. J. M., Koedinger, K. R., & Nathan, M. J. (1995). *An analysis of the task demands of algebra and the cognitive processes needed to meet them.* Paper presented at the Annual meeting of the Cognitive Science Society, Pittsburgh, PA.

Usiskin, Z. (1988). Conceptions of school algebra and uses of variables. In A. Coxford & A. Shulte (Eds.), *The ideas of algebra, K-12. 1988 yearbook* (pp. 8–19). Reston, VA: The National Council of Teachers of Mathematics.

Usiskin, Z. (1997). Reforming the third R: Changing the school mathematics curriculum: An introduction. *American Journal of Education, 106*(1), 1–4.

Van Dooren, W., De Bock, D., & Depaepe, F. (2003). The illusion of linearity: Expanding the evidence towards probabilistic reasoning. *Educational Studies in Mathematics, 53*(2), 113–138.

Van Dooren, W., Verschaffel, L., & Onghena, P. (2003). Pre-service teachers' preferred strategies for solving arithmetic and algebra word problems. *Journal of Mathematics Teacher Education, 6,* 27–52.

van Essen, G., & Hamaker, C. (1990). Using self-generated drawings to solve arithmetic word problems. *The Journal of Educational Research, 83*(6), 301–12.

Verschaffel, L., De Corte, E., & Lasure, S. (1994). Realistic considerations in mathematical modelling of school arithmetic word problems. *Learning and Instruction, 4,* 273–294.

Vicente, S., Orrantia, J., & Verschaffel, L. (2007). Influence of situational and conceptual rewording on word problem solving. *British Journal of Educational Psychology, 77,* 829–848.

Walkington, C., Sherman, M., & Petrosino, A. (2012). Playing the game of story problems: Coordinating situation-based reasoning with algebraic representation. *Journal of Mathematical Behavior, 31*(2), 174–195.

Wyndhamn, J. & Saljo, R. (1997). Word problems and mathematical reasoning: A study of children's mastery of reference and meaning in textual realities. *Learning and Instruction, 7*(4), 361–382

Ya-amphan, D., & Bishop, A. (2004). *Explaining Thai and Japanese student errors in solving equation problems: The role of the textbook.* Paper presented at the 27th annual conference of the Mathematics Education Research Group of Australasia, Townsville, Australia.

Zbiek, R. M., Heid, M. K., Blume, G., & Dick, T. P. (2007). Research on technology in mathematics education. In F. Lester (Ed.), *Second handbook of research on mathematics teaching and learning* (pp. 1169–1207). Charlotte, NC: Information Age Publishing.

Made in the USA
Middletown, DE
28 December 2018